Women, Art, and Technology

LEONARDO

Roger F. Malina, series editor

Designing Information Technology, Richard Coyne, 1995
Technoromanticism: Digital Narrative, Holism, and the Romance of the Real, Richard Coyne, 1999
Metal and Flesh: The Evolution of Man: Technology Takes Over, Ollivier Dyens, 2001
The Visual Mind, edited by Michele Emmer, 1994
The Robot in the Garden: Telerobotics and Telepistemology in the Age of the Internet, edited by Ken Goldberg, 2000
The History of Virtual Art and Its Future, Oliver Grau, 2003
Leonardo Almanac, edited by Craig Harris, 1994
In Search of Innovation: The Xerox PARC PAIR Project, edited by Craig Harris, 1999
The Digital Dialectic: New Essays on New Media, edited by Peter Lunenfeld, 1999
Women, Art, and Technology, edited by Judy Malloy, 2003
The Language of New Media, Lev Manovich, 2000
Immersed in Technology: Art and Virtual Environments, edited by Mary Anne Moser with Douglas MacLeod, 1996
Information Arts: A Survey of Art and Research at the Intersection of Art, Science, and Technology, Stephen Wilson, 2002

Women, Art, and Technology

edited by
Judy Malloy

The MIT Press
Cambridge, Massachusetts
London, England

This book was set in Garamond 3 and Bell Gothic by Achorn Graphic Services, Inc.

Printed and bound in the United States of America.

Library of Congress Cataloging-in-Publication Data

Women, art, and technology / edited by Judy Malloy.
p. cm — (Leonardo)
Includes bibliographical references and index.
ISBN 0-262-13424-1 (hc. : alk. paper)
1. Art and technology—History—20th century. 2. Technology and women—History—20th century. I. Malloy, Judy. II. Leonardo (Series) (Cambridge, Mass.)

N72.T4W66 2003
700′.1′05082—dc21

2002045178

Contents

Series Foreword

The cultural convergence of art, science, and technology provides ample opportunity for artists to challenge the very notion of how art is produced and to call into question its subject matter and its function in society. The mission of the Leonardo book series, published by the MIT Press, is to publish texts by artists, scientists, researchers, and scholars that present innovative discourse on the convergence of art, science, and technology.

Envisioned as a catalyst for enterprise, research, and creative and scholarly experimentation, the book series enables diverse intellectual communities to explore common grounds of expertise. The Leonardo book series provides a context for the discussion of contemporary practice, ideas, and frameworks in this rapidly evolving arena where art and science connect.

To find more information about Leonardo/ISAST and to order our publications, go to Leonardo Online at ⟨http://mitpress.mit.edu/Leonardo/⟩ or send e-mail to ⟨leo@mitpress.mit.edu⟩.

Joel Slayton
Chair, Leonardo Book Series

Book Series Advisory Committee: Pamela Grant-Ryan, Michael Punt, Annick Bureaud, Allen Strange, Margaret Morse, Craig Harris.

LEONARDO/INTERNATIONAL SOCIETY FOR THE ARTS, SCIENCES, AND TECHNOLOGY (ISAST)

Leonardo, the International Society for the Arts, Sciences, and Technology, and the affiliated French organization, Association Leonardo, have two simple goals:

- To document and make known the work of artists, researchers, and scholars who are interested in the ways that the contemporary arts interact with science and technology and
- To create a forum and meeting places where artists, scientists, and engineers can meet, exchange ideas, and, where appropriate, collaborate.

When the journal *Leonardo* was started some thirty-five years ago, these creative disciplines existed in segregated institutional and social networks, a situation dramatized at that time by the debates initiated by the 1959 publication of C. P. Snow's *Two Cultures.* Today we live in a time of cross-disciplinary ferment, collaboration, and intellectual confrontation enabled by new hybrid organizations, new funding sponsors, and the shared tools of computers and the Internet. Above all, new generations of artist researchers and researcher artists are now at work individually and in collaborative teams bridging the art, science, and technology disciplines. Perhaps in our lifetime we will see the emergence of "new Leonardos"—creative individuals or teams who will develop a meaningful art for our times, drive new agendas in science, and stimulate technological innovation that addresses today's human needs.

For more information on the activities of the Leonardo organizations and networks, please visit our Web site at ⟨http://mitpress.mit.edu/Leonardo⟩.

Roger F. Malina
Chair, Leonardo/ISAST

WOMEN, ART, AND TECHNOLOGY

Women, Art, and Technology is a volume of manuscripts produced by women artists, curators, critics, and theoreticians. Edited by Judy Malloy, this collection of critical commentaries and artists' documentaries serves to illuminate the important historical contribution to art-and-technology enterprises by women. *Women, Art, and Technology* provides insight into essential artistic frameworks that helped shape several generations of artists and define the field of new media.

Providing a scholarly forum for women artists, researchers, and theorists to write about their work has long been a consideration of *Leonardo.* In 1993 the *Leonardo* Women, Art, and Technology Project, under the guest editorship of Judy Malloy, focused on the achievements of women artists and functioned as a catalyst for this book. Introductory essays by critical theorists Patric Prince, Margaret Morse, Sheila Pinkel, Anna Couey, and Kathy Brew set the stage for the central collection of artists' writings that include both pioneers and contemporary artists describing their work, methods, and concepts. *Women, Art, and Technology* presents a vast terrain of creative and scholarly exploration, ranging from the experimental video and installation works of Steina to the genetic engineering systems of Nell Tenhaaf. Several generations of artists including Rebecca Allen, Diane Fenster, Monika Fleischmann, Lynn Hershman, Brenda Laurel, Pauline Oliveros, and Pamela Z (among many others) are represented. Important to their writings is the framing of contemporary discourses provided by Jaishree K. Odin, Simone Osthoff, Martha Burkle Bonecchi, Carol Stakenas, and Zoë Sofia.

The Leonardo Book Series is proud to include *Women, Art, and Technology.*

Joel Slayton

Foreword: The *Leonardo* Women, Art, and Technology Project

Patricia Bentson
Leonardo Special Project Manager

The *Leonardo* Women, Art, and Technology Project began in 1993 with the primary aim of encouraging women artists working with technology-based media to write about their work. The tremendous response to the project confirms the importance of recognizing these artists: since the start of the project, we have received proposals and papers from all over the world in response to our letters of invitation, our published calls for papers, and word of mouth. The result is a cross-section of writings about works in a variety of media, styles, and forms, created for a variety of purposes—some to explore the visual, aural, and textual possibilities of new media; some to comment on the artists' lives, communities, or on the world around them; and some to push the boundaries of the technologies themselves.

Leonardo began publication in Paris in 1968, led by founding editor and kinetic artist Frank Malina, who modeled the journal after the science journals he read and wrote for during his first career in astronautics. Malina envisioned *Leonardo* as being a vehicle of communication for artists who worked with media involving science and technology and as having a primary focus on artists' writings about their own work. Notable papers were published in *Leonardo*'s early years by art-and-technology pioneers Colette Bangert, Margaret Benyon, Ruth Leavitt, Vera Molnar, Jasia Reichardt, Sonia Landy Sheridan, and others, effectively archiving the concepts and technological bases behind a number of early groundbreaking works. The journal has come to serve as an archive documenting technology-based artworks as the technologies on which the works are based and exhibited become obsolete. As computer and technological systems and media evolve and are reinvented and as the viewing of a number of early works on original equipment is becoming impossible, papers such as these chronicle the art of the times.

Since the mid-1980s, under the editorship of Frank's son and successor Roger Malina, *Leonardo* has focused on special projects guided by guest editors working within targeted areas, often relating to specific media or research such as telecommunications art or artificial-life art. Over the years, however, we noted that women artists were not responding in large numbers to the calls for papers based on media. While the number of artists, including women, working with tech-based forms have increased steadily over the years, by the early 1990s the number of papers by women artists received at *Leonardo* still lagged far behind those by men. A notable exception was the *Art and Social Consciousness* special issue, *Leonardo* 26, no. 5 (1993), which was guest-edited by Sheila Pinkel. This topic-focused rather than medium-focused issue features a number of articles that were written by both women and men artists about their artworks that address social and political issues. Currently, through the Women, Art, and Technology Project and other special projects, *Leonardo* seeks to connect communities across gender, geographic, and disciplinary boundaries.

When the Women, Art, and Technology Project began, led by guest editor Judy Malloy, it was our hope that the project would convince many artists that writing about their work is important. Some artists find it difficult to write about their work, preferring to express themselves solely through their artworks. Rigorous writing can be challenging, even painful. However, many artists discover that the process of writing can also help them view their own work in new ways—teaching them much about their processes and aiding their future work.

Early in the project two goals were defined: (1) to increase the numbers of women artists writing and publishing in the journal *Leonardo,* thereby ensuring that the works and concepts of women artists are included in the full discussion of tech-based art (the writings published in the journal are not set aside or labeled as "women's writings") and (2) to publish a selection of the papers in a book, as part of the Leonardo Book Series, providing another context by stepping back and recognizing the roles of women artists in contemporary media at this point in time. Dozens of papers have been published in the journal since the project began; with publication of this book, both goals have been reached.

It is with great excitement that everyone at *Leonardo* greets the publication of this volume. We are indebted to Judy Malloy for her dedication to the project and the book; to Doug Sery of the MIT Press for his recognition of the importance of the topic for the Leonardo Book Series; to the California Tamarack Foundation for its early support and encouragement of the project; and to all the authors who have participated in the project by documenting their work in the book, in the journal, or on the Web. We hope that this volume will encourage additional women artists to document their work, methods, and concepts.

Preface

You/I approach this comprehensive volume as we would approach a work of interactive art—Carolyn Guyer's *Quibbling,*[1] for instance, where the narrative is unfolded in four separate parts (and you may or may not discover how they are related) or Brenda Laurel and Rachel Strickland's virtual environment *Placeholder,* which participants enter as a spider, snake, fish, or crow "and thereby experience aspects of its unique visual presentation."[2]

"Another ancient myth binds the ordering of time and space to the efficacy of the narrative string, implying gender and heteronormative processes in its production," Sue Ellen Case has written. "As the ancients told it, there was a maze."[3]

You/I, beholding or assuming the shape of its inhabiting minotaur, holding and following multiple threads, are negotiating this information maze that art has become. We navigate this volume knowing that at the turn of the century many paths emerge from the matrix where art, technology, and gender intertwine.

We (reader, artist, viewer) cannot expect to remain passive, either in experiencing and immersing ourselves in new-media artworks or in formulating answers and questions about the intersection of new media and gender at the turn of the century—questions such as the dichotomy between the evidence set forth in this book of a strong, influential, central female presence in the field of new media and the continuing male domination of the computer industry.

By its weight, this book makes clear that women's voices are now integral in new-media art practice. Therefore, this book does not need to reinforce the stereotype that women are not interested in technology or to chronicle the changing ratio of female to male voices as the online environment evolves.

Nevertheless, as Jane Yellowlees Douglas has noted, "It is still the norm for commercial interactive infotainments to be male centered. . . . The choice is, essentially, to either go along with the thing and strap on the virtual phallus, or drop out entirely."[4]

"Gender matters," Dawn Mercedes wrote in the online panel *Gender Identity in New Media.* "It is that simple and that complex. Our identities are shaped by rigid patriarchal structures that function as constant barriers to self-definition. Our technologies, in turn, evolve from androcentric, gendered perspectives. Therefore, gender, along with many other contextual factors, is an influential consideration with regard to how artists and viewers perceive, understand, and think about art in the age of new media."[5]

Conversely, in the same online panel, Helen Thorington stated that "For me the question of gender has been fading. . . . Perhaps it's a function of age, perhaps of an environment in which women function far more comfortably and with far more opportunities than they did when I was younger."[6]

Other women viewed the female position in the art and technology environment with wary optimism. "The electronic arts liberated me from the prejudice that I faced when I was a woman painter," Cynthia Rubin wrote. But she added that "In recent years, as the computer has become more accessible to more people, the old prejudices are creeping back in."[7]

Aviva Rahmani commented that "As in race issues, I also see the discrimination as more subtle and sophisticated and therefore harder to fight."[8]

"In my studies of hypertext, I have often wrestled with difference," Carolyn Guertin wrote, addressing the question of the impact of the creator's gender on the work itself. " 'Do women use hypertext differently from men?' is not a question that I have ever been able to answer, nor am I certain that it is a distinction that I want to make. . . . I too could name men who had used similar techniques in most approaches. However, I have found broad-based tendencies born largely of feminist politics and projects."[9]

"Is the tech-art made by women different from men? This is the heavy question," Annick Bureaud emphasized. "The answer is yes and no (not very helpful, I know). No, in the sense that if you list all the women in tech art and their artwork, you will not find a 'common' ground that make those artworks special or different from a gendered point of view." Those works that address feminist or lesbian issues do have a commonality, she notes, ". . . which is already identified as they precisely work into that direction. Yes, if you consider some specific artworks for which you have this strange 'knowledge' that they could not have been done by the other gender."[10]

The complexity and centrality of the issue of gender was noted by Catherine Richards and Nell Tenhaaf in the documentation for the bioapparatus seminar at the Banff Center. They referred to the bioapparatus as a "gendered territory," but rather than creating a category that focused on gender issues, they chose to integrate gender into questions of race, class, and cultural differences: "This idea encompasses gender issues that can be raised within a reexamination of historical and contemporary constructs of nature and culture, mind and body, and machine and spirit."[11]

Similarly, the powerful sense of postcolonial identity that characterizes the Internet at its best pervades both the artwork that this collection of writings describes and also the field as a whole. At the intersection between art and technology, boundaries are emphasized: "This work is about those empty hands that are on the fringe of the Web," Rejane Spitz has written about her Web work *Private Domain.* "It is about those words that cannot be translated, about those emotions that cannot be shared, and those meanings that cannot be understood by people from other cultures. It is about the richness of human beings living in their different realities, with their own systems of ideas, concepts, rules, and meanings."[12]

Boundaries are disrupted: "We have expanded a single gender identity (the feminine) to include a sexual synthesis (the hermaphrodite), which has collapsed into an in-between condition (intersex) that echoes the sexual indifferentiation attributed to angels," Maria Klonaris and Katerina Thomadaki have written about the work *Orlando-Hermaphrodite II.* Their multimedia installation centers around an excerpt from Virginia Woolf's *Orlando* in which the sleeping hero wakes up as a woman: "What should remain hidden makes itself manifest: the feminine as a disrupting force ruining gender order."[13]

And sometimes boundaries are transcended. For instance, in her paper in this book, Carol Stakenas writes that "I have annually coordinated Creative Time's *DWA Web Action* to facilitate the coming together of individuals and organizations to defy the geographical boundaries and unite publicly around the grave impact of the AIDS pandemic."

The following texts offer an array of inviting doors. Although neither a definitive gender definition nor a separation of genders is ultimately desirable, a consideration of gender identity is implicit in the opening and entering.

Judy Malloy

NOTES

1. Caroline Guyer, *Quibbling* (Cambridge, MA: Eastgate Systems, 1993).

2. Brenda Laurel and Rachel Strickland, "Placeholder," in Mary Anne Moser with Douglas MacLeod, eds., *Immersed in Technology, Art and Virtual Environments* (Cambridge, MA: MIT Press, 1996), 297–298.

3. Sue Ellen Case, "Eve's Apple or Women's Narrative Bytes," *MFS Modern Fiction Studies* 43, no. 3 (1997): 631–650.

4. Jane Yellowlees Douglas, "Virtual Intimacy™ and the Male Gaze Cubed: Interacting with Narratives on CD-ROM," *Leonardo* 29, no. 3 (1996): 207–213.

5. Dawn Mercedes, "Keynote," in Judy Malloy, ed., *Gender Identity in New Media,* an online interactive document created as a part of the Invencao Conference, Sao Paulo, Brazil, 1999. Available at ⟨http://www.judymalloy.net/identity/panel.html⟩.

6. Helen Thorington, comments posted in "Open Forum," in Malloy, *Gender Identity.*

7. Cynthia Rubin, comments posted in "Panel 1" in response to statements by Dara Birnbaum and Agnes Hegedüs, in Malloy, *Gender Identity.*

8. Aviva Rahmani, comments posted in response to Anna Couey, "Keynote," in Malloy, *Gender Identity.*

9. Carolyn Guertin, comments posted in "Panel 1" in response to statements by Dara Birnbaum and Agnes Hegedüs, in Malloy, *Gender Identity.*

10. Annick Buread, comments posted in response to Anna Couey, "Keynote," in Malloy, *Gender Identity.*

11. Catherine Richards and Nell Tenhaaf, eds., *Virtual Seminar on the Bioapparatus* (Banff, Alberta: Banff Centre for the Arts, 1991), 8.

12. Rejane Spitz, in Paul Hertz, "Colonial Ventures in Cyberspace," *Leonardo* 30, no. 4 (1997): 249–259.

13. Maria Klonaris and Katerina Thomadaki, "The Feminine, the Hermaphrodite, the Angel: Gender Mutation and Dream Cosmogonies in Multimedia Projection and Installation (1976–1994)," *Leonardo* 29, no. 4 (1996): 273–282.

Introduction

Judy Malloy

> Once in a lifetime, if you are lucky, you may have the opportunity to be present at the birth of something special, something possessing tremendous power. You know that you yourself don't have that power, but you have been vouchsafed the vision, the knowledge that somewhere very close by, somewhere you can actually get to, it lies, still helpless but huge with promise. And no matter how we manage to reinterpret its purpose later, in the pale illumination that hindsight provides, we know the power of the thing is the force that drives our daily work and us ourselves and the whole damn universe. Whether it expresses itself as speciation or linguistic diversity or the restlessness of fashion, we're looking at the same phenomenon: The power of Change. Transformation. Trans.
>
> —ALLUCQUÈRE ROSANNE STONE[1]

At the intersection of art and technology in this time of transformation, it has been a joy to compile a volume that both documents the work of women who have been working innovatively with art and technology for many decades, such as Rebecca Allen, Joan Jonas, Lynn Hershman, Brenda Laurel, Pauline Oliveros, Nancy Paterson, Sonya Rapoport, and Steina, and includes projects and voices now integral in the field, such as Dawn Stoppiello and Mark Coniglio's Troika Ranch dance collaboration; *The Day With(out) Art* telecom project produced by Carol Stakenas, and visual artist Diane Fenster who describes herself as a "modern alchemist who transforms electrical patterns into art."

In this field, there is a curator-driven push for works that integrate the fastest, newest equipment. Yet hardware, despite its deceptively solid presence, is ephemeral. Abandoned IBM ATs and Apple IIIs lie in the corners of salvage barns, the victims of upgrades or industrial support decisions, and although UNIX survives (almost smothered under the weight of commercial interfaces), other excellent operating systems on which seminal works of computer-mediated art were designed to run—Apple DOS and MS-DOS, for instance—have been left for dead or imperfectly emulated by the very institutions that produced them.

No book can fully describe an actual experience of art, but until emulators of superseded systems become widely available, extensive documentation is likely to be the final form of many computer-mediated artworks. Therefore, documentation and inclusive, comprehensive resources are essential. Given the many years needed to produce a volume of this size, this book was not intended to focus on the latest thing. Rather it documents a substantial number of the artists whose work was integral in the formation of the field, while at the same time it looks to the future.

In addition to the foreword by the current Women, Art, and Technology Project manager, *Leonardo* senior editor Pat Bentson, this source volume is organized in three parts. Introductory chapters provide an overview to the history and understanding of the field. Classic papers originally published in *Leonardo* document core work that was created as many as twenty years ago; papers written expressly for this book by women whose work has shaped/is reshaping the field are interspersed with the classic papers. To close the book, five critical essays either compliment the introductory overviews or set the stage for a future new media practice that is diverse and inclusive.

OVERVIEWS

With their individual experience of the work and their knowledge of the field as a whole, critics can significantly increase our artist/scholar/viewer understanding. In the five introductory chapters of this book, critic and curator Patric D. Prince; Margaret Morse, critic and professor of film and digital media at the University of California at Santa Cruz; Sheila Pinkel, artist, associate professor of art at Pomona College; artist and information activist Anna Couey; and artist Kathy Brew, who was director of the New York City initiative ThunderGulch, set the stage for the book's central collection of artists' documentation.

Patric Prince documents the work of Collete Bangert, Lillian Schwartz, Vera Molnar, and other computer art pioneers in "Women and the Search for Visual Intelligence." And she sets forth the evidence of computer art as an artmaking system that requires a "deeply involved understanding of the diverse elements connected to the process and the intellectual ability to control a complex procedure."

Describing the work of Christine Tamblyn, Lynn Hershman, Sara Roberts, Sonya Rapoport, and Catherine Richards, among others, in "The Poetics of Interactivity," Margaret Morse portrays a deep and persistent interactivity—"expressed not only in art, but ubiquitously in every sphere of contemporary life where chips reside, from automatic tellers and garage-door openers to computers that access discs, CD-ROMs and the World Wide Web."

In "Women, Body, Earth" Sheila Pinkel documents artists who have focused on some aspect of body or environmental representation in their work, such as Kim Abeles, Linda Benglis, Coco Fusco, Suzanne Lacy, and Cindy Sherman. "The challenge for women," she concludes, "is to continue this intensity of activity, to remain true to the interior voice that gives veracity and energy to art making, to continue to lobby for parity, and to generate innovative solutions for exhibition and change."

In "Restructuring Power: Telecommunications Works Produced by Women" Anna Couey, whose own works have involved the construction and use of communication systems as social sculpture, emphasizes the tremendous influence that communications technologies have had "in shaping individual and cultural perceptions across the planet." Through interviews with early women network artists including Lucia Grossberger Morales, Karen O'Rourke, Sherrie Rabinowitz, and Lorri Ann Two Bulls, Couey highlights and preserves an important history.

In "Through the Looking Glass" Kathy Brew, who for over twenty years has followed the intersection of multimedia and contemporary art—where as she says "the journey is the destination"—reviews the work of artists working in multimedia, including Maryanne Amacher, Laurie Anderson, Toni Dove, Beryl Korot, and Mary Lucier.

ARTISTS' PAPERS

For several decades, artist-produced documentation has been with some regularity the final form of performance and installation and other conceptually rooted art forms. The artist's paper—a hybrid of conceptual art documentation and of the technical paper—is also at the core of *Leonardo*'s publishing practice, and it has become an integral component of new media practice.

In recognition both of *Leonardo*'s integral role in hosting the artist-written paper and of the women who blazed trails in this field, the artists' papers in this book include classic papers by pioneer video artist Steina, musician-composer Pauline Oliveros, environmental artists Helen Mayer Harrison and Newton Harrison; and three artists—Sonya Rapoport, Lynn Hershman, and Nancy Paterson—whose work and ideas in interactive installation, developed over several decades, have been widely influential in the field.

Interspersed with the classic *Leonardo* papers are artists' articles that document the female creator's role in shaping new media activity at the turn of the century and that were written expressly for this book. In these chapters, twenty-six artists describe work that has shaped new media in the (sometimes intersecting) fields of video art, environmental art, computer graphics, interactive art, sound, computer music, and dance. The order of the chapters is not chronological but rather reflects the intertwining of ideas and structure in this interdisciplinary field.

The artists' papers begin with Steina's "My Love Affair with Art: Video and Installation Work," in which she describes how her current multichannel video compositions evolved from the early days of video in the late 1960s. "As soon as I had a video camera in my hand—as soon as I had that majestic flow of time in my control—I knew I had my medium," she states.

In "Transmission" Joan Jonas weaves together performance, myth, her ground-breaking video techniques, sound, time, and her evocative mirror imagery—detailing one artist's life-long, art-making quest. "The fact is, the videos still dance and make music," she concludes.

Dara Birnbaum's "The Individual Voice as a Political Voice: Critiquing and Challenging the Authority of Media" describes works such as her multimonitor *Tiananmen Square: Break-In Transmission* and documents her exploration of the fragmented relationship between viewer and televised news.

In "Small Leaps to Ascend the Apple Tree" Jo Hanson describes the making of her seminal 1974 multimedia installation *Crab Orchard Cemetery.*

Chapters 10 to 14 are classic papers, originally published in *Leonardo.* They document the work of five artists whose seminal work in the development of contemporary new media practice continues to creatively evolve.

In "Shifting Positions toward the Earth: Art and Environmental Awareness," among other works, Helen Mayer Harrison and Newton Harrison describe their *Lagoon Cycle* that takes endangered estuarial lagoons as a starting point. They state, "The *Lagoon Cycle* can be read as a story in seven parts; each part, as in a picaresque novel, is its own story. It can be read as an array of storyboards for a very unusual movie. As artists we see it as an environmental narrative, one of whose properties is to envelop the reader with its form and subject matter."

Sonya Rapoport describes her intellectual, information-based, interactive art works—*Digital Mudra,* for instance, where the mudra words you select become "a gesture-dance sequence on the monitor," and, as you watch, the printer embodies them. In describing the deep interaction at the heart of this work, she states that "the various responses of the participants created the media, i.e., the 'paint' to be mixed, manipulated, and applied with the use of the computer 'brush'—which developed into a grand finale of an integrated interactive artwork."

In Lynn Hershman's film, *Conceiving Ada,* Ada (Augusta Ada Byron, Countess of Lovelace, the first programmer in history) moves around the film in a digital environment that could not have existed without her original inventions. Hershman's paper in this book, "Touch-Sensitivity and Other Forms of Subversion: Interactive Artwork," traces the origins of her own interactive work, including the interactive videodisk *Lorna* and the computer-mediated installation *A Room of One's Own.*

Nancy Paterson documents *Bicycle TV,* an interactive installation that uses real-time, full motion, prerecorded video imagery "to place the viewer in an authentic rather than digital landscape." In her chapter, Paterson observes that the experience of producing and exhibiting *Bicycle TV* resulted in her greater awareness "of the issues and questions that will become increasingly relevant as more people are directly exposed to virtual-reality environments." She emphasizes the role of artists in not only exploring the creative potential of virtual-reality systems, but also in addressing issues of physical control and psychological manipulation in electronic applications. "If our greatest challenge thus far has been the development of the technology to make these systems possible, then our next challenge will relate to the use and commercial application of these systems," she writes.

In "Acoustic and Virtual Space as a Dynamic Element of Music" composer-musician Pauline Oliveros describes how her experiences as a performer and composer relate to the development of her interest in acoustics and technology. "Deep listening is listening in every possible way to everything possible: this means one hears all sounds, no matter what one is doing. Such intense listening includes hearing the sounds of daily life, of nature, and of one's own thoughts, as well as musical sounds. Deep listening is my life practice," she states.

In chapters 15 to 31—beginning with artist and designer Rebecca Allen's "passionate relationship with innovative technology"—seventeen artists whose work is situated in and merges computer graphics, animation, interactive art, virtual reality, communications, Internet applications, biotechnology, music, and dance—describe their work.

Interviewed by Erkki Huhtamo, Rebecca Allen discusses the evolution of her work in computer graphics and animation.

Donna Cox details an "evolving dialog with the universe" in "Algorithmic Art, Scientific Visualization and Teleimmersion."

Agnes Hegedüs documents the approaches to interactivity in her installation art, which "aims to configure a social dimension for the new digitally visual spaces of communication."

Judith Barry probes the role of technology in her installation work. "This is not a linear description of my work but more an attempt to delineate some of the questions that have interested me," she writes. "These include notions about the city and its architecture, public and private space, rural and agrarian landscapes, first- and third-world relations of capital, the flow of information and its capital accumulation. This list is not programmatic but reflects the way I have thought about how contemporary experiences and their representations might be interrogated."

Jennifer Hall and Blyth Hazen document computer-mediated work, such as Hall's *Acupuncture for Temporal Fruit*, and they describe the community-based Do While Studio laboratory.

Brenda Laurel tells the story of Purple Moon, the software company she founded: "I consider our work to have been a cultural success in the sense that it touched the lives of millions of girls and gave them fresh views of girlhood and a new portal into technology. But our business failed."

In "Imagine a Space Filled with Data" German artists and researchers Monika Fleischmann and Wolfgang Strauss set forth their collaboratively shaped communications spaces. "The old metaphors of explorer or navigator have remained the same: the language of the conqueror is apparent to everyone on the Internet," they write. "As artists, we explore aesthetic strategies with communication processes to influence and transform the development of the community of the future."

Char Davies documents her immersive and virtual-reality environments which are grounded in nature as metaphor.

French composer/musician Cécile Le Prado writes about her sound installations, such as *The Triangle of Uncertainty:* "In all these examples, we are no longer in a concert situation or in a frontal relationship to the music. Most of the time I try to immerse the listener in the middle of the sound."

Pamela Z eloquently describes her own work and the contemporary environment for women composers and musicians.

Canadian artist and theorist Nell Tenhaaf documents her art-based examinations of genetic engineering and artificial life in works such as *Apparatus for Self-organization* and *Orphaned Life-form.*

Interlaced with descriptions of her installation *Your Words, My Silent Mouth.* Allucquère Rosanne Stone examines the field as a whole. Of the task ahead, she writes: "That task, if we choose it, is to create that physical base for our work, to reach deep into the institutions that house us and change them at their deep hearts. Such change is effected not by rhetoric, not by revolution. In the postmodern prospect such change is accomplished by viruslike structures carrying information that changes the way the organism's cells self-replicate, so that over time the organism ineluctably incorporates that new information: the changed becomes the changer. In this method, there is no battle, no confrontation. Instead, change simply is."

In "Video Arte Povera: Lo-Fi Rules!" Valerie Soe emphasizes the integration of low-tech materials and social content in her works such as *Mixed Blood* and *Picturing Oriental Girls.*

In "Face Settings: An International Co-cooking and Communication Project" Kathy Rae Huffman details her work with Eva Wohlgemuth, a collaborative communications project that integrated communications issues and real-time food preparation and consumption.

With Celia Rabinovitch, Diane Fenster describes her use of digital tools "to depict a new poetry, a multidimensional world of images."

"Pigs, Barrels, and Obstinate Thrummers" by choreographer-dancer Linda Austin and composer-performer-instrument builder Leslie Ross details their evocative and collaborative uses of interactive objects and mechanical devices in live performance.

To close the artists' papers, Dawn Stopiello with Mark Coniglio describe the evolution of their computer-mediated dance collaboration that operates in their words: "At the intersection of flesh and silicon, blood and television, body and computer that our culture is in the midst of splicing together."

CONCLUDING ESSAYS

Beginning with "Embodiment and Narrative Performance," in which Jaishree Odin examines Shelley Jackson's hyperfiction *Patchwork Girl,* the book closes with a series of critical essays that complement the introductory overviews. "In politically informed narratives, hypertextual strategies are indispensable for exploring the issues of embodiment and reinscription of cultural or literary history that are pivotal in the works of women and minority writers," Odin emphasizes.

Simone Osthoff's "Women and Media Arts in Brazil" documents the pioneering work of Jocy de Oliveira, Marcia X, Lygia Pape, and many others.

Mexican critic Martha Burkle Bonecchi eloquently and persuasively reminds us that technology has "forgotten" many women in the developing world.

Carol Stakenas, who annually coordinated Creative Time's *Day With(out) Art Web Action,* describes this collaborative network project, which had at its heart "the coming together of individuals and organizations to defy the geographical boundaries and unite publicly around the grave impact of the AIDS pandemic."

To conclude the book, in the classic *Leonardo* paper "Contested Zones: Futurity and Technological Art," educator/researcher Zoë Sofia sets forth the work of Australian artists VNA Matrix, Linda Dement, and many others, stating, "I hope this chapter, by focusing interest on works by contemporary women technological artists, contributes to their success in shaping alternative futures that do not simply intensify the powers of the already strong but enlarge the influence of the values and interests of those not satisfied by the pursuit of the new as a good—or even a god—in itself."

AND THERE ARE MANY OTHERS

In one book and one associated Web site, it has been impossible to include all the women artists who are working in new media, a fact that is a testament to our central role in this field.

The work of many other women who are key players in the field—Maria Blondeel; Red Burns; M. D. Coverly (Marjorie Luesebrink); Abbe Don; Marjorie Franklin; Jacalyn Lopez Garcia; JoAnn Gillerman; Reiko Goto; Carolyn Guertin; Carolyn Guyer; Tina LaPorte; Deena Larson; Muriel Magenta; Ann Powers; Aviva Rahmani; Beverly Reiser; Cynthia Beth Rubin; Anne-Marie Schleiner; Jill Scott; Helen Thorington; Victoria Vesna—to name just a few, has been documented in *Leonardo* and/or will be documented in the Web site associated with this book. These artists are no less important than those included in this book. In many cases, their papers were published fairly recently in *Leonardo,* and in the book it seemed important to include previously unpublished papers as well as seminal, classic papers.

In the Web site associated with this book, an opportunity exists for continuing documentation. As Cathy Marshall wrote in a paper about our collaborative hypernarrative, *Forward Anywhere*, "Closure was never a goal of this piece."[2]

NOTES

1. Allucquère Rosanne Stone, "*Your Words, My Silent Mouth: Trying to Make Sense out of Nonnarrative Work* (A Brief Collection of Interviews and Polemics in the Interest of Aesthetic Quasiclarity)" (chapter 26 in this volume).

2. Judy Malloy and Cathy Marshall, "Forward Anywhere," in Lynn Cherny and Beth Weise (eds.), *Wired Women* (Seattle, WA: Seal Press, 1996), 56–70.

I

Overviews

1

Women and the Search for Visual Intelligence

Patric D. Prince

When I was first introduced to computer art in 1980, the most interesting aspect was the relationship between art and the obvious intelligence of the artist. In addition to formulating the art idea, artists using computer technology had to have the means to take it to fruition. I first appreciated the concept of visual intelligence after talking to Frank Dietrich and reading his respected early history of the movement.[1] The status of artists in the West has changed from servant or artisan to member of the intellectual elite (if Tom Wolfe is to be believed). Although artists proclaim that the creative process is a vigorous intellectual exercise and should be recognized as such, there is resistance to this belief. Nonetheless, here was an art-making system that required a deeply involved understanding of the diverse elements connected to the process and the intellectual ability to control a complex procedure. The use of a single "medium tool idea-spawner inspirational-focus" was compelling and exciting. There seemed to be endless possibilities for new forms of artistic expression. What began as picture storytelling in caves has evolved into elaborate conceptual experiences. Data visualization has expanded our understanding of the world and, coupled with new methods of communication, makes amazing changes in the way we deliver and receive information.

Women have participated in the computer art and technology movement from the first decade, learning to speak the language of the machine as well as enjoying the implementation of ideas, techniques, and experiences derived from its inherent logical involvement. Early prominent practitioners include Colette Bangert, who works with her husband Jeff with algorithmic processes; Vera Molnar, who works in heuristic techniques; Jasia Reichardt, who works as a critical thinker and curatorial innovator; and Lillian Schwartz, who is associated with appropriation art and pictorial analysis. These pioneer women contended with concepts and aesthetic challenges that have become seminal and are the foundation of current digital work.

IN AT THE BEGINNING

This adventure begins in the mid-1960s, when Jasia Reichardt sought to show the world the new directions that art was taking. She was a curator at the Institute of Contemporary Arts (ICA) in London and had just opened an exhibition on concrete poetry. According to Reichardt,

> The idea for *Cybernetic Serendipity* was born in the autumn of 1965 at the opening of another exhibition at the old ICA that I had organized, called *Between Poetry and Painting.* It was London's first international exhibition of concrete and visual poetry, and several people came from abroad to celebrate the event, including a group from Stuttgart. One of them, a great supporter of concrete poetry, . . . was Professor Max Bense. . . . [He] was a philosopher with an interest in generative aesthetics and a taste for intellectual adventures. . . . Sometime during the next few days, he asked me what I was going to do next, and I admitted that I didn't know. "Look into computers," he suggested, and my work, on what was to become *Cybernetic Serendipity,* started the following day.[2]

Immediately after meeting with Max Bense, Reichardt began work on the seminal computer exhibition *Cybernetic Serendipity,* which opened in London in 1968 (figure 1.1). It was the first important exhibition of digital art to be held since the initial exhibitions of 1965[3] and included 325 participants, whose work was viewed by 60,000 visitors. She writes that the exhibition "dealt with the relationship of the computer and the arts. It was concerned with the exploration and demonstration of connections between creativity and technology (and cybernetics in particular)—the links between scientific or mathematical approaches, intuition, and the more irrational and oblique urges associated with the making of music, art, and poetry."[4] Her selection of artists, performers, and musicians was acute: many have continued to captivate contemporary art, engaging viewers in further investigations of the mind. The catalog from the exhibition has become, with a few notable exceptions, a guidebook for most of the prominent art and technology work created to that time.

As curator, Reichardt discovered significant factors in conjunction with this early exhibition of art and technology, one of which was that each and every artwork needed to be explained to the general public. This is an aspect of mounting new-media exhibitions that has not changed substantially in the last thirty years. Even though the viewer has truly become a participant in the art-making process, explanations are necessary to overcome the reluctance and lack of experience that still exist, as well as to mediate newer aesthetic modes inherent in new experiences.

Reichardt was at the forefront of understanding the need to redefine just who was making the art experience. Was it the artist, the technologist, the viewer participant, or a combination of all three? She states that early exhibi-

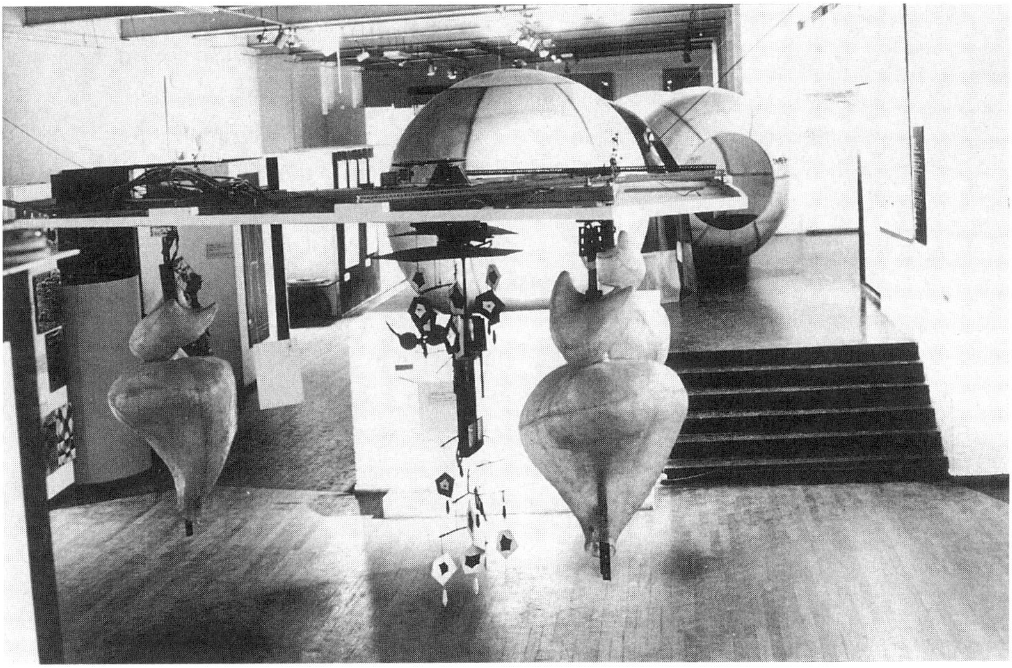

Figure 1.1
Jasia Reichardt, *Cybernetic Serendipity,* 1968, Institute of Contemporary Arts, London. The exhibition was designed by Franciszka Themerson. The site photo includes, in the foreground, Gordon Pask's *The Colloquy of Mobiles* and, behind them, the fiberglass spheres in which people could sit and listen to computer-composed and computer-generated music.

tors made no distinction between artists using science and scientists expressing art ideas. The crossover artist scientist has become part of the arena of digital art. Today, artists can choose to work in an intuitive mode; current technology allows for direct interaction. The early artists had no options; they had to learn to work directly with the language of the computer.

Reichardt has authored several books and articles. A recent curatorial effort situated in Japan, titled *electronically yours,* was about electronic portraiture. From *Cybernetic Serendipity,* the landmark curatorial experiment, computers in the arts became known to a wider population and led eventually to the interest in art and technology today.

Research facilities like Bell Labs in Murray Hill, New Jersey, had the means to support visual research. One of the best-known women working in art and technology from the 1960s is Lillian Schwartz. She went to Bell Labs in 1968 as one of a number of digital pioneers who were experimenting with creative processes and the computer. Her early work includes collaboration with Ken Knowlton, author of BEFLIX, one of the original animation programming languages. Over the years, she has worked in partnership with distinguished mathematicians and scientists to produce a remarkable range of digital artistic production, including film and animation, sculpture, and two- and three-dimensional images. For her work in film and animation, Schwartz has earned acclaim and many awards.

More recently, her work in analyzing pictorial data as well as her appropriation art has brought her worldwide recognition.[5] Schwartz's appreciation of the advantages and uncertainties associated with digital processing is vital to understanding the significance of her works such as *Big MOMA* (1984), in which she mapped images of works from the Museum of Modern Art in New York onto a simulation of a large female sculpture (also from the collection). The compilation of images as well as the placement of the works create an ironic statement that has become an influential icon in the history of digital art. Picture processing and scanning techniques led to the movement known as appropriation art, which incorporates works of art into new works and redefines the interpretation of the ownership of images: "Creation is a process that begins with the initial instinct to express oneself and finishes in the attainment of the end result. The need to create can be triggered from within oneself, from nature, from cultural surroundings, or from other works of art."[6] However, the issue of copyrights associated with digital appropriation is important today and has yet to be resolved.

Collette Bangert and her husband, Jeff, have had an unusual creative collaboration since the 1960s. They both have art degrees and "share form preferences." As a team, they concluded early in their careers that to produce credible works they would have to rethink their methodology. Instead of starting with the composition of a work, they have learned to start with the mark itself. Jeff, a mathematician as well as an artist, examines the aesthetic properties of algorithms. When they find something they believe will produce an interesting mark, they then proceed with the composition of the drawing. Collette and Jeff have been working in the digital realm since 1967, when they both participated in an early programming seminar. Their first work, *Spiral Coils* (1967), is a plotted array of apparently randomly placed spiral forms, ink on paper. They live in Kansas, the geographical heartland of the United States, where the nature of the prairie has a visual impact on all forms of Colette's and Jeff's work. Her paintings reflect the aesthetic elements that are found in her digital drawings; she says that all of her work is "symbiotic."

Another interesting aspect of the Bangerts' work revolves around the issue of support. They have owned and housed their own systems since their earliest digital production. Many of the other pioneers in computer graphics and animation have had the opportunity to experiment, from time to time, on much larger systems. As a result of the limitations of their equipment, the Bangerts have dealt with questions of scale and productivity from the beginning. To compensate for the limited range of paper sizes, many early artists transferred their output to film and then made either photographic reproductions or screenprints. The Bangerts have always treated their work as single originals. Each plotter drawing is unique. Based on abstract concepts gleaned from mathematics, their work has evolved in the last thirty years to a high degree of visual substance while maintaining a fluid graphic technique: "Computer art is in the process of humanizing our experience of technology."[7]

Colette also contributed to the field by organizing numerous art exhibits over the years. In the first decade, artists supported each other at conferences and exhibitions, and she strives to improve the perception and understanding of the work, reexamining the nature of their art through dialogs and statements: "Our discussion of our work changes. We have at times focused our interpretive thoughts on the computer, but we have said that our medium is software and math and geometry." Collette and

Jeff Bangert believe that "Algorithmic processes have contributed to our aesthetic environment and are a basic element in all digital work."[8]

Vera Molnar is the most significant European woman artist who used digital technology in the 1960s. Her art and her papers are essential to the development of digital art. She was born in Budapest in 1924 and is now living in Paris. By the time she first used a computer in 1968, she had already been working on abstract geometric images for a decade. In Paris in 1960, she cofounded Groupe de Recherche d'Art Visuel (GRAV), a group dedicated to understanding mathematical simulation and aesthetics. "After ten years of 'Machine Imaginaire' (1959–1968), I started to work with a real computer in 1968."[9] Molnar worked with a series of computers and plotters in the Centre de Calcul of the University of Paris–Sorbonne in Orsay (figure 1.2). Her style is that of a constructivist, interpreting mathematically based abstract forms that explore aesthetics and perception.

Molnar's methodology is focused on an understanding of heuristic problem-solving techniques. The most appropriate solution is selected at successive stages of discovery for use in the next step. She states that she works in a "series of small probing steps" and that each step is followed by an evaluation. By varying only one parameter at each step and "by comparing the successive picture," she has "control over the stages of development."[10] Molnar was fascinated by the process of going back to former versions of her work. She programmed in FORTRAN and later in BASIC. Recently, eye problems have led her to collaborate with a programmer. Molnar's early recognition of the computer's ability to save and rework artistic research is crucial to the history of digital art and one of the important elements of contemporary graphics. Artists need never fear change. All stages of development of work can be realized.

Freder Nake, one of the original crossover artist scientists to exhibit his computer-generated art, says that Vera Molnar was "exceptional."[11] Her precise creative abilities have emerged to become a lyrical form of geometric expression.

THE NEXT WAVE

Recognizing the potential of new media and new aesthetic experiences, numerous artists became interested in the movement during the next decade, the 1970s. Some prominent women artists from that period in-

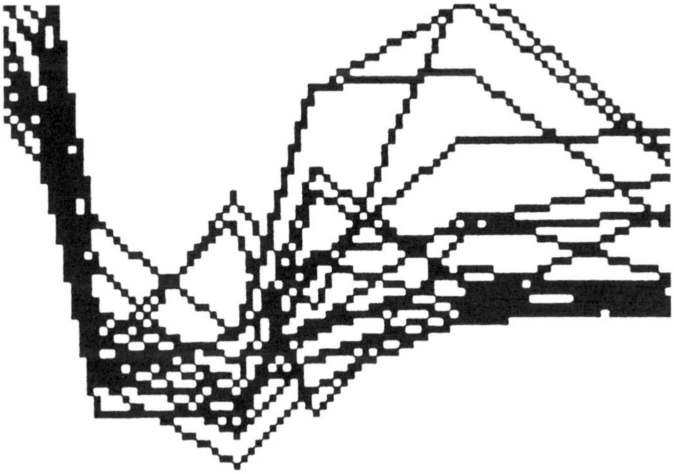

Figure 1.2
Vera Molnar, plotter drawing from the series *Variations Ste. Victoire,* 1989–1997, ink on paper: "I develop a picture by means of a series of small probing steps, and each step is followed by evaluation. In my opinion, painters should employ such a procedure, especially if they consciously wish to learn what of aesthetic importance is occurring on the canvas as the painting develops and what effect the work may have on viewers."

clude Rebecca Allen, Eudice Feder, Darcy Gerberg, Copper Giloth, Ruth Leavitt, Nadia Magnernat-Thalmann, Barbara Nessim, Sonia Landy Sheridan, Vibeke Sorensen, Joan Truckenbrod, and Jane Veeder, as well as many others represented in this work and in the online Web installation. Women were also involved in the organization of exhibitions and of computer art publishing.

Ruth Leavitt added to the literature.[12] Grace C. Hertlein curated art exhibits and founded the publication *Computer Graphics and Art* in the mid-1970s.

Examples of artists who contributed early to the movement are Vibeke Sorensen and Joan Truckenbrod. Joan Truckenbrod is one of the early adopters of the computer; she has been using electronic imagery for many years to produce a body of work centered on conveying connections within society. In her paper "Tears in the Connective Tissue" (available on the Web site associated with this book), she describes her body of work as "electronic totemism" and states that she uses "different modes of digital representation simultaneously." Truckenbrod's recent installation, titled *Torn Touch,* introduces tactile experience to produce a symbolic reality. She developed her work using a range of media from experimental large-scale two-dimensional images to interactive environments.

Vibeke Sorensen began using analog devices in her video production in 1971; she made her first digital printout in 1976. Her animations have developed into a passion for stereoscopic imagery. In addition to realistic forms, Vibeke employs her "fundamental understanding" of images and the computer to create fluid, nonobjective works that evoke internal elements that are reflected "externally." Sorensen says that she "analyzes and deconstructs how images work." Many of her three-dimensional images are designed to be animated. Actually, any image that is fully realized in the computer's memory is but a captured moment in time. Sorensen's most recent work comprises the creation of performance, music, two- and three-dimensional elements, and digital video elements in a single programming language.

ELECTRONIC CURRENTS

When critics explain that this art is all about the technology, perhaps they have forgotten that until recently (about a hundred years ago), when paint became available in tubes that could be carried outside, art making was all about the technology then available. In the medieval era, artists were hired for a particular commission because of their ability to produce certain colors and not explicitly for their style. Tempera was developed as a medium because of its availability as well as for its color-fixing properties. Oils on canvas were used and proliferated because Venetian painters couldn't employ fresco in the damp atmosphere of the Adriatic environment. The list of technology examples is endless.

Digital artists have taken different paths to create art forms; some have learned to program, building their own special voices. Others have worked with collaborators, allowing themselves time and freedom to refine alternative aspects of their work. Many of the initial works were graphical in nature because of the limitations of the devices that could translate the data into art media. We now see an end to restrictions that focused aesthetic choices on the machinelike qualities seen in the earliest works. Of course, as a graphics pioneer once told me, artists were so delighted that the technology functioned that in the early days specialists did not deeply question the aesthetics. At the exhibition titled *WYSIWYG,* the curator was challenged by several attendees to explain where computer processes came into it.[13] Computer technologies and graphic techniques have been assimilated into artistic production so that the technological aspects can either be invisible or present algorithmic foundations. What began in the early 1960s as experiments in graphical picture making and interactivity has evolved into a total art experience.

Artists have invented their own environments and sensory devices using the technology to respond to specific challenges. The interactive experience divides into four categories: active (each viewer is offered the same experience), interactive (viewers may have different experiences depending on the intervening choices they make), reactive (the presence of the viewer changes the art work), and immersive (through various devices, the viewer enters another environment).

Christa Sommerer and Laurent Mignonneau work in a reactive environment, "focusing on real-time interaction and evolutionary image processes." Their artwork is designed to be reactive and to change as the viewer comes in contact with the art. They developed the means to eliminate the mechanical devices that dominate many interactive experiences. Their revolutionary work *Interactive Plant Growing* (1992–1993), was the beginning of experimental immersive artistic production. This work was both reactive and immersive. Their newer work uses their interest in biology, artificial evolution, and interaction to produce imaginative new forms: "The artists in such interactive computer installations are only now setting the frame within which the artwork is developing itself through the interaction of the visitors and the evolutionary image process."[14] Their work presents the viewer participant with choices that relate to understanding the aesthetic potential of interactivity with human action and perception.

Another innovative artist using immersive techniques is Agnes Hegedüs. She has an unusual view of the world that involves imagination and an understanding of personal space and perception. Her work includes CD-ROM production, immersive experiences, and virtual-reality (VR) environments. She describes it as involving "chance encounters and potential relationships." Hegedüs also states, "Unlike the journeys of Alice in Wonderland, who had to change her size to accommodate the worlds that she entered, here the travel in a virtual world becomes an effortless journey by means of a simultaneous displacement of the viewer's body in both the physical and virtual dimensions." She is currently working on a VR experience involving more than one viewer participant interacting in several levels of "worlds."

REAL SYNTHETICS

As this technology matures, the emphasis changes from technical inspiration and experimental innovation to an exploration into the nature of meaning as expressed through newer experiences. Barbara Nessim first used computers in 1980. She states, "Content reveals itself on both intellectual and intuitive levels. It is emotional, delving into something deep but going out in many directions, not unlike the architecture of the Internet."[15] Nessim, like pioneer Chuck Csuri, is one of the rare contemporary artists to work consistently with elegant graphical figures. Her spare figurative style is visually powerful and always related to the narrative. Nessim has consistently explored the nature of relationships, both universal and intimate. She is experimenting with new media and is producing compounded interactive artworks that combine tactile and information elements. "Artists always need to focus on the content," says Paras Kaul, who works in virtual reality. "If the changing nature of the software doesn't limit you, it may be possible to go deeper into meaning."[16]

In an article in *Leonardo,* Jennifer Steinkamp states, "One of my greatest challenges is to create a work where complex ideas can be best experienced as works of art."[17]

TERMINAL THOUGHTS

Art making with technology will continue to expand our ability to experience ideas. I see the movement as an informed approach to artistic appreciation. The complexity of the involvement with the technology is only one factor in the art of the future. A hundred years from now, the art-historical contributions from women in computer art to a broader aesthetic experience will include algorithmic and heuristic processes, interaction, and telematic art. The media for these future experiences will change with the times, as always. When I asked Masao Komura, one of the original members of the early Japanese art group, CTG, where he thought the technology would take us, he responded, "It will become our third skin"—the first being our flesh, the second our protective clothing, and the third the new technology.[18] We will adopt whatever technology we choose to develop, incorporating it further into our lives and making it possible to communicate in newer intelligent terms.

NOTES

1. Frank Dietrich, "Visual Intelligence: The First Decade of Computer Art (1965–75)," *IEEE Computer Graphics and Applications,* 5, no. 7 (1985): 34–35.

2. Jasia Reichardt, letter and manuscript sent to the author, 13 July 1998.

3. The first exhibitions of computer-generated art took place in Germany in 1965 with the work of Georg Nees. Later in that same year, in New York City, work by A. Michael Noll and Bela Julesz was shown at the Howard Wise Gallery.

4. Jasia Reichardt, letter and manuscript sent to the author, 13 July 1998.

5. Lillian F. Schwartz, "The Mona Lisa Identification: Evidence from a computer analysis," *Visual Computer,* 5 (1988): 40–48.

6. Lillian F. Schwartz, "Computers and Appropriation Art: The Transformation of a Work or Idea for a New Creation," *Leonardo,* 29, no. 1 (1996): 48.

7. Statement written for the Di Paola/Ercius Gallery Exhibition, 1988, 1–3.

8. Ibid.

9. Vera Molnar, letter to the author, 20 July 1998.

10. Vera Molnar, "Toward Aesthetic Guidelines for Painting with the Aid of a Computer," *Leonardo,* 5, no. 8 (1975): 185.

11. Freder Nake, e-mail to author, 28 August, 1998.

12. Ruth Leavitt, *Artist and Computer* (New York: Harmony Books, 1976).

13. *WYSIWYG (What You See Is What You Get),* exhibition curated by Tony Longson, Fine Art Gallery, California State University, Los Angeles, May 1999.

14. Christa Sommerer with Laurent Mignonneau, "Art as a Living System: Interactive Computer Artworks," *Leonardo,* 32, no. 3 (1999): 165–173.

15. Barbara Nessim, telephone interview with the author, 12 September 1998. See ⟨http://www//barbaranessim.com/⟩.

16. Paras Kaul, neuro-art researcher, telephone interview with the author, 9 September 1998. See ⟨http://www.paraswest.com⟩.

17. Jennifer Steinkamp, "My Only Sunshine: Installation Art Experiments with Light, Space, Sound, and Motion," *Leonardo,* 34, no. 2 (2001): 109–112.

18. Masao Komura, founder of Computer Technique Group (CTG), 1965–1968, interview with the author in Tokyo, Japan, May 1995. For his new work, see his Web page ⟨http://www.ntticc.or.jp/special/babel/profile_e.html⟩.

REFERENCES

Bangert, C., and J. Bangert, eds. "Computer Art: Hardware and Software vs. Aesthetics." In Proceedings of the Seventh National Sculpture Conference, April 27, 28, 29, 1972, 184–189.

Bangert, C., and Charles Bangert. "Experiences in Making Drawings by Computer and by Hand." *Leonardo,* 7, no. 4 (1974): 289–296.

Benthall, J. "Anatomy of an Anomaly." *Technology and Art, Studio,* 182 (1971): 63–64.

Dietrich, F. "Visual Intelligence: The First Decade of Computer Art (1965–75)." *Leonardo,* 19, no. 2 (1986): 159–169.

Franke, H. W. *Computer Graphics, Computer Art.* Berlin: Springer-Verlag, 1971.

Goodman, C. *Digital Visions: Computers and Art.* New York: Abrams, Everson Museum of Art, 1987.

Handhardt, J. G., ed. *Video Culture: A Critical Investigation.* Salt Lake City, UT: Peregrine Smith Books, 1986.

Lovejoy, M., ed. *Postmodern Currents: Art and Artists in the Age of Electronic Media.* Ann Arbor, MI: UMI Press, 1989.

Molnar, V. "My Mother's Letters: Simulation by Computer." *Leonardo,* 28, no. 3 (1995): 167–170.

Molnar, V. "Toward Aesthetic Guidelines for Paintings with the Aid of a Computer." *Leonardo,* 8, no. 3 (1975): 185–189.

Reichardt, J., ed. *Cybernetic Serendipity: The Computer and the Arts, Studio International* special issue (1968).

Schwartz, L. F. "Computers and Appropriation Art: The Transformation of a Work or Idea for a New Creation." *Leonardo,* 29, no. 1 (1996): 43–49.

Schwartz, L. F. "The Staging of Leonardo's Last Supper: A Computer-Based Exploration of Its Perspective." *Leonardo,* Electronic art supplemental issue (1988): 89–96.

Wolfe, Tom. *The Painted Word.* New York: Farrar, Straus and Giroux, 1975.

Youngblood, G. *Expanded Cinema.* New York: Dutton, 1970.

2

The Poetics of Interactivity

Margaret Morse

A HOPELESS TASK?

Interactivity once was a useful term for distinguishing art that has been influenced and shaped by a media-saturated and computerized contemporary world from painting and sculpture. However, as Marjorie Franklin said in an interview, *interactivity* now means *too* many things. It does not comprise a genre or even many genres of art. Rather, it identifies a mode of engagement between ourselves and machines—usually but not necessarily involving communicating with a computer—that finds expression over a wide range of forms and techniques. It is expressed not only in art but ubiquitously in every sphere of contemporary life where chips reside, from automatic tellers and garage-door openers to computers that access discs, CD-ROMs, and the World Wide Web.[1] Even traditional art forms are now displayed and presented "interactively" in ways that address the gallery visitor via audio or computer, offering information at the visitor's own pace at the click of a button. Adding further to the confusion, the critical discourse on "interactivity" is ideologically loaded, even schizophrenic in its tension between pejorative connotations and utopian values and expectations. Received notions extend polarized, normative criteria for evaluating interactive art to the critic even before we as a culture are quite sure what possibilities, functions, and aesthetics could be or have been realized in such work—or, for that matter, not realized.

We have to go back in time to a fundamental break in culture that occurred in the late 1960s and the 1970s to see *interactivity* as a cultural novum. An egalitarian impetus opposed one-way and hierarchical relations in society at large. In the arts, the proscenium between performers and the public was lowered, sculptures descended to floor level, and images exited the frame and entered into everyday life. In conceptual, pop, performance, body, and video art, artists explored the ephemeral and shifting experience of the here and now. With the goal of vacating their privileged relation as authors and creators, artists invited spectators to become *participants* in art events, from happenings to closed-circuit and recorded video installations.[2] The liberatory associations of interactivity with mutuality and reciprocity owe much to the presentational and participatory arts of this era.[3]

This was also a period of struggle by women and minorities for entry into and validation by the art world. Beyond their advantages as a means of expression, new art forms without set conventions and entrenched practitioners also promised new opportunities for women. Despite revolutionary shifts in technology, there are continuities of person, themes, and practices between the live and analog art in the earlier period of media art and the digital interactivity of the 1990s to be discussed below.[4] Both second- and third-wave women in technology art have faced the difficulties of pioneers and transgressors in ideological and practical domains defined and controlled by men. Women encountering strong external obstacles and constraints are also likely to have internalized resistances and ambivalencies toward their own artistic means of production. Such psychic distancing is advantageous, for these tricksters and thieves of cool fire are more likely to produce *metainteractive* work that foregrounds the contradictions and mystifications of interactivity itself, while devising ways to animate its liberating potential.

Whatever it may be in the larger socioeconomic and cultural sphere, artists have chosen to inflect prosaic interactivity to their own expressive ends. Metainteractive aesthetic strategies—like poetry, with its rhythms, assonances, and figures—does not merely transport us to another scene or world but is itself an experience charged by semantic and formal values of expression. Interactivity is not just an instrument or a perhaps irritating interval between clicking and getting somewhere else but an event that brings corporeal and cognitive awareness to this increasingly ubiquitous feature of the contemporary world.

DEFINING INTERACTIVITY NEVERTHELESS

Reception theory tells us that the reader of a novel and the theater- or filmgoer have always *cognitively* "interacted" with the text by filling in the gaps.[5] Audience studies tell us how fans of mass-culture print, sound recordings, television, and radio have actively received, revised, and extended texts without, however, changing the text itself in real time. However, the interactive user/viewer *corporeally* influences the body of a digital text itself—that is, a database of information and its manifestation as display of symbols[6]—in real time.

Inter-—from the Latin for "among"—suggests a linking or meshing function that connects separate entities.[7] Unlike *intra-*, a prefix for connections or links within the same entity, *inter-* joins what is other or different together. That liaison between mind, body, and machine, between the physical world and the other virtual scene, requires a translator or *interface,* most often hardware that includes a keyboard (or, for instance, a motion sensor or other tracking device), a monitor, and a controller such as a mouse, as well as software programming. One interacts by touching, moving, speaking, gesturing, or another corporeal means of producing a sign that can be read and transformed into input by a computer.

The common graphical user interface (GUI) or screen display of icons or graphical symbols and menus of commands, conventionally organized by the prosaic desktop metaphor, is also, confusing enough, often known as the *interface.*[8] The communication links between hardware and software and between the user and the computer compose a layered, complex site of exchange that is virtual as well as physical in multiple dimensions. The symbols to be manipulated may be text, graphics, images, and audio. They may appear on a screen—on CD-ROM, on digital video disk (DVD), on the Internet, or on the World Wide Web. They may appear in an installation or in a fully immersive virtual reality. With distributed computing, the symbols may even consist of "wired" or computer-controlled objects in physical space. Thus, interaction occurs across an interface or cybernetic frontier between the physical and conceptual, between the human body and the machine, and between biotechnology and communications.[9]

One vision of *interactivity* considers it largely as a tool for getting "into" the other scene presented on screen or projected elsewhere. Conceived in this way, the interface and interactivity may be seen as obstacles or barriers to "immersion"—a concept that conveys the state of being totally inside a created world both virtually and emotionally, in a way comparable to a novelistic or cinematic fiction but, by implication, to a far greater degree. The wish to design an interface that is transparent, and an interaction that is "intuitive," or that demands little awareness of a user is often expressed in industrial quarters, as well as by makers of fictional texts and scenes, who aim at immersive involvement.[10]

However, there is a problem in achieving such aims of immediacy, since interactivity is a level of expression that is not likely to be wished away from conscious awareness. Rather than presenting a story that seems to tell itself or a world that arises of itself, by definition interactivity involves decision making or the active *participation* of a user. (Even the direct computer-brain thought connection of cyberpunk fantasy would need some way of translating choice-making activity.) However, awareness of mediation and its sensory material of expression does not necessarily preclude sinking into fantasy. As in the spheres of poetry and the daydream, there is a middle realm between the capacity for regression into a world of imagination and the waking capacity to select and create such a world out of metaphor.

Participation as an activity is not, however, dependent on technology; it is rather an historical elaboration and transmutation of dialogic modes of encounter, the archetype of which is face-to-face conversation. Now, however, one "interfaces" or communicates by means of a computer-mediated simulation.[11] To *interact* is a kind of doing that entails purposiveness, conclusiveness, and agency—qualities that, namely, point to a subject. One might assume that the humans involved in the roles of author, designer, programmer, and user are the subjects of interactivity and that the machine in its various technological configurations is their medium. Indeed, the capacity to involve the *receiver* or *user* in the process, if not the creation of at least second-order selection and linking or assembling of elements displayed on-screen is precisely what differentiates interactive fiction and art from the passive readers and viewers of traditional cultural forms that espouse a one-sided notion of authorship. The utopian claims for interactivity as a liberatory and nonhierarchical praxis are based on the capacity to accommodate multiple and nonlinear links between elements in narration and the potentially more egalitarian or dialogic relation between artists and their audiences.[12]

However, the computer cannot be reduced to a medium of communication between human subjects. Its very capacity to give feedback and the immediacy of its response lends the quality of "person" to what is a computational tool.[13] This responsiveness allows it (and the virtual entities it displays) to pose or function as subjects—however partial, quasi, imaginary, and virtual—who are involved in the interactive exchange. The degree of influence and control of the interacter varies by design from an immediate one-to-one response to greater complexities, delays, and permutations. Interactivity may even initiate a process that grows out of the user's control into the relative autonomy of "agents" and "artificial life." From the beeps and clicks that acknowledge our touch to its capacity to mirror the user like a second self,[14] the computer can also function like an exteriorized mind. The "interface" is then a very special mirror that not only reflects but acts on and generates the symbols that we virtually encounter, enter, and process.

In answer to my interview question, "What is interactivity?" Lynn Hershman alluded to the anthropomorphic connotations that surround the term as part of a larger sphere of biological metaphors that structure our relations with machines, especially the computer—however hard we may try to evade them.[15] Qualities of "liveness" or instantaneous responsiveness and the appearance of autonomy and purposive motion support a biological interpretation of computer events, just as the language- and symbol-manipulating and -generating capacities of the computer seem to offer the computer itself as a hazy subject and container for mind that is partly us, partly other people, partly alien machine. Thus, the interface is the consummate arena for exploration and play with the enigmas of persona—including gender—and the mysteries of life and death.

In the 1980s, a well-known model of degrees of interactivity (associated with the laser-disk player) identified three levels—minimal interaction comparable to that of a remote control for television, interaction that allows the user a choice among a set of preestablished narrative outcomes, and interaction that allows the user to alter the final form of the artwork. This ability to change the subject or alter the rules is a feature of *intersubjectivity* or a dialogic relation. Intuitively, we reserve this capacity for human-to-human interaction. Perhaps for this reason, a distinction has arisen between the "interactivity" of hypertext/hypermedia/multimedia (especially on the hard-disk storage medium of CD-ROM and DVD) and the "connectivity" of the Internet and the World Wide Web. In the first instance, our interactive partner machine presents what is ultimately a closed body of information, albeit one that can be accessed in nonlinear order. In the second instance, we tend to envision our partners as human parties who exchange e-mail, chat, MUD, or MOO with us, and design personal Web pages and the like in ways that are open-ended and subject to change. To paraphrase Julia Scher, the space of the Web is enormous, beginning in the entrails or interiority of the computer user and extending out into a virtual universe that is expanding geometrically.

However, hard, binary distinctions between human and nonhuman and between open and closed do not bear close scrutiny. The anonymous relation between the user and machine enabled by an interface allows humans, agents, bots, and simulations of humans to interact in computer-supported exchange with each other as virtual subjects, whether on the Internet or Web or, for that matter, via teleconferencing systems or satellite. Furthermore, blended forms such as the "interrom" (the Muntadas *Media-architecture* CD-ROM/Web link produced by Anne-Marie Duguet) or hybrid forms composed of interlocking media (Branda Miller's *Witness to the Future: A Call for Environmental Action*) are more and more common, suggesting that the boundaries between hard and soft are fluid. More fundamentally, one may question the "openness" of sites on the Web, when "visiting" means triggering an increment on the counter of visitors, possibly entering one's credit card number, but, in any case, leaving a data trail of one's choices or "cookies" behind that can be used as consumer research ("data mining"). As I have stated elsewhere, "Ongoing surveillance by machines is then a corollary of the feedback of data from interaction with machines."[16]

What, indeed, does *openness* mean? Consider that while interactivity allows associative rather than linear and causal links to be made between heterogeneous elements, these associations are themselves part of a symbolic system that is made up not of endless possibilities but of historically and ideologically produced constraints. While I have distinguished *intersubjectivity* from *interactivity,*[17] in truth they are not so easily identified as a set of fixed oppositions, nor are they that easily separated in the psyche. In any case, the relation between the machines and humans in question is *virtual,* as is the muddle of subjectivities involved.

The theorist Jeanne Randolph has proposed that the primary ideological assumption about technology is that it should work.[18] No wonder the term *interactivity* presupposes a fait accompli—that links in networks of connections have been successfully made. However, unintentional failures of interactive hardware and software and of the humans that design and employ them occur at every level of cybersociety—from AT&T down to the artists who toil, often collaboratively, as pioneers in labor-intensive new media. The term *interactivity* thus refers to a state that occurs after or is incognizant of painful effort and myriad unsuccessful, broken, and invalid connections and attempts to interact that simply don't work.

The result of an interaction is a change of state or condition—in this case, that of connecting, but connecting to what and to what end? The answer is not yet entirely in sight, since interactivity is a feature of a great societal and cultural transformation in progress, and, as Julia Scher said in an interview, "the directory is not complete."

ARTISTS, GENDER, AND METAINTERACTIVE ART

In the following, I briefly discuss instances of media art that foreground interactivity itself in a variety of ways. Interactive art is distinguished by its eclecticism rather than by some cohesive generic quality. I make no implicit claim to common aims or generic unity in interactive media art by women, nor could I possibly aim at comprehensiveness or completeness by singling out just a few examples from a large body of work that has a history of more than two decades. What unites these examples formally is rather the uneasy situatedness of women in the worlds of art and technology that promotes a reflexive and ambivalent relation to media and incites production that self-consciously sets its own premises in question. Informally, these pieces are all linked through my having had to come to terms with them in my own writing, making them long-term objects of reflection.

Persona

I begin with an interactive multimedia essay on CD-ROM that models ambivalence: *She Loves It, She Loves It Not: Women and Technology* (1993) by Christine Tamblyn with Marjorie Franklin and Paul Tomkins. The interface or metaphorical map of the piece is appropriately enough a daisy menu over a twelve-part zodiac.[19] The CD-ROM reader is inducted "inside" the mind of "Christine," a virtual subject and guide, by clicking on and thus entering through her mouth. "Christine's" mind is, however, inhabited by many different and contradictory voices. Two of the daisy petals or loops demonstrate her contradictory moods and attitudes.

In the "INTERACTIVITY" loop, "Christine" takes a dim view of her subject: "The much touted interactivity associated with computers often consists merely of pushing a button and getting a predetermined response. This is concomitant to pushing a button on a vending machine and having a Coke delivered into the slot." A "movie" in the same loop shows a 1950s lady in pearls opening and closing an oven while a male voice of god intones about the "freedom of individual choice." Ironically,

Tamblyn's own didactic essay is organized along those very principles of button clicking along branching pathways.

On the "LABYRINTH" loop, on the other hand, her text appreciates the "fairytale"-like mystery of "navigating" through (that is, having serial interaction with) digital space: "The reader goes on a journey during which she meets challengers, completes tasks, receives omens, and decodes signs. Movement through a computer program consists of a nomadic reading that always leads to another curious space, another confrontation with demonic agents. The work is never finished."

Her "user" or interacter may also find that some clicking brings surprises. In a "personal letter" to the reader in the "INTERACTIVITY" loop, "Christine" reminisces about a boyfriend who deposited, for example, a salamander leg in the food served up at the automat cafeteria. In another letter, she remembers not being permitted to call a boy on the telephone, offering oblique commentary on the position of the interactive artist as a smitten girl in a teen romance, waiting for a phone to ring. The robot guide is yet another "objective" voice, who deposits the well-known cliché on screen: "In cyberspace, the disembodied consciousness can zoom along data highways with unparalleled mobility. It is a realm in which the mind is freed from physical limitations and thoughts are omnipotent." Elsewhere the robot is an unapologetic mouthpiece for feminism: "In 1980, a United Nations report concluded that while women constitute half of the world's population, they perform nearly two-thirds of its work hours, receive one-tenth of the world's income, and own less than one-hundredth of the world's property."

As a storage medium, the CD-ROM accommodates Tamblyn's simultaneous deployment of mixed-media and found fragments from popular culture, multiple personas and voices, and various modes of data, fiction, and speculation in moods from serious to ironic and allows her to play them off against each other. As retrieval glitches and lags, she welcomed even the "noise" of short loops and jerky animation that fragmented the flow of images as Brecht-like distancing that creates a space for reflection. Nonlinear access also permits a cyclical form that, unlike narrative, allows the end of each complete exchange or segment to return to the center. Thus, paradoxically, the very limits of the medium serve a decentering of the subject "or an identity that is multiply locatable."[20]

On the other hand, Tamblyn regarded working in the medium as onerous and even agonizing: "The torment of accommodation to preprogrammed routines" was part and parcel of this "Faustian pact we are succumbing to in the rush to embrace prefabricated, if multiple, identities in cyberspace."[21] What is a persona, and what does it mean when personality (whose?) is invested in a computer? Tamblyn's next CD-ROM builds a "lexicon of received ideas about identity."[22] *Mistaken Identities* (1995) maps ten women's biographies—among them those of Catherine the Great, Isadora Duncan, Marie Curie, and Josephine Baker—into a system or paradigm of the strategies that women adopt "to survive and prosper in environments that were not particularly receptive to their precedent-setting endeavors."[23] As the CD's title suggests, these are not canonical or idealistic biographical treatments but self-contradictory and transmutational personalities. Tamblyn's own inevitably subjective role in selecting and editing these lives is emphasized in the "Morphologies" section, which blends her image into one standard portrait pose of each of her subjects. Her final CD-ROM, a synoptic view of her own lifework (*Archival Quality,* completed posthumously in 1999), "made no attempt to separate the personal from the political from the autobiographical from the theoretical. I tried to show they [all these personas] were all closely intertwined in a big scheme."[24] Tamblyn claimed all these hybrid and multiple personas. We discover that womanhood is not an identity, just as interactivity is not one apparatus or process.

The play with persona that has become a major theme of anonymous virtual encounters on the Web has precedents in performance-art masterpieces such as *Roberta Breitmore,* an alternate identity with a driver's license and a checking account who inhabited her creator, Lynn Hershman, and multiple other women's bodies from 1971 to 1978. As they pawed through the detritus of Roberta's private life—canceled checks, letters, clothes, trash—in her vacated room at the Dante Hotel in San Francisco in 1973, visitors to the exhibition become uncomfortably aware of their own voyeurism. Participation is thus a lure that turns the spectator's gaze back on itself. In Hershman's *Lorna: The First Interactive Laser Artdisk* (1980–1984), visitors change the channels on the "television" of an agoraphobic middle-aged woman, exploring the branching pathways of her life by remote control. In *Deep Contact* (1985–1991), with Sarah Roberts, viewers interact with a luscious "Marion" at her invitation by touching her on-screen body, each body part unleashing a different narrative path. Encountering the shadow of the camera in the story world causes the visitor's own image to displace what is on-screen. The gaze seeks its target inside a dollhouse of miniatures sets of *A Room of One's Own,* an interactive videodisc installation with Sarah Roberts (1990–1993), only to be unexpectedly confronted with a surveillance image of the visitor's own eye. The theme finds further transmutation as visitors see and interact on the Internet through the eyes of CybeRoberta in *Tillie the Telerobotic Doll* (1995–1998). Thus, participation foregrounds formerly tacit and apparently innocuous features of "passive viewing," while making visitors painfully aware that surveillance is part and parcel of the two-way nature of "interactivity." Throughout this series, the visitor is displaced from intimate physical space to virtual space and from interaction with a virtual persona to inhabiting the virtual persona/avatar of a camera-eyed doll, seeing worlds telematically through the Internet.

The limitations of communicating with and by means of a machine preprogrammed with network of choices can also serve to reveal psychic automatisms and reflexes that limit human-to-human exchange. In Sara Roberts's *Early Programming* (1988), for instance, the computer poses as a bossy virtual "Mom" as expert system with no face but the monitor itself and a neutral voice produced by a DECtalk voice synthesizer. The visitor's interactions with a virtual mom through a series of childhood scenes are governed by an "emotional engine" that cycles through various affects, influenced to some degree by the user's responses to "Mom's" admonishments and cautionary tales. For instance, a fork advances toward the viewer with an enormous Brussels sprout that is "good for you." The visitor selects from a range of potential stock responses to "Mom's" classic line, "Think of all the little children of the world who would be happy to have what's on your plate." Other scenes of the struggle between discipline and desire include a view from floor level of a messy room, littered with socks and toys, as a vacuum roars in the background, small hands practicing at a piano, and a swimming pool that beckons as the virtual child is told, "Don't you go into the water. You just ate." "Mom's" attitude can vary from hilarity through sarcasm and huffiness to raving anger when telling her "child" to go outside on a beautiful, sunny day. Her mood meter is a rectangle in the middle of the screen that shifts from a small, dark block of rage to a bigger, lighter shape.

While these exchanges are associated with the repressive and controlling force of the reality principle, much about them evokes a poetics of childhood and lyrical sense memories like poking one's hand out of the car window to play with the wind. Far from a dialog based on mutual recognition, communication, and reflection, as Roberts explains, interaction with "Mom" is "an exchange of emotional tokens, not ideas." Thus, the limits of interactivity are perfectly suited to such automatic if not entirely predictable responses.

In fact, such automatisms set some of the more utopian premises about interactivity in question. Though hypertext and hypermedia are often considered subversive in putting different elements—print, sound, and image—together nonhierarchically, are such associative relationships actually subversive or synesthetic? Do they actually succeed in blurring the distinction between writer and reader? Organizing interactivity nonhierarchically through associative relationships may not be so liberating once one considers those associations themselves as a symbolic domain that is socially and culturally constructed.

Links

Collective entities built of electric charges, connections, and rhythmic patterns that shape lives more or less distant in space and even time together are a miraculous feature of the Internet and Web. These creations of the Net deserve their own lengthy and independent exposition. However, interactivity depends on the very act of linking. That act made self-conscious can be the stuff of art.

The ability to choose among linking pathways in a network is the defining feature of hypertext and hypermedia. Such choice is considered liberating, especially when it furthers a path of associations to be formed that allow different stories to be told and different causalities to operate. However, "free" associations are themselves creations of culture and, as Freud's life project demonstrated, of the unconscious. It seems to me that Sonya Rapoport's art is based on strategies of linking that are both playfully nonsensical and revealing about Cartesian and other scientific modes of interaction. In the process, she brings the act of associating itself into awareness. For example, the *Shoe-Field* (1986) interactive installation ultimately results in photos of audience members' shoes laid out into a grid and labeled "Which SHOES would you like to interact with?" The "association" work of *Brutal Myth: Are Women the Guilty Daughters of Eve* (1990), a Web site with a "Bitter Herb" menu display inspired by the *Malleus Maleficarum or Witches' Hammer* is a more obvious clue to the artist's project.

Interactivity in Rapoport's *Brutal Myth* extends beyond the computer. According to a given prescription (antique cures), the participant would print and cut out a curative herb described in the artwork, soak it in wine, and then drink.

The Transgenic Bagel, an interactive computer-assisted artwork, invites the visitor to "splice a trait of your choice." *Make Me a Man* and *Make Me a Jewish Man* invite further click-and-paste creations that suggest the absurdity as well as pleasure of such couplings. Another limit on linking as free association is the reach and composition of the system itself and the problem of who has access to the network. Even persons who are part of the global system of an information economy as workers may not be able to participate in it as subjects. It may be a work of art to bring suppressed or disconnected links to conscious awareness. Coco Fusco, among others, has been active in making the association between women workers in the *maquiladoras* strung along the Mexican side of the border with the United States and the goods that compose the material side of a virtual life.

Immersion

The last feature to be addressed in relation to interactivity is immersion. As suggested earlier, immersion as sinking into a fantasy world is often held to be incompatible with interactivity brought to conscious awareness. However, the very notion of a poetics of media art suggests that immersion and interactivity are far from incompatible or inversely proportional. Here, three examples of self-consciously immersive art will be addressed.

Of course, the most "immersive" form of interactivity occurs when the subject navigates within a virtual world. However, the visitor to Agnes Hegedüs's *Handsight* (1993) submerges at most a forearm into an utterly empty transparent globe. *Handsight* breaks elements that are conflated or condensed together in virtual reality into their component parts.[25] All three elements in this interactive installation are metaphors for the eye:

a hand-held "eyeball" interface or round Polhemus sensor that tracks hand position; a transparent sphere with an irislike opening for the hand; and, beyond it on the wall, a round projection of an eye.

Once a hand penetrates the empty transparent globe with the hand-held "eye," the projected eye on screen opens into a virtual world. Hand motions are translated instantly (in "real time") into a moving point of view like an "endoscopic eye" (*endo-* = "within," capable of penetrating an interior space) exploring a virtual world, the three-dimensional anamorphic representation modeled on the inside of a "passion jar." Nearby, a narrow-necked bottle of nineteenth-century Hungarian folk art contains—as if miraculously—the tableau of a crucifixion scene. These material objects immersed in the jar have been transformed into brightly colored, spherically distorted geometric shapes "inside" the eye or virtual world. Anamorphic perspective produces precipitous spatial relationships, quite unlike "real" space.

While the late nineteenth-century viewer of the spiritual realm in the passion jar had to be content with looking through the glass to foster imaginary access to the sacred scene, the distinctive feature of late twentieth-century virtual environments is the penetration of the gaze inside "worlds that are otherwise inaccessible by virtue of their two-dimensionality, scale, solidity, immateriality, or imaginariness." What is an ordinarily inaccessible interiority or psychic space of transcendence has been transformed into an externalized virtual space that can be penetrated and explored. In this interactive environment, the "endoscopic eye" or virtual camera is inducted symbolically into the mind itself, not to reveal some objective reality but to display parts of a symbolic system laden with historical and personal resonance. Once entered, this virtual space becomes larger than its apparent container, seemingly monumental, offering a subjective point of view onto precipitous declines and vertiginous shifts of position governed by the visitor's own hand.

The layout of the installation makes the interaction—between the sensor or "interface" (the sphere that the sensor is active in) and the on-screen events—transparent in a way that an "immersive" virtual environment that collapses the three eyes into one point of view obfuscates. (With the head-mounted display of virtual reality, for instance, the position of the sensor, the active area, and the virtual world all overlap.) Using the hand to "see" leaves the emptiness of the material sphere that is the counterpart of the on-screen world open to view. The eye within an eye within an eye is like a *mise-en-abyme* or metaphor for an irrational space based on incorporation. *Handsight* offers a metaphor for perception of a virtual realm that is not matched to the physical world but rather is a view of the "mind's eye" or of externalized imagination. At the same time, it exposes the logic of this construction and does not participate in the illusion. In this way, the piece both offers and deconstructs interactivity.

In contrast to this analytic and deconstructive piece, Char Davies's two virtual-reality pieces suggest the experience of diving into a fully immersive virtual world. A harness interface allows the visitor use her breath to navigate up or down in virtual space. Used in conjunction with a head-mounted display, the interface evokes diving gear. I have discussed my experience in a virtual pond within *Osmose* (1993) at some length in *Virtualities.* I was amazed at how different the experience of being inside Davies's subsequent virtual world, *Ephémère* (1998), would be. The virtual worlds to be navigated in *Osmose* were distinguished by an aesthetic of transparency and their delicate figurative rendering of the natural world. In contrast, *Ephémère* was like entering a virtual abstract painting that suggested variously a landscape or the interior of a human body or some microscopic view of plant life. Navigating this abstract space had none of the qualities of ground-covering speed I had experienced in *Osmose.* Because there was no secure index of scale, I experienced virtual space as a more contemplative temporal flow. The interface and user of *Éphémère* was visible to spectators in a darkened theater as a shadow on a translucent screen. Not only could every navigational act be followed like a shadow play; a monocular view of the virtual world in the head-mounted display was projected on-screen in real time. Thus, it was possible to discover differences in navigational style that reflected different movement personalities and that consequently revealed different territories within the virtual realm. In this way, the difference between these two pieces was as significant and revealing as the pieces seen individually.

In a similar way, two of Catherine Richards's pieces are meant to be understood as linked together as contrasting situations. However, in this instance, immersion does not require computerized assistance. We are all immersed in a world that requires no electronic equipment whatsoever to be "plugged in" to electromagnetic waves that bombard the earth. Picking up a glass heart in Richards's interactive installation *Charged Hearts* (1997) plugs and closes the circuit with the miniature northern lights in a Terrella nearby, and the heart glows with mysterious blue light. One can communicate with another heart with a winking connection created with one's own hands. In fact, to isolate oneself from any link or connection in interaction with the world, the visitor to the *Curiosity Cabinet, at the End of the Millennium* (1995) must enter a Faraday cage of grounded copper wire. The experience inside is one of an absence achieved through the great effort to become unplugged.

NOTES

1. Some interactive programs run completely without human input—through machine-to-machine communication.

2. In *Virtualities: Television, Media Art, and Cyberculture* (Bloomington: Indiana University Press, 1998), I propose this kind of presentational relation of the artwork to the public as a *fiction of presence* to be distinguished from the fictions of the past—of theater, novel, film, painting, sculpture, and other representational arts.

3. In his "Computer-Mediated Interactivity: A Social Semiotic Perspective," *Convergence: A Journal of Research into New Media Technologies* 4, no. 3 (Autumn 1998): 40–58, Paul A. Mayer emphasizes the social ramifications of "interactivity" in enabling the user to enter into the "change and development of social structures in conjunction with new communication media" by offering a medium as "object to think with" and involving the user in forms of social reflexivity.

4. Some artists in new-media art have their roots in second-wave feminist performance and video art, while others are the third-wave students and collaborators of these second-wave feminists. Some were involved with digital technologies from the first, while other women came to interactivity from traditional, noninteractive arts. While women in performance art and video were marginalized in both art and academia in the 1970s and 1980s, after the muddled subjectivities of performance became digital, women could be found setting up and staffing the computer labs in universities and art schools in the 1990s.

5. The common assumption of audiovisual narrators and game designers—that the greater the elaboration of the fictional world, the deeper the immersion into it—may be misguided if reception theory applies here.

6. Timothy Binkley's "Refiguring Culture," in Philip Hayward and Tana Wollen, eds., *Future Visions: New Technologies of the Screen* (London: bfi, 1993), 92–122, offers a lucid explanation of the two-leveled digital text.

7. According to the *Oxford English Dictionary* (2nd ed. 1989), the term *interactive* appears in 1774 as "the short performance between two acts" of a play, in 1832 as "reciprocally active, acting upon or influencing each other," and in 1967 as "pertaining to or being a computer or other electronic device that allows a two-way flow of information between it and a user, responding immediately to the latter's output."

8. Steven Johnson's *Interface Culture: How New Technology Transforms the Way We Create and Communicate* (San Francisco: HarperCollins, 1997) offers a history of the computer interface, especially the GUI.

9. The exhibit *Interface: Encounters with New Technology* at the Museum of Contemporary Photography in Ottawa in fall 1998, curated by Carol Payne, provocatively explored this frontier.

10. In *Hamlet on the Holodeck: The Future of Narrative in Cyberspace* (New York: Free Press, 1997), for instance, Janet Murray espouses the ultimate goal of transparency (26).

11. In *Virtualities* (11–14), I explain why even face-to-face conversation in physical space involves simulation.

12. Jaishree K. Odin's "Embodiment and Narrative Performance" (chapter 32 in this book), for instance, describes the implications of nonlinear hypertext for postcolonial and other critical literatures.

13. Mayer, "Computer-Mediated Interactivity," 52, emphasizes this communicative aspect between partners or poles that are not just subjects but both subjects and objects. Each interaction involves a negotiation of the status of the subjects and the meaning of the communication. Thus, interactivity is inherently metacommunication: "Figures like 'the computer,' 'he/she/it' (the application, or a character avatar within a game world) are *experienced* as interlocutors . . . based on the immediate nature of the responsiveness experience by the user." Byron Reeves and Clifford Nass address these anthropomorphizing tendencies in their *Media Equation: How People Treat Computers, Television, and New Media Like Real People and Places* (New York: Cambridge University Press, 1996).

14. Sherry Turkle's *The Second Self: Computers and the Human Spirit* (New York: Simon & Schuster, 1984), is the most famous expression of this view. Allucquère Rosanne Stone's *The*

War of Desire and Technology at the Close of the Mechanical Age (Cambridge, MA: MIT Press, 1995) also addresses identity and masquerade in cybernetic communication.

15. In her e-mail communication, Hershman elaborated on this idea: "Net culture's model is life. It is biocybernetic because we've assigned our own history to it rather than wait until it has time to develop one itself. Most certainly, the Net is alive, and like the universe it is expanding. The language that is used on the Net is usually positive, animated, and abbreviated. It bristles with energy. It is active and volatile with words of Navigate, Link, Browse, Capture, Refresh, and Save."

16. Morse, *Virtualities,* 7.

17. Ibid.

18. In a paper delivered at the International Symposium of Electronic Arts (ISEA 96) in Rotterdam.

19. The petals address the themes of ideology, power, communication, violence, interactivity, representation, control, labyrinth, the other, homunculus, memory—and the credits make twelve. She loves it not.

20. Christine Tamblyn, "*She Loves It, She Loves It Not*: Women and Technology: An Interactive CD-ROM," *Leonardo,* 28, no. 2 (1995): 99–104.

21. Christine Tamblyn, "Remote Control: The Electronic Transference," in *Processed Lives: Gender and Technology,* ed. Jennifer Terry and Melodie Calvert (London: Routledge, 1997), p. 430.

22. Ibid., 265.

23. Ibid., 261.

24. Christine Tamblyn, interview with the author, 23 December 1997, printed in the insert to Tamblyn's *Archival Quality* (Los Angeles: Framework Press, 1998) distributed by Video Data Bank, Chicago.

25. This description of *Handsight* was written for *Hardware, Software, Artware* (Ostfildern, Germany: Cantz Verlag and Karlesruhe: ZKM, 1997), 40–41.

3

Women, Body, Earth

Sheila Pinkel

> One of art's functions is to recall that which is absent—whether it is history, or the unconscious or form, or social justice.
> —LUCY LIPPARD[1]

Once women claimed the right to their own voice, they identified new subject areas to explore. The artists included in this chapter have focused on aspects of body or environmental representation in their work. Since the 1960s, especially in the United States, an avalanche of excellent work has investigated new territories of signification. These artists represent the intensity and diversity of work that has been done by women throughout the world during the last forty years.

EARLY HISTORY

The social and political upheaval of the 1960s led to protests for free speech and against the Vietnam War, civil disobedience during the Democratic National Convention, the assassinations of John and Robert Kennedy and Martin Luther King Jr., national movements in support of civil rights, women's rights, gay rights, hippies and flower children. In this highly charged social environment artists explored new art forms, sowing the seeds of postmodernism.

Initially reacting against modernism's narrowly prescribed conventions about content, artists soon found broader subject areas for their work. They began to make art about politics, autobiography, sexuality, nature, history, and mythologies, both personal and social. The transgressive imagination, then, became the terrain for frontiers of meaning, even though in the late 1960s artists had few places to exhibit these works. Groups of young artists, marginalized by the prevailing art system, searched for alternative venues for exhibition. As a result, alternative artist-run spaces—operating on a shoestring but open to exhibit work of the artist's choice—emerged across the country. In addition, performance and installation artists began using public spaces as venues for their work.

In the 1970s, women began to find strength in community and dialog. For instance, in Los Angeles the Women's Building became a center for arts activity, providing art and performance classes and exhibition and meeting spaces. The Women's Building also offered printing opportunities to women who wanted to make books rather than galleries the container for their work. During this period books emerged as an affordable and portable alternative gallery that democratized exhibition and ownership of artworks, and allowed the addition of text.

EARLY ELECTRONIC IMAGING TECHNOLOGIES

During the 1960s and 1970s, artists Joan Lyons, Sonia Landy Sheridan, and Barbara T. Smith discovered that duplicating machines provided new possibilities for making art. In cities across the United States, from Rochester, New York, to Los Angeles, California, these women separately began to photocopy all manner of things, including their own bodies—"because that was what was most available." For instance, Lyons used one of the first halide processes, in which imaging and fusing were done in separate steps. As a result she could "paint" with the toner on the plate before the final image was fused. These artists were less interested in static, one-of-a-kind images than in transformative possibilities of duplicating machines that produced images at the speed artists were able to imagine them.

The concept of art as integrated life activity was part of their pioneering work. For instance, Joan Lyons has spent thirty years running the artists' book wing of the Visual Studies Workshop, helping hundreds of other artists manifest their ideas in book format. Lyons draws little distinction between running a book-making workshop and doing her own artwork. Smith has devoted much of her creative energy to organizing performance events, sometimes in nontraditional places like the Los Angeles River, which showcase the work of other artists and address social and political issues. And Sheridan has taught students at the School of the Art Institute in Chicago to integrate art making into the fabric of their lives, to understand that the spiritual and the political are related, and to infuse work with the power of consciousness and imagination.

EARTH, BODY, AND THE EROTIC IMAGINATION

In many cases, artists have identified sexuality as the territory for power relations and have used the taboo subject of female erotic experience as a way of claiming their right to this territory. Mary-Charlotte Domandi clearly articulated the dilemma for women when she wrote, "Women have been the leaders in freeing and redefining the body. . . . In Western culture, women's bodies and minds have traditionally been stringently regulated through social conditioning, which has taught them to be attractive and available, passive and obedient, and through physical violence, threat, and law, much of which has attempted to stifle what might be "unruly" sexuality."[2] "We are in an era in which our survival is contingent on our ability to deal with our sexuality," says installation artist Dorit Cypis, as a way of positioning the emphasis on images of her own body in her installation work.[3]

In early 1963, Carolee Schneemann did a performance called *Eye Body* in which she appeared nude with live snakes crawling on her body. She was one of the first artists to use her nude body in performance to make a statement about rebirth and fertility, using symbols from Christian creation mythology to reclaim that story for women. In other works, she wove imagery from female-centered cultures into semiautobiographical works to connect the personal and the universal. She adopted the symbol of the inverted triangle (also used by Judy Chicago in *The Dinner Party*) to symbolize woman. Her use of her own nude body as theater for expanded content gave subsequent artists the courage to adopt whatever form was necessary to convey meaning. She also questioned Western taboos against female nudity when women instead of men author those images and in so doing challenged conventional power relations between men and women in Western culture.

When Linda Benglis donned a large penis in an issue of *ArtForum* in the mid-1970s, she was joining the legions of women who at that time were asserting their right to their own self-determination by making images of their dreams and desires that often overtly challenged patriarchal representations. Other artists, including Annie Sprinkle, Karen Finley, and Hanna Wilke, have used performance as a way to address issues of sexuality and power.

In the late 1960s, Barbara T. Smith began doing performances in alternative spaces intertwining autobiography and sexuality, as a way to claim her right to her evolving erotic imagination. She viewed herself as a guide for expanding the knowledge of sexual experience and in so doing going beyond the confines of sexual practice in the Judeo-Christian tradition. For instance, in *Birthdaze* (1981), Smith began the performance by juxtaposing the mind and body erotic experience that is symptomatic of Western dualistic thinking and ended it by enacting a tantric sexual practice, which constructs the sexual experience as a journey toward higher consciousness. In her work, Smith takes risks to stay true to her own evolving consciousness, and as a result the form of her work continues to be innovative as well.[4]

In 1996, when Diane Buckler started placing images of unclothed women on the edge of simulations of Greek and Roman vessels, she was addressing the right of women to their erotic selves by joining classical symbols of power with images of women freely enjoying their own bodies. Traditionally, such vessels show women clothed and men unclothed. She is creating a new vernacular that is transgressive in terms of traditional gender power relations and subordination of female sexual experience.

DEVELOPMENT AND DESTRUCTION OF THE EARTH

In the 1960s, some women started researching primordial forms and cultures of ancient peoples, especially matriarchal cultures, because they were searching for deeper, more enduring symbolic meanings for their work than those found in contemporary culture. Rituals of ancient peoples and a concept of deep time gave these artists a frame of reference that was outside the prevailing patriarchal culture and renewed their relationship to natural processes. The periodicity of the sun and moon, the relationship of their own biological rhythms to these external forces, and an emphasis on relatedness rather than uniqueness and aloneness became the basis on which many women artists constructed their worldview and spiritual foundation. Building on this research, artists such as Agnes Denes, Mary Beth Edelson, Ana Mendieta, Donna Henes, and Nancy Holt began constructing their own installations, performances, and drawings to forge a healing connection between people and natural processes.

Since the mid-1960s, as people have become increasingly aware of the fragility of biological systems, many artists have focused on their love of nature and concern about its destruction. For the last thirty years, Helen and Newton Harrison have proposed imaginative solutions for ecological problems in the form of installations and books. *The Lagoon Cycle,* done in the 1970s, was a monumental installation in which they attempted to shift the focus from land-centered to water-centered concerns and to demonstrate the interrelatedness of geography, history, economics, and mythology that is necessary to make enlightened decisions about development. The Harrisons have been invited, as artists, to participate in water-reclamation projects throughout the world because they add a broad perspective to the activity of problem solving.

Since 1982, concerns about the toll taken by human activity on living systems and the future of the planet have been the basis for the performance work of Rachel Rosenthal. The focus of many of her performances has been the relationship of humans to other animals and living systems. In *Foodchain* (1985), for instance, she questions whether eating meat results in predatory behavior on the part of people toward one another. In other works, she explores the tautology of technological solutions that result in alienation and of alienation that produces the desire for more technological solutions. Rosenthal believes that the way that people relate to and treat other animals is symptomatic of the degree to which they have evolved. In her universe, a relationship with other species is a sublime experience resulting in healing for the individual and the larger biodynamic system.

During the 1980s and 1990s, artist Jacki Apple generated a number of performance and audio works, including *The Garden Planet Revisited, The Culture of Disappearance,* and *You Don't Need a Weatherman* to explore the imbalance between the materialism inherent in economic and technological priorities of industrial and postindustrial culture and environment and ecology of the planet. In all of these works, she juxtaposes political and philosophical belief systems with catastrophic events such as species extinction, environmental and social destruction, and apocalyptic weather conditions. In 2001 Apple reflected that "What we think—that is, conscious thought—generates an energy field that manifests itself in specific forms. If those thoughts are in balance and value the interrelatedness of nature and organic and the technological, then the social constructs, practices, and institutions they produce will reflect and manifest these values. When a culture does not value the principle of interconnectedness and has lost its perspective on power and accumulation, then the entire system goes out of balance. That is what we are faced with at the end of this century."

The destruction of the atmosphere that we have to breathe every day has come to symbolize the imbalance in the corporate industrial state. When artist Kim Abeles conceived of the work *Mountain Wedge* (1985–1988), she was trying to know something about the extent to which smog in downtown Los Angeles was interrupting her ability to see the mountains. For 274 days in 1985 and 1986, she photographed the mountains north of downtown Los Angeles, which were obscured by smog most of the time. On 10 September 1987, she made a pilgrimage from her studio in downtown Los Angeles to the base of the mountains to be able to know how far the smog-obscured mountains are. She walked in a straight line toward the mountains, no matter what was in the way, climbing over fences, through houses and backyards, and over cars until she reached the base of the mountain. By insisting on her own firsthand bodily experience in order to know how far the besmogged mountains were from her studio, Abeles was asserting the importance of primary experience in a culture that is usually satisfied by surrogate experience of quantification and media representation.

MEDIA MYTHOLOGIES

During the early 1970s, many feminist media theorists were addressing the constructed nature of representation of women in print media and cinema as a primary source of control in the culture. At about this same time, many women artists started to do artworks deconstructing advertising and media imaging of women in the culture. When Barbara Kruger made the works *Your gaze hits the side of my face* and *We won't play nature to your culture,* she was questioning the nature-culture duality that was embedded in constructed representations of women in media and advertising and the benefit to patriarchy of these constructions. The latter image shows a woman lying on the grass with a leaf covering her mouth. Once women began to view images as constructs and recognized their controlling function, they began looking for an alternative language to deconstruct prevailing images and name their own experience.

Cindy Sherman's early series of self-portraits, *Untitled Film Stills,* begun in the mid-1970s, succeeded in deconstructing the conventions of female representation in the cinematic images. By placing herself both behind and in front of the camera, Sherman named the constructed nature of female representation in films and film stills and in so doing interrupted their conventional reading. She also questioned who is being looked at and who is doing the looking, a language so embedded in photographic imaging that without deconstruction the image looks natural, like looking through a window.

Using human-sized medical mannequins to construct quasi-human forms, Sherman also names the cyber-and-human dilemma, the contemporary version of the human-and-machine relationship, which has been of concern for the last five hundred years. Donna Haraway has written that "late twentieth-century machines have made thoroughly ambiguous the difference between natural and artificial, mind and body, self-developing and externally designed."[5] According to Judith Fryer Davidov, "Technology, in her [Sherman's] version, shows us the impurity of intermixtures; history, in her version, is multisubjective, power-laden, and incongruent, a carnivalesque world of ruptures and fissures."[6] And Fryer finally asks, "How, indeed, do we untwist the strands of hybridization, acknowledge that there are other strands, and attempt to recover subversions without falling into the dialectical pattern of binary thinking? We can only begin by acknowledging the complexity and multiplicity that underlie all social constructions."[7]

For the last thirty years, artists have been exploring possibilities for representation reflecting sexual practices that fall outside of the Judeo-Christian heterosexual nuclear-family paradigm. In the 1980s, Deborah Bright inserted herself into film stills, employing a strategy similar to that of Cindy Sherman, except that Bright used actual film stills to make visible the heterosexual construction of romantic relationships in the cinema. By positioning herself as the "other woman" in film stills, she sought to challenge media assumptions about desire and gender relations. Other artists, such as Kathie Opie, Sue Maberry and Sherrie Galke, and Kaucyila Brooke, have photographed themselves and others in nontraditional sexual roles to articulate lesbian experience and relationships. And Brooke's video *Dry Kisses Only* is a collection of sequences from the history of Hollywood films, which make visible normally transparent homoerotic relationships embedded in film history.

The discussion of women's uses of self-representation would not be complete without mentioning the recent work of the San Francisco–based artist Laurie Long. Realizing that she had assimilated the character of Nancy Drew, girl detective, into her personality, in the mid-1990s Long did a series of fifteen tableaus entitled *Becoming Nancy Drew* in which she photographed herself to simulate the illustrations in Nancy Drew books (figure 3.1). In each of the tableaus, she included a shot looking at the action, a pinhole shot from her own frame of reference, and a quote from the Drew text. In the image shown here, we see a beautiful bound woman looking much like a Sherman film still. We look at her; she does not look at us, and she is obviously posed for the camera-viewer. The vernacular for women's representation, which Sherman made visible in the 1970s, was alive and well in 1930s in popular children's literature when this book series first started to be produced. And when we look further back in history, we find this vernacular operating as well.

During the generation before Long's, little girls were drawing their role models from movies and then television. By locating the source of her identity in literature, Long opens up the terrain of identity formation for girls and foregrounds the obvious question, "How do narratives affect personality development?" and then the next obvious question, "What are the new narratives that will embody greater self-determination for women?"

While most postmodern representations deal with naming the dilemmas of representation of women in various media, more recently women have been attempting to develop new narratives. For instance, Beverly Naidus, in her collaged book *One Size Does Not Fit All* (1993) (figure 3.2), challenges the conventions of thinness as the measure of beauty and honors the uniqueness of each person's body. This idea of self-acceptance, from Naidus's perspective, is key to a sense of self and relatedness. Text accompanies the images, which provide an alternative voice to representations of women in the media: "She learned to accept compliments gracefully," "She got fed up with the fashion industry and developed her own sense of style," "She practiced loving herself by enjoying the diversity of body styles she saw in the sauna," "She celebrated her health and vitality."

In the early 1990s, Brenda Laurel, via Purple Moon, attempted to generate new narratives for teenage girls using electronic media as a vehicle. Her Web site was one of the most popular of its class, indicating that there is an enormous audience for new feminist narratives. One of the challenges for women will be to successfully participate in all levels of creative property activity, including distribution, so that works can reach target audiences. There is much to be learned from Laurel's valiant pioneering experiences.

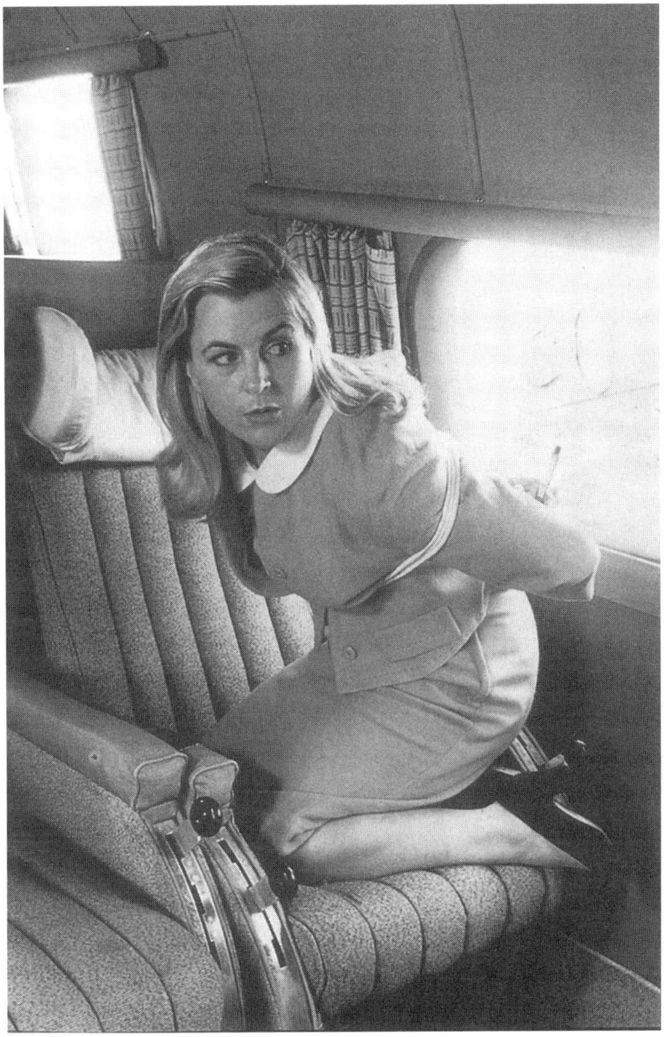

Figure 3.1
Laurie Long, from the *Becoming Nancy Drew* series, 1996, courtesy of the artist.

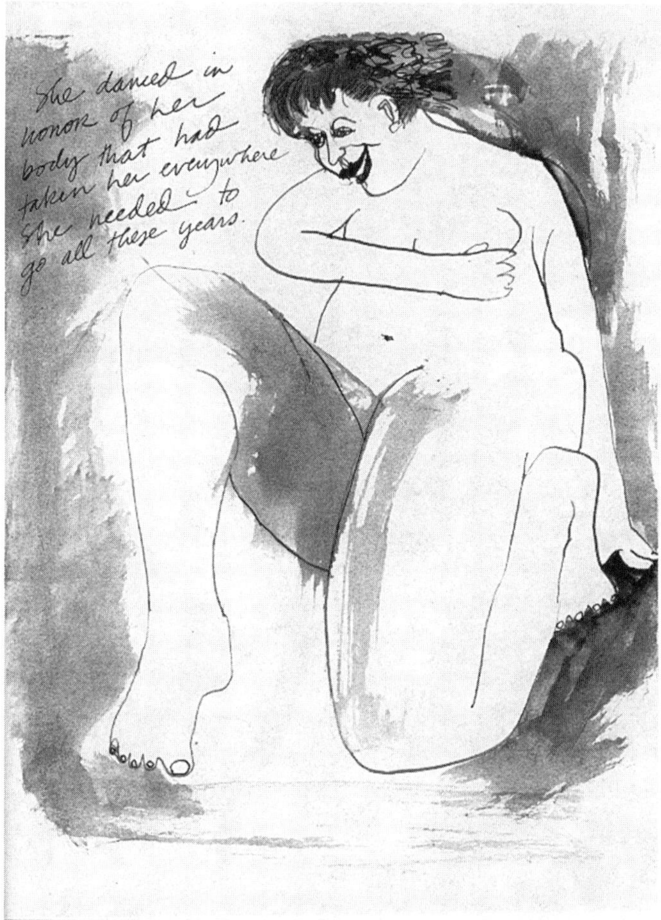

Figure 3.2
Beverly Naidus, *One Size Does Not Fit All,* 1992, courtesy of the artist.

VIOLENCE AND MEDIA

Since the 1970s, there has been a growing recognition that media representations of women in advertising, television, film, and print media were perpetuating violence toward women. Some women artists have attempted to make that violence visible. For instance, for years photographer Donna Ferrato traveled throughout the United States photographing women who have suffered domestic abuse. In so doing, she articulated the unseen world of battered women and raised questions about the nature and degree of violence toward women in this culture.

In the early 1970s, Yoko Ono and Maria Abramovitz did performances that were meant to make violence toward women visible. They did separate performances in which they included a series of objects such as handcuffs and knives and instructed visitors to their performances to do anything to them that the visitors wanted to do. Ono found that visitors progressively cut the clothing from her body, piece by piece, until she was entirely naked. Abramovitz found that men moved her into compromised positions and at one point she had to employ guards because people feared for her physical safety.

Many artists, such as Suzanne Lacy, Leslie Labowitz, and the team of Elizabeth Sisco and Louis Hoch, started making work that required television coverage to be successful. They understood that the mechanism of control lay in the hands of network executives, mainly men, who determine what is seen on television. They strategized about how to become a voice in that system and represent the point of view of women. In December 1977, Labowitz and Lacy produced the seminal performance *In Mourning and in Rage* on the steps of Los Angeles City Hall to protest the growing violence toward women in the media as symbolized by the sensationalized coverage of the murders of ten Los Angeles women by the Hillside Strangler. They held this performance at a time that gave them the best news coverage. Lacy went on to perfect her ability to attract media intervention at numerous feminist social projects. And Labowitz coordinated *The Incest Awareness Project,* which was cosponsored by Ariadne and the Gay and Lesbian Community Service Center in Los Angeles.

The issue of violence toward women and the importance of media intervention was the focus of the work *NHI (No Humans Involved),* done by artists Louis Hock and Elizabeth Sisco. While reading a San Diego newspaper, Hock and Sisco discovered that if no relatives or friends came forward insisting on an investigation when prostitutes were murdered in San Diego County, then the police would classify the case as "NHI (no humans involved)" and close it without a proper investigation. During the mid-1980s, forty-seven women were so classified. Hock and Sisco created an exhibition composed of portraits of these women and text that provided details about their lives. They also created an NHI billboard in the city of San Diego, adjacent to a busy freeway, to call attention to this issue. And they parlayed the issue into a media event in which newscasters on various local channels debated the idea of whether this work could be considered "art." From Hock and Sisco's point of view, they had accomplished their goal of infiltrating prime-time media to encourage people to think and care about this situation.

Robin Lasser has used the billboard venue to make visible the ways that women experience cultural repression about their bodies. One image with two pots aflame says, "Fear of fat eats us alive. Some women don't just diet. They die. Anorexia Nervosa" (1998). In another, entitled *How's my mothering?* (1996), Lasser presents women without faces to show that some pregnant women feel they are perceived as invisible in public and devoid of sexuality, intellect, and personality. The torsos are representative portraits of the types of cultural constructs that objectify mothers and distill their individuality to walking, ever-growing public vessels. By showing us various symbolic images of pregnancy in larger-than-life representations on a billboard outside the white cube, she is able to reach a much larger audience to consider how stereotyping of pregnancy via images functions in the culture.

Other artists, seeking to have a visible voice, have also used interventionist strategies on billboards to make cultural critiques about representations of women. In her book *Spray It Loud,* Jill Posner gives numerous examples from around the world of the alteration of billboards by women activists as a way of claiming their voice and their bodies in the public domain. Normally billboard space is expensive, financially accessible only by corporate America and outside the financial range of most artists. In addition, the idea that there is public space where people can make visible their ideas and concerns is an illusion.

Deedee Halleck has chosen to assert her voice by founding Paper Tiger Television and Deep Dish T.V. Through her own production company,

Halleck has facilitated hundreds of cultural critiques that have been done by men and women and that would not be aired on commercial television. And by creating a link with satellite broadcasting, Halleck transmits to people throughout the United States and Latin America, circumventing commercial broadcasting systems. For years, Anne Bray has headed an alternative arts organization called L.A. Freewaves, which finds venues other than commercial television for artists' videos. Her main dream is to reinscribe the vacant urban landscape with human activity that generates dialog and gives texture to the urban experience.

RECLAMATION OF DOMESTIC EXPERIENCE

Recognizing that domestic experiences and interpersonal dynamics between mothers and families had been omitted from artistic representations, since the 1970s many women artists have explored the importance of domestic space. When Mary Kelly did her seminal work *Post Partum Document* from 1973 to 1979, she documented the daily rituals and interactions between herself and her infant son to better understand the nature of that process. In so doing, she was also investigating assertions made by Freud and Lacan, the prevailing patriarchy in the world of psychoanalytic theory, about gender and personality formation during the early stages of life.

The Los Angeles–based team of Kerr and Malley did collaborative installations, both in galleries and the public domain, about reproductive rights of women: "Reproductive rights have nothing to do with morality: they have to do with business, economics, medicine, and who really controls it all."[8] In the mid-1980s, they did an installation about violence that is committed against women by people who consider abortion to be immoral. They took testimonies written by women in the nineteenth century an hour before they died describing the nature of their botched abortion and posted them on the verso side of freestanding images of women's bodies and lined the surrounding walls with black and white snapshots of abortion clinics in southern California. Their emphasis on reproductive rights highlights the increasingly active anti-abortion efforts of conservative members of Congress and the religious right. And it is testimony to the place of abortion rights in a culture in which doctors are not taught in medical schools how to perform safe abortions.

Judith Crawley, in her book *Giving Birth Is Just the Beginning: Women Speak About Mothering,* represents the relationship between mothers and children using the vernacular of snapshots of daily life to reclaim the territory of motherhood as worthy of photographic representation. And in the scholarly book entitled *Pregnant Pictures,* Sandra Matthews and Laura Wexler reproduce and discuss the work of numerous women artists who deal with various aspects of pregnancy to overturn cultural taboos against picturing the pregnant body and make visible the wide range of meanings for the pregnant body in the United States.

RECLAMATION OF HISTORY

From 1973 to 1979, Judy Chicago, in collaboration with hundreds of women, made the monumental *Dinner Party,* which exhumed the history of seminal women in Western civilization. She chose the shape of a triangle, a symbol of female power, as the table on which to display thirty-nine place settings devoted to women in history. She also created a time line of women and their accomplishments throughout Western history, which accompanied the table installation. In so doing, she was attempting to realign history to include stories and symbols from deep time as a way of challenging conventional histories and economic and social systems.

Multicultural and working-class perspectives have been an important part of the work done since 1970. A growing number of women artists have been giving voice to the realities and dilemmas that are part of their lives. Judy Bacca, Delilah Montoya, Nobi Nagasawa, Bettye Saar, Lorna Simpson, Clarissa Sligh, May Sun, Hulleah Tsinhnahjinnie, Patsi Valdez, Carrie Mae Weems, Pat Ward Williams, to name a few, have done ground-breaking work that addresses the politics and the points of view of nonwhite artists who have found their voice. Much of their work unmasks the racism embedded in cultural institutions; some of it remembers histories that have never been told, and some attempts to heal by providing images of affirmation. Along with this new generation of artists has emerged a generation of art historians, such as Whitney Chadwick, Shifra Goldman, Phyllis Jackson, Lucy Lippard, Frances Pohl, and Deborah Willis, who are recovering and strengthening the history of women artists through symposia, exhibitions, and publications.

In 1992 Coco Fusco and Guillermo Gomez-Pena did a series of performances entitled *The Year of the White Bear* about ethnocentrism and the constructed dichotomy of first-world and third-world peoples. Fusco dressed as the prototypic fetishized female "primitive" who behaved according to the Western stereotypes of pretechnological people. During the performance, she encouraged the audience to interact with her and in so doing to confront their own assumptions and biases about developing-world women. Her more recent work has evolved into installations and performances about the plight of women working in the *maquiladores* south of the U.S. border. Her larger life concern is in unmasking these biases and realities to make visible the mechanisms for controlling poor often nonwhite women there and throughout the world.

The work of Chicana artist Christina Fernandez is important to include here because it reflects her struggle to understand and connect with her Latina sisters in Mexico. She has used her own familial history or stories of women she knew or interviewed as a vehicle to begin to tell the stories of the strength and hardships of Mexicanas and Chicanas of the past and present. In *Ruin* (1999), Fernandez rephotographed images of indigenous Mexican women in the background of photographs by well-known male Mexican image makers and overlaid these pictures with her own image. In her view, most Chicanas have a nostalgic connection to indigenous women, and despite the widening cultural gap, they identify with the working-class and poor people of Mexico. Dispersion and metamorphosis of herself in these images as well as the enlargement of the pixels of reproduction are almost a metaphor for estrangement and the failure of photography to enable some type of meaningful connection between Fernandez and the women depicted in the photograph. The ultimate question posed by the work is "What is left of the self, and the knowledge, traditions, and beliefs of one's ancestors, after cultural estrangement, geographic relocation, and social displacement—after the cultural and natural sites that condition our identities are reduced to the broken stones of what once was?"[9]

For Barbara Jo Revelle, reclamation of people in history has formed the focus of her work. In the late 1980s, Revelle received a Colorado Council on the Arts commission to select eighty-seven men and eighty-seven women who had made an important contribution to the history of the state of Colorado and include their portraits in a series of mosaic murals on the front of the Denver Convention Center. The project proceeded smoothly until the question of inclusion arose. Revelle, who is an intrepid researcher, wanted to acknowledge people from the multicultural history of Colorado. The committee awarding the grant wanted to see a canonical version of history that celebrated pioneer heroics and the white male power elite rather than a history that foregrounded indigenous and working-class people. Revelle fought to include Black Panther leaders, labor activists, artists, feminists, aid workers, and leaders in the Chicano movement. Ultimately, the debate was about economic and social class. Wrangling over whose history should be included held up the project for over a year. After contentious media battles and city council hearings, Revelle's version of history prevailed. Currently, Revelle is working on a miner's history in Lafayette, Colorado, and the history of Fort Myers, Florida.

In the mid-1980s, Esther Parada did a large computer-generated piece entitled *The Monroe Doctrine* (figure 3.3) about U.S. intervention in Latin America, a subject mostly absent from U.S. media and art. In this work, she juxtaposed gritty and highly stylized images of U.S. and Latin American military personnel with extremely finegrained photo representations of women and children being subsumed by the military. The accompanying text that was integrated into the work gives verbal testimony to domination by the military of women and children. This theme of the historical domination of women was also central to the stereo transform work that Parada did at the California Museum of Photography. This work deconstructs the Christopher Columbus statues that appear in nearly every major city and town square throughout Latin America and that often include a scantily clothed woman "native" reaching up to honor Columbus. More recently, Parada has represented familial generations and constructions of family that fall outside of the conventional nuclear-family mode.

The final artist mentioned in this essay is Meridel Rubenstein, who, in works such as *Critical Mass* (1994), a collaboration with Ellen Zweig, and *Joan's Arc: Vietnam* (1999), attempts to find the healing intersection between seemingly disparate worlds at war. In *Critical Mass,* she deals with the intersection of Native Americans and the nuclear industry in New Mexico and the role played by Edith Warner in creating a "safe

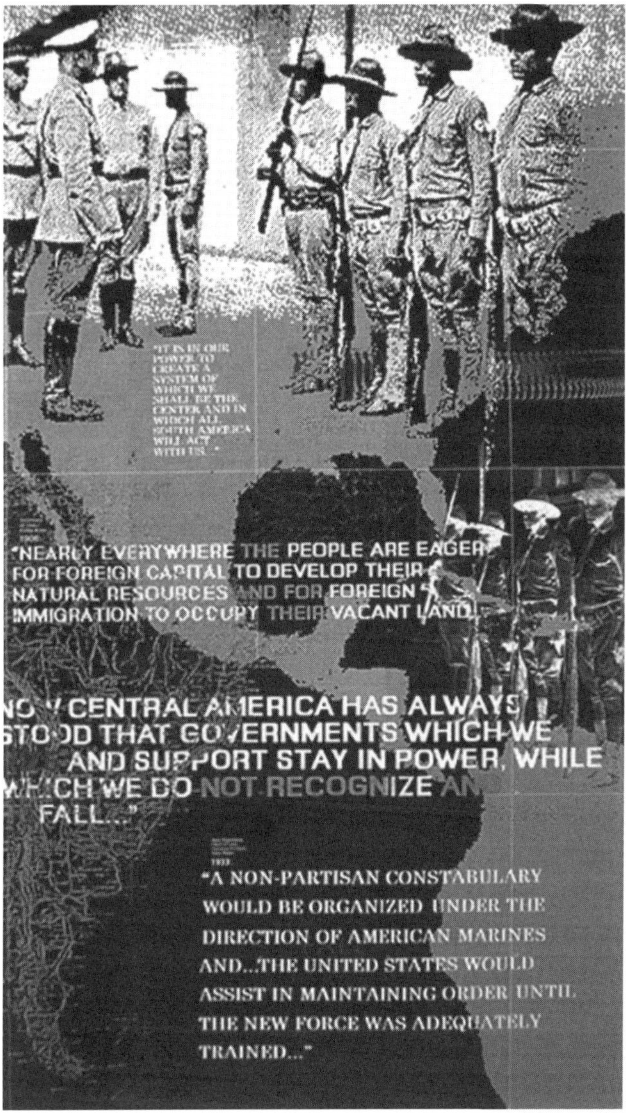

Figure 3.3
Esther Parada, *The Monroe Doctrine,* 1987, courtesy of the artist.

spot" for the scientists at Los Alamos, as symbolized by Warner's image on the side of Fat Man, the atomic bomb dropped on Nagasaki during World War II. For Rubenstein, the writing of Elaine Scarry has been important in understanding physical pain and its relationship to psychic pain. Scarry's locating the cause of war as pain within the body became a source for Rubenstein's recent work. In *Joan's Arc: Vietnam,* the warrior in a woman's body and the vanquished peoples of Vietnam are interfaced to make a work that creates a zone of communality, not just one of guilt and suffering. By naming the pain connected with war, Rubenstein investigated the forces that threatened our sense of rootedness at the end of the century and attempted to create healing through understanding.

CONCLUSION

For over five hundred years, women have actively participated in the art-making practices of their time but have been excluded from history and the system that acknowledges or communicates their ideas. In the last thirty years, however, women artists have managed to find opportunities to work together to expand the possibilities of content and for exhibition. They have seen their works enter the dialog of their time. The challenge for women is to continue this intensity of activity, to remain true to the interior voice that gives veracity and energy to art making, to continue to lobby for parity, and to generate innovative solutions for exhibition and change.

NOTES

1. Lucy Lippard, *Overlay* (New York: Pantheon Books, 1982), 4.

2. Mary-Charlotte Domandi, "The Body in Question," *Aperture,* 121 (1990): 67.

3. Ibid, 67.

4. Throughout the latter part of the twentieth century, issues of censorship plagued artists. In 1986, in an effort to curb child pornography, the National Obscenity Enforcement Unit was created by former U.S. Attorney General Edwin Meese to train local and state law enforcement officers to make obscenity-related arrests. Artists Andres Serrano, Robert Mapplethorpe, Alice Sims, and Jock Sturges are a few of the artists who have grappled with obscenity laws.

On 25 June 1998, the U.S. Supreme Court voted eight to one to uphold a provision that required the National Endowment for the Arts (NEA) to take into account "general standards of decency and respect for the diverse beliefs and values of the American public" when making grants. The decision came in a case originally brought against the NEA by performance artists Karen Finley, Tim Miller, Holly Hughes, and John Fleck, who were denied NEA individual artist grants because of the content of their work. Finley uses her body in her performances, sometimes appearing nude and covered only with chocolate, to make a statement about the abuse of women: "My work was taken out of context and eroticized by Helms on the Senate floor. . . . Who is going to be the decider of what's decent and what isn't. . . . It will be a free-for-all. The witch hunt can happen anywhere. . . . Only Justice David H. Souter, in a lone dissenting opinion, interpreted the provision . . . as violating the First Amendment. A statute disfavoring speech that fails to respect Americans' "diverse beliefs and values" is the very model of viewpoint discrimination,' Justice Souter said." Linda Greenhouse, "Justices Uphold Decency Test in Awarding Arts Grants, Backing Subjective Judgments," *New York Times,* 26 June 1998, A17.

5. Donna Haraway, *Simians, Cyborgs, and Women: The Reinvention of Nature* (New York: Routledge, 1991), 152.

6. Judith Fryer Davidov, *Women's Camera Work: Self/Body/Other in American Visual Culture* (Durham, NC: Duke University Press, 1998), 19.

7. Ibid., 44.

8. Research, "Angry Women," in *Angry Women,* ed. Andrea Juno and V. Vale (San Francisco, CA: Re/Search, 1991), 158.

9. Laura E. Perez, *Ruin* catalog essay (1999).

4

Restructuring Power: Telecommunication Works Produced by Women

Anna Couey

> We are trying to bring disparate worlds together, not so that we can all get along, but so we can see out of the "me" into "us."
> —ANNA DEAVERE SMITH[1]

Telecommunication art is the art of electronic communication networks. Its medium is communication—the structure of the distribution of ideas and meaning in a networked world. Emerging from critiques of centralized mass media (including radio, television, newspapers, and the historic role of select "visionary" artists to define our contemporary consciousness), telecommunication art often takes the form of nonhierarchical many-to-many communications—conversation.

To engage in the construction of communication structures is an ancient cultural and political practice. Communication theorist Armand Mattelart writes, "Who should control the circulation of information, the installation and functioning of long-distance communication networks—the state or the private sector? Who should be authorized to use the new services? These questions predate the arrival of the manual telegraph. They were posed during the long history of postal institutions."[2] The answers to these questions in any particular period of time reveal the power structures at the core of human relationships and cultural identity.

In the twentieth century, network communication technologies have had a tremendous influence in shaping individual and cultural perceptions across the planet. Each new technology has engaged social, cultural, and economic forces in all countries to establish (or not) "appropriate" mass communication structures. According to media historian and critic Robert McChesney,[3] corporations have increasingly seized control of new communication technologies and in the process eroded the potential for a democratic electronic public sphere. Furthermore, the costs of implementing new communication technologies have encouraged disparities of access between rich and poor nations and rich and poor people, with resulting increases in economic inequity and cultural domination by the wealthy.

Artists have engaged in questioning, envisioning, building, and using telecommunication structures since at least the early days of radio. In 1932, Bertolt Brecht proposed a restructuring of radio to "change this apparatus over from distribution to communication. The radio would be the finest possible communication apparatus in public life . . . if it knew how to receive as well as transmit, how to let the listener speak as well as hear, how to bring him into a relationship instead of isolating him."[4] Ironically, radio was initially a send and receive communication system. The proliferation of television, too, has unleashed a multitude of artistic critiques and experimentation since the 1960s, including video art, television art, satellite art, public-access broadcasting, and alternative networks. A vibrant international art exchange has developed through the fax network. The international mail art network, known as the Eternal Network, utilizes one of the most accessible and global communication networks—the postal system—to engage in democratic cultural communication.

For women artists, new communication technologies pose a particular set of issues. Mass media has firmly established itself as an immense seat of power dominated by corporations run by men. Not surprisingly, "information" disseminated over corporate mass-media networks reinforces oppressive female stereotypes, encouraging women to find liberation through consumption: "The mass media molds everyone into more passive roles, into roles of more frantic consuming, into human beings with fragmented views of society. But what it does to everyone, it does to women even more. The traditional societal role of women is already a passive one, already one of a consumer, already one of an emotional nonintellectual who isn't supposed to think or act beyond the confines of her home. The mass media reinforces all these traits. . . . Women are said to make 75 percent of all family consumption decisions. For advertisers, that is why women exist."[5]

Even though (in the United States at least) we now see more "professional" women on television programs than we did in 1970, women in the 1990s were still disproportionately exploited by electronic communication technologies. Interdisciplinary artist Coco Fusco makes the connection between the rhetoric of the Internet as a technology of liberation and the abusive production system that creates the tools that enable connectivity: "I have been conducting research on women *maquiladora* workers along the U.S.-Mexico border and in the Caribbean. Though these women have virtually no access to the Internet, they are a crucial component of the global information circuit. Not only do they assemble much of the digital revolution's hardware, but their low wages maximize multinational profits and facilitate accelerated consumption of electronic media for the virtual class. In Tijuana alone they produce more televisions than anywhere else in the world."[6]

Feminist art practice is grounded in issues of voice—talking among ourselves to understand what gender oppression is and how to transcend it; voicing our identities to destroy stereotypes that destroy us; and bringing our experiences and perspectives into the fabric of our communities as authentic voices in shaping our cultures. Women artists who build electronic communication environments, connect people to each other, share stories, and develop communication tools are taking steps toward altering social communication at a systemic level—as art. In so doing, our work not only provides a platform for our own voices but opens channels of communication for others who have been denied their voices—bringing disparate worlds together to create a world that reflects and respects all.

Women's telecommunication art practices take a variety of forms and draw from diverse influences. Access to tools has impacted the demographics of practitioners—and despite our networked communications, the field is fragmented and poorly documented. I am indebted to the women who have written about their work, to the women I have collaborated with over the years, and to the women who found time to respond to my questions for this chapter, including Isabella Bordoni (Italy), Nancy Buchanan (United States), Sarah Dickenson (United States), Anne Fallis (United States), Anne Focke (United States), Lucia Grossberger Morales (United States), Heidi Grundmann (Austria), Carolyn Guyer (United States), Jennifer Hall (United States), Sue Harris (Australia), Jacalyn Lopez Garcia (United States), Judy Malloy (United States), Aida Mancillas (United States), Cathy Marshall (United States), Karen O'Rourke (France), Sherrie Rabinowitz (United States), Elisabeth Schimana (Austria), Janet Silk (United States), Nina Sobell and Emily Hartzell (United States), Andrea Sodomka (Austria), Carol Stakenas (United States), Lorri Ann Two Bulls (Oglala Lakota), VNS Matrix (Australia), and Eva Wohlgemuth and Kathy Rae Huffman (Austria and United States). Still, this story is not complete. I hope that you, who know a different herstory than I, will add your story to this narrative.

CREATING PORTALS

In 1980, Sherrie Rabinowitz and Kit Galloway used technology to punch a *Hole in Space,* employing a satellite feed and large-scale projections to open a portal between pedestrians walking past Lincoln Center in New

York and the Broadway department store at Century City in Los Angeles: "People would be walking by, and they'd look up, and they'd see this screen and these images, and they discovered that the people that they were seeing and hearing were in fact 3,000 miles away."[7] *Hole in Space* was unannounced to the public and lasted three days: "There was the first evening of surprise discovery; the second evening was populated by word-of-mouth and long-distance telephone calls; and after the television coverage of the second evening, the third was like a mass televisual migration of families and transcontinental loved ones, some of whom had not seen each other for over twenty years."[8] *Hole in Space* collapsed geographical distance, bringing into being a window between two physical places, through which passers-by at each site could encounter each other visually in real time. The project was an expression of Rabinowitz and Galloway's concept of "image as place"—using technology to meld the artistic practice of image creation with the architectural practice of creating an environment for human contact. As a functional and public media space, *Hole in Space* articulated the connections between dispersed people in a mass media culture—and brought people together in a shared media space. The shift from artist as producer of content to artist as producer of an "image" in which the public produced content was profound.

Other early experiments with visual telecommunication technologies involving women artists include *Satellites Art Project* (1977), Sherrie Rabinowitz and Kit Galloway's investigation into "the image as place" in which dancers physically located in Maryland and California danced together in the space created by the technology;[9] *Send/Receive Satellite Network* (1977), a bicoastal program of artists' performances organized by Liza Bear, Keith Sonnier, and Carl Loeffler that was cable-cast in New York and San Francisco—extending artists' works not only across the country but into a mass-media system; Red Burns's two-way television project for senior citizens during the 1970s and 1980s; Sarah Dickenson's work with the Communicationsphere Group at the Massachusetts Institute of Technology (1977–1985) that utilized slow-scan television, two-way cable TV, and computer networking to enable artists in Japan, Amsterdam, New York, Australia, British Columbia, and Cambridge, Massachusetts, to "exchange across cultures their ideas, concepts, and work";[10] and *Electronic Café* (1984), an interactive communication link between diverse cultural communities in Los Angeles set in local eateries developed by Sherrie Rabinowitz and Kit Galloway.

While the specific investigations that these artists undertook were quite different, each focused on the creation and use of electronic communication structures to connect people to each other across physical distance, rather than utilizing new media to create visual images. As Sarah Dickenson describes it, "Because of this interaction among artists over the electronic media channels, a specific form of art began to emerge, rooted in the fine arts but shaped by the electronic media themselves. In effect, we began to see the development of a new visual and spatial language that bridged the arts and technology, for it was not concerned with the art object but rather with art as communication."[11] Another concept that formed a cornerstone of many subsequent telecommunication arts projects was the involvement of the public as collaborators in making art.

MAKING INFORMATION

While communication arts experiments investigate the potential of horizontal communication as a social construct, women artists have also invented alternative models for the production and dissemination of information through telecommunication networks.

In 1986, Judy Malloy began to gather information for *Bad Information Base No. 2,* a collaboratively produced information artwork "which was conceived as a database of wrong, misleading, inappropriate information and was meant to question our reliance on the veracity of computer-delivered information."[12] She opened a discussion topic on the Art Com Electronic Network (ACEN) on the Whole Earth 'Lectronic Link (WELL), a computer-based conferencing system, inviting users to post bad information for the *Bad Information* database. The information-collection stage lasted for well over a year, as artists, computer programmers, futurists, and other members of the WELL community shared over four hundred pieces of bad information on such topics as advice, health, relationships, football, politics, religion, technology, body odor, sex, computers, foreigners, television, horoscopes, miracle cures, the WELL, happiness, music, toilet paper, environment, pets, transportation, food, success, salt, and work. Malloy catalogued and organized the information "in a definitive database"[13] that is today accessible on the Web, complete with bad technical support.[14] As an art project, *Bad Information* engaged people who do not consider themselves artists in consciously creating art content. As an information project, it redefined the public as experts in information production.

Nancy Buchanan organizes evidence to develop actionable "portraits" of social conditions that are carefully kept out of corporate-dominated mass media: "I hope to counter the passivity of television and also to use video to challenge the accepted status quo 'truths' of various social clichés. I also want to demystify the media, challenging the notion that important work can be done only by (industry) 'experts'."[15] Buchanan's work draws from community information sources and experiences to weave a portrait of current events, often involving community members themselves in the production of information artworks. Working in and with communities in Los Angeles since the mid-1980s, Buchanan has utilized video art, public-access broadcasting, computer-based information technologies, including the Internet, as well as art galleries as vehicles for disseminating her work. Her recent CD-ROM/Web work *Developing: The Idea of Home*[16] expands the idea of home as a traditionally domestic female territory into its broader social context: "As a resident of southern California, my own home area carries with it many other considerations such as water procurement and use, destruction of fast-disappearing habitat, covenants prohibiting certain lifestyle choices within an area, location of toxic waste (and 'greenwashing' campaigns to hide them), inflated values and risky bank practices, tenant organizing techniques, alternative home ownership schemes, etc."[17] Using associative reading techniques, *Developing: The Idea of Home* incorporates the Web as a means of maintaining an updated information resource on the issues and groups contained in the CD-ROM—linking living information and real-world activism into an artwork that evolves as the issues it presents change over time.

CONSTRUCTING COMMUNICATION NETWORKS

The construction of communication networks—perpetual communication systems—as art is an extension of the idea that telecommunication events are communication sculptures. As a sculptural form, communication networks address issues of dissemination of ideas and of community building (for networks exist only to the extent of their use). When successful, communication networks blend into daily living, becoming a part of the social landscape—organic sculptures that take on a life of their own. Similar ideas have been applied in the artist space movement and art publishing. Artists working with a variety of new media and mass media technologies have developed continuous communication networks.

Following *Electronic Café,* Sherrie Rabinowitz and Kit Galloway realized they needed to establish a continuous telearts venue to continue their investigation of "image as place." Their early works stand as "models of what is possible. After *Electronic Café* in 1984, we realized that we needed to create a permanent facility. . . . We wanted to build a permanent public telecommunication lab where we could connect with other people and build an international network."[18] By 1995, Electronic Café International, based in Los Angeles, had approximately thirty network affiliates in Brazil, Denmark, Israel, New York, Toronto, and other locations around the world. Since its founding, Electronic Café International has produced an extensive range of live multipoint performances, involving virtual-reality technology, three-dimensional multiuser navigable online worlds, telerobotic devices, the Internet, and hybrids—such as mapping the movements of a live performer over the Internet to activate an avatar that performed with a video-rendered performer (1997).

Kunstradio-Radiokunst,[19] established by Heidi Grundmann in Vienna, Austria, in 1987 as a forum for original artworks for radio, continues more than a decade later as a vital space for networked sound projects that travel both radio networks and the Internet. As producer and curator of the program, her work included the production of *Realtime* (1993), a live interactive work for radio and television, and *Horizontal Radio,* a live twenty-four-hour multimedia radio project. In 1999, she curated *Sound Drifting,* an interdependent temporary system of international remote subprojects that used a wide range of methods and approaches to the generation, processing, and presentation of data, sounds, and images to form an innovative nine-day-long continuous online, on-site, on-the-air sound installation. Currently, Elisabeth Zimmerman is the producer for Kunstradio, and Grundmann consults on projects and developments. While Grundmann does not consider herself an artist, she has established a space for art communication experiments within mass-media systems.

Nancy Buchanan's collaboration with community activist Michael Zinzun to produce the monthly cable show *Message to the Grass Roots* at

Pasadena Community Access Corporation (PCAC) is one example of artists' efforts to build an alternative, community-based voice using mass-media systems. The fact that the program was produced continuously (1988–1998) rather than as a one-time event established it as a forum for community empowerment within a social structure that overwhelmingly disempowers community experience. *Message to the Grass Roots* still airs on PCAC; its programs on police brutality, racism, South African liberation, and other issues remain relevant. Community-based media production workshops are also a component of Buchanan and Zinzun's collaboration and continue through the Coalition against Police Abuse and Community in Support of the Gang Truce, further strengthening connections between local, national, and international communities and issues.

Prior to the widespread availability of Internet access in advanced capitalist countries, a number of women artists and cultural workers actively participated in the construction of computer-based communication networks. These include the Art Com Electronic Network (ACEN), which was launched on the Whole Earth 'Lectronic Link (WELL) conferencing system in 1986 as an online art publishing venue and was soon transformed into a "virtual artists' space" for creative collaborations between artists and the cyberspace public. Although Carl Eugene Loeffler and Fred Truck conceived of ACEN, women, including me, Nancy Frank, Donna Hall, Darlene Tong, and Lorna Truck, played an important role in its design and implementation. Artists Judy Malloy and Abbe Don also played active roles as ACEN community members. In the early 1990s, Sue Harris and Phillip Bannigan formed ArtsNet on Pegasus, the Australian node of the Association for Progressive Communications (an international network of conferencing systems for the social justice community). In South Dakota, Anne Fallis operated the Dakota BBS, a project of American Indian Telecommunications that provided online information services to rural and tribal communities and served as a distribution medium for Native American clip art. Anne Focke took the lead in laying the groundwork for Arts Wire,[20] a United States–based online system for artists and arts organizations that began in 1992 on the Meta Network, and I was the program's first network coordinator, continuing to build Arts Wire's culturally diverse community base.

ArtsNet and Arts Wire were designed as broad-based networks that would connect artists and arts organizations across cultural, racial, class, institutional, and discipline barriers. The idea was to construct a democratic communication system that would enable the arts community (both were conceptualized as national networks) to share ideas, exchange information, and facilitate arts advocacy (particularly crucial for the U.S. arts community as it struggled against the Christian right–instigated culture wars in the early 1990s). American Indian Telecommunications focused on addressing the cultural and economic implications of new communication technologies for Native Americans. As Jim May (Cherokee) pointed out in testimony to Congress: "Not only are we not publishing materials about ourselves, but we also do not have adequate access to reliable information from the outside world. This is a serious problem since it affects our health, our economic development, our education, and almost all those aspects of our daily lives that we have in common with all people. We miss out on opportunities to improve our lot by not being connected to electronic resources."[21]

As Internet access has become more commonplace in the advanced capitalist countries, the methodology for building cultural communication systems has changed. Rather than connecting computers and telephone networks to build a network, artists tend to use existing Internet infrastructure. Specialized electronic mailing lists and listservs serving all kinds of cultural communities and communities of interest proliferate and often become so active that their members lack the time to participate in more general communication systems. The vision of a new mass-communication paradigm—horizontal communication among millions of people—has in practice become more focused and exists beside the commercial, broadcast models that mass media and other corporations have brought to the World Wide Web.

Face Settings,[22] initiated by Kathy Rae Huffman and Eva Wohlgemuth in 1996, is a network project that combines dinner performances with offline and online communication and community building, connecting women in Belgrade, St. Petersburg, Bilbao, Glasgow, and Vienna. The project grew out of Wohlgemuth and Huffman's observations about the status of women in Russia, women's communication styles, and the lack of support for European women to work with new communication technologies: "The objective was to join real groups of women in networks and to engage women in discussions of importance on local, regional, national, and international levels."[23] Over several years of dinner performances in different cities, *Face Settings* explored how communication practices differ across cultures, how women communicate differently than men on and offline. The project also sought to encourage women to participate in the formation of "online culture." FACES,[24] an online mailing list that grew out of the project, has become a community-building mechanism of its own that "connects women from areas that border on the fringe territories (and concerns) of the European media centers."[25]

WEAVING NETWORKS

Communication networks form an invisible geography that intersects the geography of physical place but is defined by political, economic, and cultural systems. The interconnections between communication networks and places enables a kind of conceptual weaving—the opportunity to map the world according to different sensibilities and to form reciprocal communications across geographic, political, perceptual, and temporal borders. These weavings are steps toward an art that connects diverse modalities of living into language and experience that articulate the whole. It becomes a kind of cubism of social space that reflects simultaneous diversities and initiates relationships between them in an art/life process.

Jennifer Hall, Susan Imholtz, and Joan Shafran conceived *Netdrama* in 1985 as a "telecommunication project that develops the conceptual parameters of electronic spaces for artistic use, defines the performance aspects of creating online theater, and tracks and documents the electronic lives of characters created within the electronic networking environment."[26] Written and performed on bulletin board systems (BBSs) throughout the east coast of North America in collaboration with online communities, *Netdrama* flattened the hierarchy between audience and artist and enabled a theater work to play a daily role in the lives of online communities. Although framed as art—*Netdrama* producers posted announcements on participating BBSs—the characters performed on the same virtual stages as community members. "The audience becomes gradually less aware that this character is simply part of a performance and begins to treat her more like a real member of their electronic community," Hall writes about the character Mindy.[27] The art/life hybrid took on a life of its own, independent of its initial creators: two years after the event appeared to have ended, the producer discovered that "characters are now self-propelled personas. Without the aid of designers, writers, or producers, the actors keep their characters alive by establishing them on additional networks and inviting more people to participate in the drama. Designers are found setting up scenarios for active audiences on four separate networking systems. The two characters in love have gone on to begin a new event where they now live the electronic lives of their choice."[28]

Cultures in Cyberspace, a project I organized in 1992, took the structural form of an open panel that developed within five online communities and was subsequently shared among them to foster a grassroots cross-cultural discussion on an issue that affected all participants:

> As with multinational corporations, computer networks are drawing new lines of social organization. . . . This technology would seem to incur a new social order—one based on reciprocity and interaction rather than imperialist domination. . . . The catch, of course, is that computer networks are not accessible to everyone. . . . What will happen to cultural groups that remain offline? Will cultural groups that do access cyberspace lose their distinct identities through a process of interaction? And, if so, is such an occurrence cultural evolution or homogenization?[29]

I asked community members at each online system to introduce and facilitate "local" discussion: George Baldwin, Anne Fallis, and Randy Ross on Dakota BBS (a project of American Indian Telecommunications, South Dakota), Sue Harris and Phillip Bannigan on ArtsNet (Australia), Joe Matuzak on Arts Wire (United States), John Quarterman on ⟨alt.cyberspace⟩ (USENET), and Judy Malloy and Eric Theise in the Virtual Communities conference on the WELL. The participating communities were culturally and geographically diverse and represented different types of online communities—local, national, and international, communities of place or culture, as well as distributed communities of interest. In realizing the project, we encountered cultural differences and technical difficulties—our proposed USENET newsgroup that was to carry cross-system discussion threads was denied distribution; one of the coordinators could not get online from Italy to facilitate the ⟨alt. cyberspace⟩ discussion; discussion on several systems was quite active, while on other systems there was little or no activity—reflecting either the nature of the online communities, the coordinator's available time, or the degree of relevance the project had for the communities it sought to engage. The work itself revealed very real social and technical borders in developing cross-cultural communications for discussion of communication policy issues at the grassroots.

In 1993, Aida Mancillas created a digital manifestation of *Project Artnet* in the form of an electronic artists' book. Developed in collaboration with Lynn Susholtz, *Project Artnet* was designed "to bring together artists, community members, social service agencies, and local arts organizations as part of a collaborative effort to foster neighborhood pride and cross-cultural respect."[30] The project engaged children who lived in an ethnically diverse inner-city San Diego neighborhood in learning about, documenting, and sharing their family histories through interviewing, drawing, and writing poetry. The telecommunication component of *Project Artnet* connected children with people outside their immediate neighborhood. Using an interactive conference on Arts Wire, *Project Artnet* displayed poems and stories written by the children. An interactive discussion item enabled Arts Wire members and attendees of SIGGRAPH 1993 in Anaheim, California, to share their own stories with the children. (*Project Artnet* was exhibited at SIGGRAPH 1993 as part of *Matrix: Women Networking,* organized by Lucia Grossberger Morales and myself.) As part of the project, the children went online—still a rarity in 1993. Mancillas described the impact of the children's online experience as expanding their sense of space—the world outside their immediate neighborhood became alive, something they could participate in.[31]

Isabella Bordoni weaves communication systems and perceptual structures. In 1985, she cofounded Giardini Pensili in Rimini, Italy, with Roberto Paci Dalò as a theatrical ensemble that focuses on "investigation of acoustical and visual perception, of language and communication systems, new technologies and their relationships with memory and history."[32] For the collaborative work *Realtime* (1993), a Kunstradio-Radiokunst performance, Bordoni took care of the text and its dramaturgy. The artists involved in *Realtime* were both authors and performers; together they developed a shared telematic stage for actors and the public in three Austrian radio and television stations. As images, sounds, and Internet data transmissions fed into the network, robots and sound generators enabled artists in multiple locations to modify and reciprocally control the development of events in real time. For Bordoni, who brings to telecommunication art a background in writing and theater, "media art offers the possibility of activating perceptual structures and exploring models of associations across disciplines, across diverse areas of knowledge."[33]

Like many telecommunication artists, Andrea Sodomka prefers a collaborative approach to making art—in any medium: "Many of the structures of telecommunication art, like the question of authorship or simultaneous events, existed in my work long before I started to deal with new communication technologies."[34] *State of Transition* (1994), a project conceived by Sodomka, Martin Breindl, Norbert Mather, and Gerfried Stocker, utilized radio and the World Wide Web to explore "forms of migration, highways, immigration rates, transit spaces, crossing borderlines, transitional stages of all kinds [that] were subject and structure of this live event."[35] The virtual paths of the audience through the work wove into the fabric of the event: when a visitor accessed the project's Web server, a trace-route routine began to analyze the route of that individual connection, resulting in an "acoustical map" that marked their journey. Sodomka views the Web as having "different aesthetic and conceptual structures than known before. A new language had to be learned. And creating this new language was my main interest."[36]

Karen O'Rourke started the project *Paris-Réseau* in collaboration with the art group Art-Réseaux in 1994 "to realize an imaginary portrait of Paris using the combined experiences of people who live there and others elsewhere, who picture, each from his own particular vantage point, travel in that city."[37] In gathering data to map the portrait, Art-Réseaux "characters" described former habitual itineraries, which were then documented by "implacable photographers" who retraced the routes. With their bodies, traveling from the center of Paris outward, six "reporters" left the Vidéotheque de Paris, taking photographs and making videorecordings of their journey. On the Internet, Art-Réseaux posted an information request, asking "informers" who believed that they have met one of the "characters" to describe the encounter. Making connections through history, the artists recorded the itineraries of "the ancestors" of Paris—from the time of Saint Denis to Andre Breton. The resulting data was meticulously catalogued on an initial CD-ROM in a parody of a guidebook, for public perusal: "The would-be traveler will be taken aback by the sheer quantity of all this rather useless information catalogued with such precision. . . . Nowhere will the traveler find the history of the Louvre or even the price of a decent hotel room."[38] *Paris-Réseau* maps the city as an overlapping accumulation of physical networks (streets, subway routes, telephone lines), temporal networks (linkages across generations that have lived in the same place), and memory (lived experience, memories of lived experience, inventions, news).

Over several years, *Paris-Réseau* has continued to expand and shift. The artists' initial concern with process developed into a desire to connect the fragments of collected data into a whole. O'Rourke developed a *Paris-Réseau* Web site that organized the data in juxtaposed fragments, with linkages that at times seemed significant and sometimes not. *Paris-Réseau* was redesigned and released in 2000 as a CD-ROM that provides four points of view from which to explore the city, four different systems of navigation—perceiving, imagining, exploring, and transforming the objects on the screen. Data fragments may reappear in different contexts, changing their meaning. The project continues, as O'Rourke describes it, as a "(net)work in progress," with a concurrent investigation into how to meaningfully archive or represent a living information system.[39]

In *Local 411,* Janet Silk and Ian Pollock used San Francisco's public phone system to expose the impact of a new downtown arts center on the lives of the members of the community displaced by it: "We want to talk about histories that can exist in the present and the psychological dimension of the telephone network that speaks of vanished spaces that remain in memory."[40] The artists used several strategies to gather together the scattered memories of the vanished neighborhood. They constructed fictional vignettes based on research about the area and its former residents, recorded them in English, Spanish, Mandarin, and Tagalog, and made them accessible on a voice-mail system: "one might start by hearing a story in English and then be exposed to another language, much like how one might come across different languages while walking down a busy street in San Francisco."[41] The voice-mail system also solicited listeners' stories, which were added to the pool of memories and presented to subsequent callers. Lastly, the artists gathered a group of performers to develop characters created from the history of the place: "people looking for friends or lovers, individuals looking for buildings that were no longer there, and ghosts haunting the phones."[42]

The characters placed calls to public telephones in the Yerba Buena area and engaged the person who answered in a conversation about the changes in the neighborhood: "For example, one of the performers would ask to speak to a specific person by name ('Is Jeff there?'). Then when the person who answered the call would report that the person was not there ('Hey man, this is a park'), the response was a lead into conversation about the neighborhood and people that used to live there."[43] Silk and Pollock report that these one-on-one performances enabled audience participants to explore the issues of dislocation and gentrification in depth. *Local 411* reveals the multiple layers of an urban place and the complex issues of urban development—the destruction of one neighborhood for another, particularly questioning the role of art in the gentrification process.

Day Without Art Web Action[44] engages artists and AIDS activists together in shaping a public place on the Web for people to mourn the people who have died, to find critical medical information, to "seek out services, find community, to become an electronic activist."[45] Organized annually by Carol Stakenas at Creative Time in New York since 1995, *DWA Web Action* coincides with the Day Without Art held on December 1. Carol works with artists to design Web works for each year's event and has accumulated an extensive collection of links to AIDS information resources and activist Web sites. One recent art project, *The Wish Machine* (see figure 35.2*)*, developed by Chrysanne Stathacos, invites the public to "submit a wish to activate the powerful energy of imagination and hope."[46] The wishes are posted to the site, articulations of a communal ritual that brings people together and encourages action: "To wish, or desire, becomes especially resonant against the broad continuum of how the HIV/AIDS pandemic continues to effect our lives and our culture from the promising news of a possible vaccine to the ominous certainty that this fatal virus is continuing to spread."[47] The project also facilitates political action across the Web: Stakenas invites other Web site producers to participate in the Day Without Art by adding the Day Without Art logo and link to their home pages for the day: "Several sites employed a more radical approach that echoed the classic DWA shrouding ritual.

They disabled entry to their Web site and simply displayed the linked DWA logo on a black background for the day. Other sites opted for an elegant solution that didn't deny access to their site by placing the DWA logo on a special buffer page for the user to 'pass through' and yet held their attention for a moment to consider the special focus of the day."[48]

In 1996, Austrian sound artist Elisabeth Schimana[49] created *the fugue,* a score for four remote conductors whose commands are interpreted live in a concert that occurs simultaneously on air, online, and on-site, with collaborators Michael Moser (musical concept and violoncello), Ludwig Zeininger (sampler), Peter Machajdik (electric guitar), Fuchs/Eckermann, Andrew Bentley, Karl Petermichl and Paul Modler (conductors), Martin Leitner (technical director), August Black (Web design), and Martin Schitter (score programming).

Describing *the fugue,* Schimana writes:

The Scene
is the place—where the river
March joins the Danube, forming
the political border between
Austria and Slovakia. On the
Austrian side is a small belt of the
Au. Opposite this, on the
Slovakian side is the limestone hill
of Devin telling the story of
permanent settlement since the
stone age.

The Bridge
a boat—as a connecting point, will
be located on the mouth of the
March. Here, 4 musicians will play
and receive commands via Internet
from 4 separate conductors.

The Sounds
environmental sounds, cello,
sampler, vocals, electric guitar

The Musical Structure
the fugue—the fleeing of voices from each other. The theme of the fugue is the break, followed by the instrumental answer to the place.

The Sound Projections
to the radio, the Internet, and the promenade of speakers along the river bank of Devin[50]

The performance is a dialog between the sounds of the place and the musicians' response. Schimana writes, "Each voice in this piece has its own theme. These sounds are taken live from the acoustic environment of the place of our performance via microphones. There are sounds from the ship (engine room), the AU (birds, trees, wind), the water (the river March), and Devin (the audience). The musicians had to find an instrumental answer to the place: the five models. These complex sound structures are together with the themes the basic sound materials of the piece."[51]

IDENTITY

In a networked, mediated world, what happens to the identities that we have formed across history, identities shaped by our physical, local communities and cultures? While broadcast media, controlled by powerful corporate interests, reflects a narrow spectrum of human experience and perspectives, the relatively open territory of the Internet has held a promise for expanding that spectrum. Conversely, as a media environment in which identity is often articulated through language rather than physical presence, the Internet has also provided a place to explore new identities and to expose the ways in which identity is constructed. Relationship to identity—historically rooted in physical place, defined by physical appearance, and politically impacted by physical manifestations of power and aggression—is a cultural battlefield in cyberspace. Will cyberspace simply mirror our "real-world" legacy of identity marginalization, or can it be made a place to transform culture erasure and ensure a way of viewing that embraces diverse histories and systems of knowledge?

"I am Oglala Lakota and come from a culture that is rich in philosophy and thought, and I was raised up in the 'old way' where women are considered to be the backbone of the family. My work simply reflects my life and self-expression. I consider this feminist movement as belonging more to the white man," writes Lorri Ann Two Bulls.[52] In 1993, she created over 100 digital drawings using NAPLPS, a computer graphics program designed for the videotext industry. At that time, NAPLPS was perhaps the only graphics tool that allowed images to be viewed online and was not widely used: English text was the dominant communication medium on computer networks. Two Bulls' digital drawings depicted Native American themes—"drums, dancers, bead work, traditional costumes"[53]—and were produced to be sold, as part of a Native American clip art catalog that Anne Fallis was developing. The drawings were distributed as share art (view for free, download for $25) on Dakota BBS. While other image formats of the time allowed computer graphics to be transported over computer networks for downloading and viewing offline, NAPLPS viewing software (distributed as shareware), allowed people to see the digital images drawn online right before their eyes, establishing an immediacy of visual communication. The drawings by Two Bulls and other Native American artists not only explored new methodologies of cultural exchange but established a culturally appropriate communication system emphasizing visual rather than text-based communication.[54] As Two Bulls points out, "In hindsight, Anne was probably at the cutting edge of what was to come in computer clip art and saw ahead of her time because now Native American clip art is widely sought after."[55] Unfortunately, Two Bulls experienced the down side of online art distribution: her works were copied and sold by others without permission and with no remuneration to her. In 1993, the digital medium was too new for her to mount an effective legal battle against copyright infringement, and since then, she has focused on creating computer art offline (protected under traditional copyright law) and her handmade painted jewelry business.

Jacalyn Lopez Garcia writes, "As we crossed the Mexican border, the border patrol would ask me my citizenship. I would reply 'American' because my parents taught me to say that. But in California, people would ask me 'What are you?' I guess they didn't quite know how to ask 'Are you American?' I would proudly reply, 'Mexican.' It wasn't until I became a teenager that I claimed I was Mexican-American. . . . It wasn't until I became a reentry student at UC Riverside (in the 90s) that I developed a Chicana consciousness."[56] Lopez Garcia developed her work *Glass Houses*[57] in 1997 as a self-portrait that would tie her personal experiences and issues of identity with broader issues of "race, class, acculturation, and nationalism."[58] Yet she struggled with doubts about using her family history as the basis of the work and with the ethics of revealing painful family memories for public viewing. Ultimately deciding to proceed with the work, Lopez Garcia designed a Web environment that is mapped from the floor plan of her own home. By leaving a house key under the doormat and inviting members of the public to enter as house guests who are free to wander through her home, Lopez Garcia creates a friendly and intimate space in which to experience her stories. *Glass Houses* is also a conversation space: in the kitchen, house guests can leave their own comments for her and other visitors to read. The power of the work becomes clear in the comments from house guests, many of whom reveal their own struggles with similar issues. For Lopez Garcia, the decision to tell her own stories publicly is validated in these messages, which she reads with her mother "out loud as tears roll down our faces."[59]

Mother Millennia[60] is a collaborative work conceived by Carolyn Guyer that weaves personal stories into a communal, planetary portrait of mother. "Remembering our mothers," Guyer writes, "or hearing older relatives remember their mothers is a human commonality that will never sound the same twice but will always resonate with the memory and desire we all use to create ourselves."[61] For Guyer, an important key to the work as a successful cross-cultural experiment is that it holds two viewpoints simultaneously: that of specific individual experiences and that of a vast composite that forms the whole—our global conception of mother. To underscore this point, she uses NASA photographs of earth viewed from space as the visual context for the work, a reminder "of that process of combining two viewpoints and that, in an ancient sense, we all share the same mother."[62] *Mother Millennia* invites visitors to the Web site to contribute stories about their mothers—images, texts, multimedia, multilingual, essays, fictions, poems; the form is limited only by the Web itself. But the project also links to the earth; Guyer seeks stories from people who do not have access to the Internet. The planetary portrait of mother has multiple points of entry. One may find a story by going to the story index—where each story is alphabetized and listed with author, language, medium, type (essay, poem, etc.)—or by looking up the list of authors. Alternatively, one might meander through the stories thematically or geographically or by following what Guyer calls "idiosyncratic" links of her own choosing. Started in 1997, the project continues. While Guyer serves as project editor, participants also shape editorial direction. For example, *Mother Millennia* includes a thematic thread of stories about fathers because contributors wrote stories about their fathers and insisted that fathers be included.

In 1994, Nina Sobell and Emily Hartzell created a virtual *ParkBench*[63] that provides people in New York City with open access to the Internet as well as the ability to see and communicate with each other using videoconferencing and a collaborative drawing space: "In a city where strangers rarely talk to one another on real park benches, *ParkBench* would be a safe place to congregate in cyberspace."[64] *ParkBench* also provides a platform for the public to remote-access an online performance series called *ArTisTheater*. With the use of a telerobotic video camera, Sobell and Hartzell have developed a number of performances that explore identity, voyeurism, and power in digital space. In *Alice Sat Here* (1995), a telerobotic video camera engages on-site and online audiences in riding a wireless vehicle named VirtuAlice. The online audience, using remote control, determines the direction the vehicle will take. The on-site audience drives the vehicle, acting as "chauffeur" for the remote user. During a Web performance in 1997, Web viewers watched Sobell and Hartzell through an active female gaze: "The camera moves from our eyes to our mouths to our hands to the work, as Nina sculpts Emily drawing Nina. The female gaze is perceived as observation in the art-making process. The cameras establish a rhythm with their movement; they record the physical process of perception and representation. Eyes move to observe and record, mouths move involuntarily, hands move to coax form out of media, and the work records the materialization of the process. Through observing one another we discover ourselves, and as the piece progresses, each artist appears on the opposite screen, in the hands of the other."[65]

VNS Matrix was a group of Australian cyberfeminist artists that was active from 1991 to 1997. As VNS Matrix, group members Virginia Barratt, Francesca da Rimini, Julianne Pierce, and Josephine Starrs presented several installations, events, and public art-works involving new media, photography, sound, and video. The impetus of the group was to "investigate and decipher the narratives of domination and control that surround high-technological culture and explore the construction of social space, identity, and sexuality in cyberspace."[66] The project that the group pursued was one of debunking the masculinist myths that might alienate women from technological devices and their cultural products. The artists believed that women who hijack the tools of domination and control can introduce a rupture into a highly systematized culture by infecting the machines with radical thought, diverting them from their inherent purpose of linear topdown mastery.

VNS Matrix's *Cybermanifesto for the 21st Century* articulates a cyberfeminist identity and subverts the language of capitalist technoculture:

> we are the modern cunt
> positive anti reason
> unbounded unleashed unforgiving
> we see art with our cunt we make art with our cunt
> we believe in jouissance madness holiness and poetry
> we are the virus of the new world disorder
> rupturing the symbolic from within
> saboteurs of big daddy mainframe
> the clitoris is a direct line to the matrix
> VNS Matrix
> terminators of the moral code
> mercenaries of slime
> go down on the altar of abjection
> probing the visceral temple we speak in tongues
> infiltrating disrupting disseminating
> corrupting the discourse
> we are the future cunt[67]

Along with Sadie Plant, VNS Matrix coined the term "cyberfeminism" in the early 1990s. Their Internet-based works include "corpusfantasticaMOO," an interactive text-based environment designed as the interior of a body. Visitors enter and traverse the body at will, using it as a locus point for engaging in discussion about gender and cyberspace.

"corpusfantasticaMOO" is a colonized body, where entities without number meet. You may not understand some of the language you encounter in this body, and it would be advisable to familiarize yourself with other methods of constructing meaning. Never assume that you are speaking to a member of a privileged class, race, gender, or species. We provide mindnet access for entities with particular needs. What resident or guest entities say or do may not always be to your liking. Beware—there is no moral code in this "place."[68]

STRUCTURING LANGUAGE

Telecommunication systems are designed environments. While many of the tools are conceptualized with utility in mind, they reflect the cultural and political biases of their designers and developers. The Internet is an information communication space that is deterritorialized in the sense that it is nonphysical and crosses geographic borders. Yet the concept of deterritorialization obscures the Internet's extreme territoriality: access is heavily dependent on wealth, class, and technical literacy. Furthermore, the structure of the system is consciously driven by corporations seeking to build new methodologies of gathering and using demographic information to target product sales, to entertain and seize "eyeballs" (as they disparagingly describe visitors to their Web sites), and to be more "efficient" by developing information-delivery systems that displace workers. The communication exchange that evolves from such goals is often manipulative and frequently covert, a linguistic structure that turns people into consumption targets. While women's telecommunication works in general challenge the structure of corporate-designed communication systems, some women artists work in particular on designing new media tools and linguistic systems that have a nonhierarchical agenda.

Judy Malloy's work extends from developing text-based artworks that mimic human thought processes to developing linguistic structures for the Web in collaboration with Cathy Marshall. According to Malloy: "For many years, I had been working on a series of artists books that attempted to simulate our fragmented, random, repetitious, nonsequential human memory patterns—using card catalog containers or electromechanical address books. I saw that—in tandem with interactive, community-based publication—computer-mediated telecommunication made possible these nonsequential or simultaneously parallel narrative structures that

I sought. *Uncle Roger,* begun on ACEN in 1986, used a database linking structure similar to what is now called hyperfiction."[69] *Uncle Roger*[70] is a narrative told by Jenny, who views the technoculture of Silicon Valley from her experience as a baby sitter and clerical worker. The narrative is comprised of three "files," each a compilation of story fragments. Files 1 and 2 are accessible by keywords—Jenny's experiences with particular characters or places. The reader thus chooses a path through the story, at times, like Jenny's memory, calling up a repeated fragment. File 3 uses a random number generator to recall Jenny's memories much as she remembers them herself. This method of storytelling is for Judy a consciously feminist approach toward writing: "[My] hypernarratives *Uncle Roger, Its Name Was Penelope, The Yellow Bowl, Forward/Anywhere* (with Cathy Marshall), *lOveOne,* and *The Roar of Destiny Emanated from the Refrigerator* are based on the feminist approaches of making the woman the subject rather than the object of the work, of connecting the reader with women's lives and thoughts; of validating daily and personal experience as a way of understanding and expressing a culture; of creating a collaborative and/or interactive environment."[71]

Forward/Anywhere,[72] a collaboration between Judy Malloy and Cathy Marshall, began as a communication exchange between an artist and a researcher who had just become acquainted with each other. The project was a response to Xerox PARC's artist-in-residence program, which was designed to build bridges between artists and research scientists. Malloy and Marshall exchanged "lexia" or screens via e-mail over two and a half years, drawing from each participant's life experiences "with new content arising through association."[73] When they sought to design the exchange for a Web environment, they looked for ways to represent for the reader their own experiences of the stories unfolding, as well as to offer a more active role for reading the text than typical Web hypertext links. The Web implementation that Malloy and Marshall designed uses a "Forward" link to enable the reader to follow the associative process that the authors employed. An "Anywhere" link employs a random number generator to allow the reader to jump around in the text, providing a reading experience that approximates the repetitive way that memory works. "Lines" gives readers the option of entering a term of their own choice, retrieving all references to that term as one-line text fragments, and linking to the stories in which they appear. As Marshall writes, "Clicking, changing channels, and—in a most unholy appropriation of verbs—surfing have, then, become the common modes of interaction with texts. Choice equates with interaction: 'I click, therefore I am' may well be the slogan of the Pepsi generation. But link following can be (and often is) a very passive form of engagement."[74] *Forward/Anywhere* shapes an active reading methodology that restructures the power dynamics between reader and writer.

Do While Studio—directed by Jennifer Hall in collaboration with Blyth Hazen, education coordinator, Joan Shafran, principal, and others—provides an organizational framework to nurture collaborations between artists and industry as a way of shaping communication tools and providing artists a means of financial sustainability. "Do While Studio is a small and focused community offering an alternative to the way technology is assimilated in day-to-day art practice," write Hall and Hazen.[75] Rather than exploiting new technological tools for their artistic potential, Do While takes the tactic of consciously designing the tools that will define how we communicate online. One artist-industry collaboration developed video-conferencing tools to create a *World Wide Simultaneous Dance* on the Internet. Laura Knott, the coordinating artist-in-residence at Do While Studio, worked with Tim Dorcey of BoxTop Interactive, who was leading the development of iVisit, a video-conferencing package. With the software still in beta, Knott and Dorcey were able to discuss and reconstruct the underlying values and philosophies built into the tool. In describing the framework for Do While Studio's work, Hall writes about the social role played by artists in an information age: "Participation in the new global dialog demands the ability to navigate through massive quantities of information while reassessing notions of space and time. We find ourselves immersed in a tremendous volume of disconnected ideas. Artists have responded to the information complex by developing a new working paradigm: connecting chunks of data. We have become concerned not with a specific piece of information or a particular image but with the changing relationships of elements within an overall structure, with customized views, and with the coherent transmission of ideas from one place and time to another. The data that are collected, as well as the information systems that explore and reveal the data, shape our social perception."[76] In the context of current global Internet technology and policy developments, Do While's work represents an important strategy for constructing a socially beneficial communication system from a creative, culturally grounded perspective.

LOOKING BACK, MOVING FORWARD

The women artists who are shaping the undefined and shifting territory that comprises our contemporary media landscape have taken significant steps in redefining ways that mass communication systems can be designed, as well as redefining the hierarchical standard in Western art production. Their works embody reciprocity with communities and people, charting an important model for cultural and political power—one that relies on inclusion, on multiple parts forming a whole. In developing this work, women artists often step outside the traditional territories of art making and collaborate with community-based organizations, social justice activists, scientific researchers, industry, and the education system. Their art is frequently woven into the fabric of daily living.

While one can look at these works as being feminist in their approach to voice and power, not all of the artists define their work as feminist in content or intent. Some identify more closely with their culture than with gender, viewing feminism as a "white" influence. Others explore nonhierarchical communication concepts—collaboration, networked communication, dispersed authorship, public art—without attaching a feminist significance to their work. And there are still others who consider their work to be feminist, whether or not its content specifically explores gender issues.

Telecommunication art has been accused, with some validity, of being an elite art form due to the cost of tools and access to cyberspace. Our work, open as the Internet is today, is still not open to all. Those of us who work with electronic telecommunication tools wrestle with the tension of trying to shape democratic communication systems in an environment that excludes on the basis of wealth, knowing that women are the majority of the poor. Our work has not yet impeded the increasing centralization of media corporations or the privatization of communication networks—the advance, as VNS Matrix calls it, of Big Daddy Mainframe.

As Karen O'Rourke points out,[77] there remains a tension in collaborative telecommunication works: they still often have an individual artist's name attached to them (whether initiated by men or women). What is the role of the public that we invite to be artists with us? Lorri Ann Two Bulls' experience with having her online work sold without remuneration is a painful reminder that we have not moved beyond cultural exploitation. As artists seek ways to restructure power and to collaborate with others in the creative process, we must be mindful of the rights of our collaborators.

What women telecommunication artists have achieved is to create spaces for horizontal communication, understanding that this implies a new language and a new set of relationships. We have connected our work to other people, extending beyond our selves, drawing from many individuals' lived experiences as the basis for making art that authentically reflects the multiplicity of all of us.

NOTES

1. Anna Deavere Smith, quoted in Sara Boxer, "Enter, the Audience: Trying to Gather Everyone In," *New York Times,* 29 August 1998: A13.

2. Armand Mattelart, *Mapping World Communication: War, Progress, Culture,* trans. Susan Emanuel and James A. Cohen (Minneapolis: University of Minnesota Press, 1994), 6.

3. See Edward S. Herman and Robert McChesney, *The Global Media: The New Missionaries of Global Capitalism* (London: Cassell, 1997), and McChesney, "Towards a Democratic Media System: Interview with Robert McChesney," *Corporate Watch,* available at ⟨http://www.corpwatch.org/trac/internet/corpspeech/mcchesney.html⟩.

4. Bertolt Brecht, "The Radio as an Apparatus of Communication," trans. John Willett, in John G. Hanhardt, ed., *Video Culture: A Critical Investigation* (New York: Peregrine Smith Books/Visual Studies Workshop Press, 1986), 53.

5. Alice Embree, "Media Images 1: Madison Avenue/Brainwashing—The Facts," in Robin Morgan, ed., *Sisterhood Is Powerful: An Anthology of Writings from the Women's Liberation Movement* (New York: Vintage, 1970), 181, 183.

6. Coco Fusco, "At Your Service: Latinas in the Global Information Network," keynote lecture at the International Symposium on Art, Science, and Technology, 1998, available at ⟨http://www.hkw.de/forum/forum1/doc/text/fusco-isea98.html⟩.

7. Sherrie Rabinowitz, interview with the author, 1999.

8. Electronic Café International, "Telecollabrative Art Projects of ECI Founders Galloway and Rabinowitz, 1977 to Present," available at ⟨http://www.ecafe.com/getty/table.html⟩.

9. Sherrie Rabinowitz, e-mail to the author, 2001.

10. Sarah Dickenson and Marilyn Schaffer, "Art, Images, Communications, and Children," in Roy Ascott and Carl Eugene Loeffler, eds., *Connectivity: Art and Interactive Telecommunications, Leonardo* (special issue), 24, no. 2 (1991): 189.

11. Ibid., 190.

12. Judy Malloy, interview with the author, 1999.

13. Judy Malloy, "Bad Information," available at ⟨http://www.well.com/user/jmalloy/bad.html⟩.

14. Ibid.

15. Nancy Buchanan, interview with the author, 1999.

16. The Web component of *Developing: The Idea of Home,* is available at ⟨http://cmp1.ucr.edu/buchanan/Index.html⟩.

17. Nancy Buchanan, interview with the author, 1999.

18. Sherrie Rabinowitz, interview with the author, 1999.

19. Kunstradio-Radiokunst, available at ⟨http://kunstradio.at⟩.

20. Arts Wire, available at ⟨http://www.artswire.org⟩.

21. Jim May, American Indian Library Association Testimony before Congress, quoted in George Baldwin, "Networking the Nations: Information Policy and the Emerging Indian Network Marketplace," *Journal of Navajo Education,* 9, no. 2 (Winter 1992): 48.

22. Kathy Rae Huffman and Eva Wohlgemuth, "Face Settings," available at ⟨http://thing.at/face⟩.

23. Kathy Rae Huffman, "Face Settings: An International Co-cooking and Communication Project by Eva Wohlgemuth and Kathy Rae Huffman."

24. FACES, available at ⟨http://www.faces.vis-med.ac.at⟩.

25. Eva Ursprung, quoted in Huffman, "Face Settings."

26. Jennifer Hall, "*Netdrama:* An Online Environmental Scheme," in Ascott and Loeffler, *Connectivity,* 193.

27. Ibid., 194.

28. Ibid.

29. Anna Couey, *Cultures in Cyberspace* invitation to participate, 1992.

30. Aida Mancillas and Lynn Susholtz, "*Project Artnet:* Building Community through Shared Histories," project description, 1993.

31. Aida Mancillas, conversation with the author following the event, 1993.

32. Isabella Bordoni, interview with the author, 1999.

33. Ibid., translated from Italian by the author, 1999: "Credo che un grande vantaggio dell'arte dei media sia la possibilità che ha di attivare strutture della percezione, esplorare modelli di associazione tra discipline, tra aree diverse del sapere."

34. Andrea Sodomka, interview with the author, 1999.

35. Ibid.

36. Ibid.

37. Karen O'Rourke, "*Paris Réseau:* Paris Network," *Leonardo,* 29, no. 1 (1996): 51.

38. Ibid., 55.

39. Karen O'Rourke, "*Paris Réseau,* Paris Network," *Artificial Intelligence and Society* (2000).

40. Ian Pollock and Janet Silk, "*Local 411:* Private Conversations in Public Space," in *Words on Works* (San Francisco: ISAST, 1998), 296. Also available at ⟨http://mitpress.mit.edu/e-journals/Leonardo/isast/wow/wow303/local.html⟩.

41. Ibid.

42. Ibid.

43. Ibid.

44. *Day Without Art Web Action,* available at ⟨http://www.creativetime.org/dwa/⟩.

45. Carol Stakenas, "Crossing the Threshold: Examining the Public Space of the Web through *Day Without Art Web Action,*" 1998 (chapter 35 in this volume).

46. Ibid.

47. Ibid.

48. Ibid.

49. Elisabeth Schimana, available at ⟨http://elise.at⟩.

50. Elisabeth Schimana, *the fugue,* available at ⟨http://thing.at/orfkunstradio/RIV_BRI/PROJECTS/schimana/openfrm.html⟩.

51. Ibid., available at ⟨http://thing.at/orfkunstradio/RIV_BRI/PROJECTS/schimana/score.html⟩.

52. Lorri Ann Two Bulls, interview with the author, 1999.

53. Ibid.

54. Randy Ross (Oglala Sioux) described the importance of visual communication for Native Americans as follows: "I remember listening and watching my grandfather speak in his Native language to his peers and of their responses to him. There was a lot of use of hands, arms, facial expressions, intonations, his eyes, that made the language come alive, a[nd] as a result a very few words had a lot of meaning and depth. Even in the music he sang, the words were short, but the song meant quite a bit about a particular situation. For American Indians who still speak their language, I don't know of any computer pro-

gram that can enhance the totality of language and music through the keyboard. Even the translation of Indian music to euro-standards on a keyboard does not work! Perhaps there is some hope in the new graphics programs that can be used online, such as art graphics, pictorials, the creative of native languages online, etc." In *Cultures in Cyberspace,* organized by Anna Couey, 1992.

55. Lorri Ann Two Bulls, interview with the author, 1999.

56. Jacalyn Lopez Garcia, "Pushing the Boundaries of the Internet: *Glass Houses,*" *Leonardo* 33, no. 4 (2000): 263.

57. Jacalyn Lopez Garcia, *Glass Houses,* available at ⟨http://cmp.ucr.edu/students/glasshouses⟩.

58. Lopez Garcia, "Pushing the Boundaries," 1998.

59. Ibid.

60. Carolyn Guyer, *Mother Millennia,* available at ⟨http://www.mothermillennia.org⟩.

61. Carolyn Guyer, "More about *Mother Millennia,*" available at ⟨http://www.mothermillennia.org/moreabout.html⟩.

62. Ibid.

63. Nina Sobell and Emily Hartzell, *ParkBench,* available at ⟨http://www.cat.nyu.edu/parkbench/⟩.

64. Nina Sobell and Emily Hartzell, "Sculpting in Time and Space: Interactive Work," *Leonardo* 34, no. 2 (2001).

65. Nina Sobell and Emily Hartzell, artists' statement, 1997.

66. VNS Matrix, "Dirty Work for Slimey Girls," available at ⟨http://sysx.org/vns/⟩.

67. VNS Matrix, *Cybermanifesto for the 21st Century,* available at ⟨http://sysx.org/vns/manifesto.html⟩.

68. VNS Matrix, "c o r p u s f a n t a s t i c a M O O," available at ⟨http://sysx.org/vns⟩.

69. Judy Malloy, interview with the author, 1999.

70. Judy Malloy's *Uncle Roger* is being adapted for the Web and is available at ⟨http://www.well.com/user/jmalloy/party.html⟩.

71. Judy Malloy, interview with the author, 1999. The works she refers to are available at ⟨http://www.well.com/user/jmalloy/awquilt.html⟩.

72. Judy Malloy and Cathy Marshall, *Forward/Anywhere,* available at ⟨http://www.csdl.tamu.edu/~malloy/html/beginning.html⟩.

73. Judy Malloy, e-mail to the author, 2001.

74. Judy Malloy and Cathy Marshall, "Notes on an Exchange Between Intersecting Lives," in *In Search of Innovation,* ed. Craig Harris (Cambridge, MA: MIT Press, 2000).

75. Jennifer Hall and Blyth Hazen, "The Community of Do While Studio," 1998 (chapter 19 in this volume).

76. Ibid.

77. O'Rourke, "*Paris Réseau:* Paris Network," 56–57.

5

Through the Looking Glass

Kathy Brew

> What I would like to evoke, about Image in general (the media-image, the technological image), is the perversity of the relationship between the image and its referent, the "supposed real"; it is the virtual and irreversible confusion of images and of the sphere of a reality, whose principle we can grasp less and less.
> —JEAN BAUDRILLARD[1]

Is it live or Memorex? Is it a real or a virtual experience? As Jean Baudrillard noted more than a decade ago, the boundary between the real and the simulated, the simulacrum, is less and less distinct and is becoming more and more semipermeable, just like the mirror in Alice's looking-glass world. Artists have historically been at the forefront, blurring boundaries and crossing borders, penetrating the looking-glass, pushing the envelope with the R&D of their creative explorations utilizing new technologies, and in the process redefining the very notions of both art and artist. In our current postmodern Oz, a cyber-Toto as avatar yanks the curtain and reveals the wizard at the screen, confronting viewers with questions of authorship, physicality, identity, and space. In many of these cases, the art experience is no longer passive but requires some sort of (inter)active participation. Yet with all the hype and seduction accompanied by the dazzle of all the latest bells and whistles, many of these multimedia artists remind us that in spite of this technological revolution that we are living and breathing, a tool is after all just a tool. And the tools continue to change at hyperspeed. It's really the thought, the concept, that counts; here the journey *is* the destination where multimedia conveys the message. The artists presented in this chapter, and the multitude of those not included, reflect just how *multi* multimedia is.

When I looked up the word *multimedia* in my somewhat dated dictionary, I was sent off on a branching exercise, a book-flipping experience as analog precursor to the digital labyrinthine branching navigated on today's World Wide Web. First, the word is simply defined by a cross-reference: "n. *same as* MIXED MEDIA." So I moved on to the entry for *mixed media,* which in my dictionary is defined as follows: "1. the simultaneous presentation of a series of effects in more than two media, as by combining acting, flashing colored lights, tape recordings, etc."[2] Sounds a bit retro, like a sound and light show from the 1960s. Today we've certainly come a long way from that dictionary definition of 1984. In fact, multimedia artists have been ahead of the curve of the original definition for quite some time, and the exponential technological advances that have occurred over the last thirty years have forever changed most of our definitions.

I have been following the intersection of (multi)media and contemporary art for the past twenty-plus years. My focus has been on the nonstatic (time-based) arts, following artists working in multiple disciplines, working cross-platform, creating new forms from hybrid combinations, defying the labels of definition, and inventing new languages. As a result, I have had the great privilege of knowing and working with many of the artists working in these *multiple* forms. My career has also been *multi,* or hybrid, in nature, and I have often thought of myself as a "Jill of all trades," wearing many different hats at different points in time, sometimes juggling more than one at a time. The hats have included video producer, writer, arts administrator, curator, performance artist, radio interviewer, publicist, fundraiser, video artist, collaborator/accomplice, and rabble-rouser. At the root of all this work has been the role of communication, of expanding audiences, of helping more people become aware of certain ideas and issues that often aren't given adequate (if any) coverage in the mass media, which really could be perceived as monomedia. And this has often involved helping to present the independent, cutting-edge artist's voice and sensibility to a larger audience to expand the dialog beyond the cognoscenti.

First, let's rewind for a brief flashback by way of introduction to my own particular entry into multimedia. I graduated from a New England liberal arts college in 1975, where I received a B.A. in sociology and anthropology with a minor in art history and the added bonus of being certified to teach elementary school. (After all, what was a girl growing up in the fifties going to be when she grew up? My options, other than mother, seemed to be either teacher or nurse. And given that I was a doctor's daughter, I opted for the nonmedical route.) Ironically enough, my first job out of college was in multimedia as an audiovisual producer for an educational publishing company, producing filmstrips, one of the prehistoric, now basically defunct, forms of multimedia. I certainly remembered this primitive form from my own elementary school days, where frames were advanced manually by the teacher with the cue of an audio beep. By 1976, the technology had advanced enough so that the filmstrips we were producing could advance a bit more fluidly to the next frame with "inaudible beeps" propelling the filmstrip mechanically forward. Communicating something by freezing motion into static images that were then accompanied by sound soon seemed so archaic that I found my way to moving images and video.[3]

But it took another hybrid or *bi* experience, moving to the west coast in 1981, before I really began to synthesize and fuse my interest in contemporary art and media. I thought that California would be a two-year experiment and that I would soon dash back to the hub of the universe, a worldview illustrated by the Saul Steinberg *New Yorker* cover that places New York as the only place on the planet. Instead, I soon discovered there were other hubs; I stayed in San Francisco for fourteen years, where I became aware of a reversal or coastal shift of Manifest Destiny that tipped the axis between the coasts in the other direction. After all, Silicon Valley did precede Silicon Alley, even if New York still claims to be the content capital of the world. As Gertrude Stein noted, there is a there there, and the gold rush mentality of the new technology industry began there and has migrated eastward.

I first worked for the San Francisco Public Broadcasting System television station, working on many arts-related documentaries and championing the contemporary arts as a "fifth column" within the conservative programming philosophy there. While I was on a trip to New York, I learned of the upcoming New Year's Day 1984 broadcast work by Nam June Paik called *Good Morning, Mr. Orwell,* a live simulcast featuring real-time intercontinental interactive performance works by multimedia artists from around the world. I returned to San Francisco and tried unsuccessfully to convince PBS executives to participate in broadcast history with the live satellite link-up; instead they decided to air reruns of *Sesame Street* to babysit the children of the previous evening's New Year's revelers and to run a taped version of *Mr. Orwell* later. Clearly, mass media, even via supposed enlightened public television, was not open enough or experimental enough for my taste, and so I soon defected to the art camp.

Before long, I found myself as the second director of the Capp Street Project, the unique site-specific artist-in-residence program begun by Ann Hatch in 1984 that graduated in 1998, merging with the California College of Arts and Crafts. During my tenure at Capp Street, from 1985 to 1987, my real education and work with multimedia artists began. For me, this experience was like going to art school, having the amazing opportunity to work with many of the country's leading contemporary multimedia artists, facilitating and being part of their behind-the-scenes process of creation. Maryanne Amacher, Mary Lucier, Liz Phillips, and Elizabeth Diller (of Diller + Scofidio) were all Capp Street artists from this time who could easily be contributors to this book (or in a sequel, as one recognizes the impossibility of being totally comprehensive in one volume). I would like to acknowledge some of the contributions of these artists (and others whose papers are not included in this book) to the multimedia realm.

Along with Judith Barry, Dara Birnbaum, Joan Jonas, Beryl Korot, and Steina (all included in this book), Mary Lucier is from the "first wave" of women artists working in video, creating both single-channel tapes and video installations. Lucier became involved in video in the early 1970s, after nearly a decade of working in more traditional art forms such as sculpture and photography. Since 1973, she has concentrated primarily on video installation. Her early work, such as *Dawn Burn* (1976), was concerned with properties intrinsic to the medium itself. But she soon began to focus on pictorial qualities and became known for her form of *pictorial narrative* video installations, where she structures images using diverse editing strategies—slow motion, reversal, hyperspeed, and audio processing, sometimes in relation to a constructed environment. More recently Lucier completed work on a commemorative flood commission for the North Dakota Museum of Art entitled *Floodsongs,* which premiered in Grand Forks in the fall of 1998 and was shown at New York's Museum of Modern Art in 1999. Through the lens of modern technology, much of her work addresses the paradoxical relationship between "progress" and its impact on the forces of nature. Artists like Lucier are attempting to awaken us to the roar of technology while there is still time. Her current work involves several new projects concerned with issues of technology, obsolescence, and false notions of utopia.[4]

Joan Logue (an early Capp Street resident, prior to my tenure there) is another artist from this early wave of women video artists who worked in both single-channel and video installation. With a background in painting and photography, Logue has been a video innovator since she first discovered the medium in the mid-1960s, when she extended her interest in portraiture through video's real-time imagery. She is perhaps best known for her *30 second spots* on American-based contemporary artists, what she refers to as *TV Commercials for Artists.* She collaborates with her subjects to create thirty-second, single-channel "video portraits" and utilizes state-of-the-art techniques of the medium (when appropriate) to capture the essence of each person's artistic expression. She began working on the spots in 1979 and has collaborated with artists in New York, San Francisco, Boston, Paris, Berlin, and Warsaw. In fall 1998, she presented her video portraits as the kick-off for a new series—*TV Dinner*—at the Kitchen in New York and also exhibited a new video installation there, *Self-Portrait 1973–83,* where she deconstructed her long- and short-term memories and then reconstructed them in a new multiple video format, creating a multilayered portrait.[5]

Maryanne Amacher, another former Capp Street resident who is featured as one of Joan Logue's portraits, has been exploring musical language and environmental sound in terms of acoustical and architectural space for over thirty years. Like the worlds of science and science fiction, where she draws much inspiration, Amacher's work is focused on the future, on exploring situations of boundary and perception. She was an early pioneer in using audio telemetry (the wiring together of different sites, generally for concert purposes) and created several "long-distance music" works utilizing what she has referred to as the CITY-LINKS format, which occurred as early as 1976. This involved placing microphones at distant locations and transmitting the "live" sound to mixing facilities, either at her studio or at installation or performance sites. More recently, she has created several multimedia installations, including her own original form of "music theater," which she refers to as MUSIC FOR SOUND JOINED ROOMS and which uses the architecture of a site to stage the sound and evolve "stories" that are dramatized through the music and visuals. For these, she brings recorded material to a site, which she considers a script, and then proceeds to test the material on speakers placed at different locations throughout the space. The final sound is created in live performances, where Amacher modulates the original sources through a synthesizer and other technology. Unlike the conventional environment for musical presentations—a proscenium concert setting where the audience sits and listens—Amacher's form of music theater creates a living atmosphere that immediately engages the audience in a perceptually different manner; the work explores alternative listening strategies, unusual acoustic environments, expanded technologies, and new relationships with audiences. In 1997, Amacher received the Prix Ars Electronica Golden Nica Distinction for her artistic achievements in computer music for her work *The Levi-Montalcini Variations.*[6]

Amacher had been a fellow at the Massachusetts Institute of Technology's Center for Advanced Visual Studies in the 1970s, and through her connections with colleagues from that time I was able to join her in a field trip down to the National Aeronautics and Space Administration (NASA) Ames Research Center in Palo Alto for my first virtual-reality experience in 1985, donning a VR helmet and becoming immersed in a wire-frame environment. Of course, the applications here involved militaristic flight simulations. Not for ten more years, back in New York, would I immerse myself in an artistic virtual-reality experience like Char Davies's *Osmose* (in an exhibition entitled *Code,* presented at Ricco Maresca Gallery in New York City). Prior to that experience, I had seen Patrice Caire's virtual-reality project—*C.A.I.R.E. '94—(Cyberhead . . . Am I Really Existing?),* an artwork presented at San Francisco's Yerba Buena Center for the Arts that allowed viewers to enter into a reconstructed model of the artist's head in a fictionalized, five-minute vertiginous journey by looking through a stereoscopic viewing device. The work raised issues about the interconnectedness of the body and the human creations of technology and information processing.

Like Maryanne Amacher, many other artists, such as Pauline Oliveros, Frances Dyson, Pamela Z, and Cécile Le Prado, are also involved with creating unusual aesthetic experiences with sound and technology. During the time of this writing, I had the opportunity to experience Pauline Oliveros and her Deep Listening Band kick off a season-long decade celebration in a concert called *Suspended Music* presented at Columbia University. The evening included composer Ellen Fullman's Long String Instrument Band, involving a one-of-a-kind musical instrument with strings nearly 100 feet long, stretching across the public space of the site, as well as the newest incarnation of the Expanded Instrument System (EIS), the computer-driven musical machine that is part of the ongoing sonic evolutions of the Pauline Oliveros Foundation.

Also occurring at the time of this writing, and from a very different multimedia perspective, is Beryl Korot's collaboration with composer Steve Reich, *Hindenburg,* "a documentary video opera," which was presented in its New York premiere as part of the *Next Wave Festival* at the Brooklyn Academy of Music. Korot and Reich first collaborated in 1993 with *The Cave,* which coupled Korot's multichannel video imagery and Reich's rhythmic structures. Continuing here with her notion of palimpsest, Korot utilizes images, including historical footage of the iconic aircraft and taped interviews, creating a visual analog to the twentieth century's collective memory to explore the impact of new technology on twentieth-century society.

The staged and performative aspect of multimedia work is something that has an ongoing thread, beginning with the early work of Steina (who likens the early days of video to playing an instrument in an improvisational and spontaneous way) and of Joan Jonas (who cites an interest in links of high tech with the original gesture). Steina opened the 1997 New York State Council on the Arts Governor's Conference on Arts and Technology, performing *Violin Power,* "a piece about connections." Taking center stage, she played a violin that was connected to a MIDI interface that both transformed the sound of the instrument and allowed her playing to influence and activate the images on three large screens behind her, creating something akin to a virtual duet between image and live performer. Her staccato bow strokes affected the images and their movement; in essence, the violin was powering the visuals.

From one violin to another, the next generation of those working in multimedia performance art include artists like Laurie Anderson and Julia Heyward, where the work centers around the orchestration of music, image, and language in the forms of video and live performance. In the mid-1970s, Heyward stopped painting, began working in video and multimedia performance art, and soon was touring Europe and the United States. In 1978, she began working on a long-form music video album, which premiered at the Kitchen in 1981, a year before the arrival of MTV. She has continued to create multimedia performances, "visual musicals." Her *Miracles in Reverse* (1996) premiered in Potsdam, Germany, and had its American premiere at the Kitchen in New York City in 1997. An accomplished artist in her own right, she recently went to Rensselaer Polytechnic Institute's graduate program iEAR (Integrated Electronic Arts at Rensselaer), getting her Master of Fine Arts degree in electronic arts so she can have the "proper" credentials for teaching while continuing her own work.[7]

Laurie Anderson, on the other hand, is one of multimedia art's great crossover stories. Today her multimedia performance art—combining voice, performance, film, technology, sculpture, and other media—has become part of the culture and is often credited with having pioneered new artistic territories, laying the foundation for explorations of a younger generation of artists. She was one of the first artists to combine media in a transdisciplinary approach. In fall 1998, Artists Space in New York celebrated her historic influence by opening its twenty-fifth season with an exhibition presenting a selection of Anderson's work from the 1970s and 1980s and *Whirlwind,* a recent large-scale viewer-activated audio installation never before presented in the United States. For the opening reception, Anderson re-created *As:If,* her first public performance, which was presented at Artists Space as part of her first exhibition in 1974. At that initial time, she had performed on the violin with her feet in roller skates encased in blocks. For the reprise, she came out in iced blocks and then proceeded to play electric violin linked to a commercial digital sound manipulator that repeats, delays, speeds up, or slows down tonal input. The exhibition underscored the early roots of Anderson's working with technology and how it continues to evolve and transform in her current work of today.[8]

Pamela Z and Laetitia Sonami are two multimedia artists whose performative work involves technology combined with a very strong presence of the physical body. Sonami performs with a virtual-reality dataglove with multiple controllers and uses the movement of her hand and arm to control MIDI signals to various digital outputs. In essence, her movements become music.

Toni Dove, on the other hand, invites viewers to physically interact with her multimedia work; she is interested in an immersive experience that triggers feelings of engagement that can be used to intensify the narrative experience. Her work *Artificial Changelings,* an interactive movie installation that premiered at the 1998 Rotterdam Film Festival, is a romance thriller about shopping that follows the life of Arathusa, a kleptomaniac in nineteenth-century Paris during the rise of the department store, who is dreaming about Zilith, an encryption hacker with a mission from the future. The installation uses video motion sensing to track the location and movement of a viewer standing in front of a screen. Viewers can take turns either as participants or spectators. There are four "zones" to this narrative, and physical movement within each zone alters the video and sound in a variety of ways: it may generate speech, change the emotional tone of the sound environment, or move a character's body. The viewer can move back and forth between centuries in a cinematic landscape with no linear direction, can enter into a character's head, or can enter other sequences by means of various physical behaviors. The installation was recently included in an exhibition titled *Body Mechanique: Artistic Explorations of Digital Realms* at the Wexner Center for the Arts in Columbus, Ohio.[9]

Some multimedia artists have come to the *multi* part more through the channel of one particular medium. Beth B is one such example. Although probably better known for her films, B has also brought her filmmaking background into creating multimedia installations where some kind of moving image component is combined with other sculptural, sound, and created environments. The interplay between media is apparent, and B is adept at setting a dramatic stage to invoke her narratives, which often explore taboo but socially relevant topics. *Out of Sight Out of Mind* is one such multimedia installation revolving around the notions of madness, the media, and the power of propaganda. It was presented in New York in 1995 and then traveled to the American Center in Paris in 1996. It includes an elaborate sculptural construction—a re-creation of a "rotary machine," an 1820s device used in psychiatric hospitals to "cure madness." Visitors could strap themselves in, push a button, and spin at speeds of up to 100 revolutions per minute. Another section involved padded cells with a sound installation, a barrage of words and music comprised of a *Who's Who* of celebrities who committed suicide or were diagnosed as mad—Marilyn Monroe, Antonin Artaud, Kurt Cobain, Vincent van Gogh, and others. The final chamber of the installation consisted of a projected video (which can also stand alone as a single-channel piece) that included archival footage of acts of daring juxtaposed with media coverage of a juvenile homicide case, further questioning the institutional treatment of insanity.[10]

Fast forward to the more recent present. From March 1997 until November 2001, I was involved with artists working with multimedia and new technologies as the first director of Thundergulch, the Lower Manhattan Cultural Council's (LMCC) new media arts initiative that seeks to provide new forms of interaction between artists, audiences, and new technologies. This nonprofit art-tech initiative was first located in the financial district, across from the New York Stock Exchange, in an office in the New York Information Technology Center, one of New York's first "totally wired" buildings with high-speed connectivity and video teleconferencing, where many technological businesses such as Sun Microsystems and IBM have their headquarters (as well as start-up tech companies, many of them now defunct, after the fall-out of the dot-com frenzy five years later).

In the fall of 1998, Thundergulch moved into LMCC's offices at 5 World Trade Center, so for almost three years more, Thundergulch was headquartered at 5 World Trade Center. On September 11, 2001, all of that changed. In addition to the devastating loss of lives, the offices and five years' worth of research and program files and materials were destroyed. Given a rather dramatic reason to morph, I concluded my tenure as Director of Thundergulch at LMCC at the end of November 2001.

Flash back to 1997: As a way to introduce artists working in some of these newer forms, Thundergulch began by presenting a series showcasing artists' multimedia work (video, CD-ROMS, and Web sites) on a fourteen-foot video wall located in the lobby of the New York Information Technology Center. Perhaps it was the video producer in me that recognized that this large video wall was hardly being used except for corporate advertising and as wallpaper cycling through the building's Web sites. I knew it would be an ideal large presentation format for artists' work, even if in a lobby of an office building.

So Thundergulch began infiltrating the business and the tech-industry culture, showcasing artists' work in the middle of the lobby during lunchtime, when people were flowing in and out of the building—a new form of public interactive art. Some people intentionally came to the presentations and sat in the provided chairs for the hour; others hovered on the edges, on their way in or out of the building, but stayed long enough to get a hit. Artists and/or curators/presenters were on hand in an informal salon-type presentation, navigating through their Web site or CD-ROM and available to address questions from the audience, questions that ranged from technological to aesthetic to conceptual. With all the hype of interactivity and the ubiquity of cyberspace, there seems to be a hunger for human interactivity in real time and real space, for physical linkages. The series subsequently migrated to other venues, and

Thundergulch continued to provide a forum in both formal and informal ways, where when possible the artist was actually interacting with the audience in a dialog about the work. The presentations are meant to be introductions to these artists' works, where the viewer must then go engage with the work in a more direct manner (either at a performance, an exhibition, or the computer). In my tenure as director, Thundergulch showcased the work of over *two* hundred artists, and many of these have been women working in some of the newer directions of multimedia. Again, the list is too long to be all-inclusive, but a mention of some of these women artists demonstrates just how *multi* the concept of multimedia is.

Adrianne Wortzel's installations and theoretical writings explore the possibilities of new electronic media in relation to traditional art forms such as opera and the novel and the use of the Internet as a performance medium. Her production *Globe Theater: Sayonara Diorama* was a multiple-site electronic media performance that featured a repertory company of robots and actors and online participants, utilizing CU See Me.

Helen Thorington has been involved with Internet performance events since initiating the *Adrift* project for the 1997 Ars Electronica Festival. *Adrift* was a collaboration with other artists; each artist worked from different geographical locations and with different computer environments (text, sound, and Virtual Reality Modeling Language or three-dimensional graphics). *Adrift* is archived on the Turbulence Web site. As director of New Radio and Performing Arts, Thorington initiated the Turbulence art site, which commissions artists to create new work, especially for the Web medium. Permanent works shown there include Diane Bertolo's *FT2K (Frontier Town 2000),* Annette Weintraub's *Pedestrian,* and Marianne Petit's *The Grimm Tale.*[11]

Melanie Printup Hope's work is an exploration of her Native American identity and ancestry. She conveys her personal experiences of cultural and spiritual growth through drawing, traditional beadwork, sculpture, computer-generated images, animation, digitized sound, video, and installation. *The Prayer for Thanksgiving* was presented in many of these multiple forms—as a CD-ROM that was incorporated into a sculptural installation that also had a Web component.[12]

Tennessee Rice Dixon and Zoe Beloff, both recipients of the New York Foundation for the Arts first-ever computer arts fellowships in 1997, are two other artists working in CD-ROM, as well as other digital forms. Dixon's *ScruTiny in the Great Round,* an internationally award-winning CD-ROM cocreated with Jim Gasperini, is analogous to an animated picture book, where pages come to life in animation and sound as the viewer moves over the imagery and through the story with a changing cursor. Various sequences and narratives are initiated by moving the cursor and clicking on "active" spots. Beloff's CD-ROM, *Beyond,* operates in a playful spirit of philosophical inquiry exploring the paradoxes of technology, desire, and the paranormal. *Beyond* won first prize in Apple's 1998 QuickTime VR competition (best multimedia, best of show). More recently Beloff completed another QuickTime movie serial— *Where, Where, There, There, Where*—with the Wooster Group.[13]

Kristin Lucas, a younger-generation artist, comes to multimedia from a video background. She sets up virtual interactions with mediated devices, such as automated tellers, public-access television, computer games, and the World Wide Web. She was commissioned to create a work for ISEA 1998 (presented in Manchester, England) and also a Web site, *Between a Rock and a Hard Drive,* for the Dia Center for the Arts. Two of her installations were presented for several months in the Thundergulch offices. *Ground Control* is an audio-video installation that comments on sexual metaphors inherent in the language of machine maintenance manuals. *Installation* is a sound installation where patterns on a wall are derivative of constellations, and the sounds are constellated from the barrage of audio information overload—fax tones, modem hook-ups, phones ringing.[14]

Because I had the luxury and opportunity to witness so many of these artists presenting their work, I began to realize that it might be important for others to have this experience, and so I began interviewing the artists after their presentations *@ the wall,* to create an archive, an oral history of the range of thinking about the nexus of art and technology at the moment it was occurring. At the time I wrote this paper, the archive was "under construction" to become a dynamic part of the Thundergulch Web site so that the presentations could have both a local and global presence and impact. However, as a result of the terrorist attacks on September 11, 2001, all this material was lost.

"Mirror, Mirror on the wall, who's the fairest of them all?" When I was interviewed for the position at Thundergulch, one of the questions

I was asked was, "Where do you see yourself in five years?" I explained that I was unable to answer such questions, that my life didn't work like that, and that when I thought I could predict where things might lead, often the very opposite could occur. (Who could have ever projected to the events of September 11th and its impact on so many lives?) I am not a psychic or prognosticator. There is no crystal ball here. And there is no monolithic art, no monolithic technology, but rather a plethora, almost an overload, a *multi* of everything. It does seem that we are in the infancy of some new art forms that are taking their very first baby steps. And as with baby steps, things can be a bit wobbly in the beginning. There is a need for trial and error, for falling down and getting up again, for patience and faith. Much of the new multimedia work seems to be in this early "baby steps" phase, an improvisational, jamming phase not dissimilar from the early days of video art, where artists are experimenting as they create new art forms and new languages with new tools, which are in a constant, exponential rate of change. The language and the tools are not yet fully developed. The tadpole has not yet turned into the frog that turns into the prince who is kissed by the princess. Besides, the tadpole is really a baby chameleon. And like the chameleon, art and the tools for creating it—which technology has become an integral part of—will keep changing colors. In the future, we won't just be reading a book such as this. We'll be viewing some multimedia presentation of artists' work that will be almost as good as "being there" in some yet unknown form. Ars longa, vita brevis.

NOTES

1. Jean Baudrillard, "Beyond Right and Wrong, or the Mischievous Genius of Image," in *Resolution: A Critique of Video Art* (Los Angeles: LACE, 1986).

2. *Webster's New World Dictionary of the American Language,* Second College Edition (New York: Simon & Schuster, 1984).

3. My own video, *Mixed Messages* (1990), is an experimental documentary collage that incorporates found footage, performance, interviews with young girls, documentary, animation, images from advertising and television, and a dream narrative in a work that examines gender stereotyping in popular culture, concluding with a postmodern version of the Pandora myth.

4. Mary Lucier's installation *Noah's Raven* was recently acquired by ZKM/Museum für Neue Kunst, Karlsruhe, Germany, and was on view there in the summer of 1999. She has been the recipient of numerous grants and was most recently awarded a $25,000 grant from Anonymous Was a Woman, one of ten women artists selected each year by nomination. Her work since 1971 is profiled in a book in the *Performing Arts Journal* series PAJ Books: Art + Performance. The book is *Mary Lucier,* edited by Melinda Barlow (Baltimore, MD: Johns Hopkins University Press, 2000).

5. Joan Logue most recently presented *Digital Limnings: Miniature Memories*—tiny wearable video screens that depict moving portraits—in an exhibition entitled *ID/entity: Portraits in the 21st Century,* which was presented at the MIT Media Lab and at the Kitchen in 2001.

6. Maryanne Amacher received a John Simon Guggenheim Memorial Foundation fellowship in 1997 for her sound installation works. Recent projects include the creation of major works: a string quartet with an electroacoustic installation commissioned by the Kronos String Quartet and the Lila Wallace–Reader's Digest Fund and two new installation works produced in 1998 for the Kunstmuseum Bern *Taktalos Festival* and for *Tunnel Vision* in the three-story Maastunnel, Rotterdam, the Netherlands. Several CD recordings have recently been released: *Sound Characters (Making the Third Ear)* (Tzadik, 1999) and works on the Asphodel Sombient Triology—*The Storm of Drones* (1996), *The Swarm of Drones* (1995), and *The Throne of Drones* (1995). Amacher was included in the 2002 Whitney Biennial.

7. Julia Heyward is currently engaged in the production and programming of an interactive digital video disk, which is the audiovisual "album" version of a multimedia performance piece presented at the Kitchen in 1997 entitled *Miracles in Reverse,* for which Heyward wrote the script and music and created the visuals. She received a Guggenheim fellowship in 1999 and was awarded a Rockefeller Foundation film/video/multimedia fellowship in 2001.

8. Laurie Anderson presented the New York premiere of *Songs and Stories from Moby Dick* at the 1999 Brooklyn Academy of Music's *Next Wave* series and a new solo piece *Happiness* at the Lincoln Center Festival 2002.

9. Toni Dove's *Spectropia* is the second project in a triology of interactive fictions on the unconscious of consumer economics. A research fellowship from the Institute for Studies in the Arts at Arizona State University provided the programming and engineering resources to develop the technology prototype.

10. Beth B's most recent documentary film is about the intergenerational trauma from the Vietnam war. She currently has a feature film in development with Good Machine based on a novel by Athol Fugard. She had two exhibitions of her visual work in New York in the fall of 1999, one at Smack Melon Studio and the other at Visionaire Gallery in conjunction with Deitch Projects.

11. Helen Thorington participated in an entirely new *Adrift* performance at the Walker Art Center in March 2000, with performance locations also in New York and Los Angeles, in addition to Minneapolis. *Solitaire,* created with Marianne Petit and John Neilson, was selected as one of ten top hypertext works in the trAce/Alt-X International Hypertext Competition and was reviewed in the sixth edition of CIAC's *Electronic Art Magazine.* She is the cowriter with Jacki Apple of *Breaking the Broadcast Barrier: Radio Art in America, 1980–1994* (forthcoming).

12. Melanie Printup Hope's video and installation work has been shown throughout the United States and Canada. She has received a Rockefeller Foundation intercultural film/video/multimedia fellowship and other fellowships from the New York Foundation for the Arts, the National Endowment for the Arts, the Andy Warhol Foundation for the Visual Arts, the Jerome Foundation, and the Lyn Blumenthal Memorial Fund.

13. Zoe Beloff produced *Illusions,* a Web site that was sponsored by a Turbulence commission funded by the Jerome Foundation. She has created a live film, stereo slide, sound performance in collaboration with sound artist Ken Montgomery entitled *A Mechanical Medium,* that was first performed at the San Francisco Cinematheque with subsequent presentations in New York and at the Virginia Film Festival in 1999. She recently completed a new film shot in 3D entitled *Shadow Land or Light from the Other Side,* based on the 1897 autobiography of Elizabeth D'Esperance, a materialization medium who could produce full-body apparitions.

14. Kristin Lucas presented an interactive video performance, *Drag and Drop,* at the RPI Performance Series at Renssalear Polytechnic Institute in Troy, New York, in 1999. In fall 1999, she participated in the ARCUS artist-in-residence program in Moriya-machi, Ibaraki, Japan. She created a work commissioned by O.K. Center for Contemporary Art in Linz, Austria, where she had a solo exhibition of her work in May 2000. *Screening Room,* an interactive video installation, was commissioned by the Foundation for Art and Creative Technology for ISEA 1998.

II

Artists' Papers

6

My Love Affair with Art: Video and Installation Work

Steina

My love affair with art was all-consuming from the time I was eight or nine years old until my late teens. I lived by it. I went to every concert, play, opera, and gallery show I could. Nothing else in life made any sense to me. I never chose to be an artist. I just knew I would not work in a bank or wait on tables. I loved playing my violin, but when faced with the prospect of being a professional musician, I realized I had made a dreadful mistake. I found myself in New York in the mid-1960s going from gig to gig, wondering if there was not more to life than black dresses and meager fees.

DISCOVERING VIDEO

I had met Woody Vasulka in the early 1960s in Prague, where I was studying music at the time, and by the mid-1960s we had moved to New York. Woody was a filmmaker, and through his film contacts he came across video in 1969. Both of our lives were changed forever. Woody introduced me to his new discovery. What a rush! It was like falling in love. I never looked back. As soon as I had a video camera in my hand—as soon as I had that majestic flow of time in my control—I knew I had my medium.

We already owned a two-track audiotape recorder, which allowed delays and speed changes. We immediately proceeded to process and manipulate videotape along the same principles we had applied to audiotape. During the same period, we were taking the portable video equipment to New York's cultural playgrounds—WBAI Free Music Store, Judson Church, La Mama, Automation House, the Village Vanguard, Fillmore East, Blue Dom, Max's Kansas City.

After those outings, everyone would gather in our loft to look at the instant playback—something that most people at that time had never

---

This paper originally published in *Leonardo,* 28, no. 1 (1995) documents the life and work of video pioneer Steina. Since this paper was written Steina has produced many other works including *Orka* (1997), a three-channel video environment and the video installation . . . *of the North* (2001).

experienced before. Even the word *video* was a brand-new addition to the vocabulary. By 1971, there was so much traffic in our loft that, when a friend told us he had found a large space in an abandoned kitchen in the old Broadway Central Hotel, we were ready. The space was intended to serve the artists, not the audience. We therefore named it the Kitchen-LATL (Live Audience Test Laboratory).

ENVIRONMENTS

In the early days of video, everything was an installation or an "environment" as we used to call it. In the first generation of half-inch reel-to-reel video, editing was accomplished by cutting and gluing, as was done with film and audiotape. Our environments, therefore, consisted of either "live" camera or "live" switching of tapes. Woody and I preferred to use multiple screens—typically, a stack of monitors and several players. One of our first installations involved the horizontal drifting of images from one monitor to the next. After we started the Kitchen, we had plenty of opportunities to do environments and live video performances. Later, when electronic editing became technically feasible, everyone became infatuated with editing, and installation work disappeared for a while—to be reinvented later by the art world.

In the 1970s, I did a series of environments entitled *Machine Vision* (figure 6.1) the first variation of which was called *Allvision.* For *Allvision,* I put a bar across a turntable and mounted two video cameras at the center. The cameras were back to back, and each pointed at a small mirrored half-sphere placed at each end of the bar. Each camera viewed 180 degrees of the space and displayed the results on four pairs of monitors placed in the corners of the room. As the table slowly turned, the cameras captured the entire room, including viewers, monitors, and the turning machinery itself. In a later version, I put a large sphere in the middle of the turntable with the cameras at each end pointing in. Another *Machine Vision* variation used a motorized moving mirror placed in front of a camera. Depending on the horizontal or vertical positioning of the mirror, the video monitor would display a continuous back-and-forth pan or up-and-down tilt of the room. A third variation implemented a continuous rotation of the image through a turning prism, while still another involved a zoom lens in constant motion, zooming in and out.

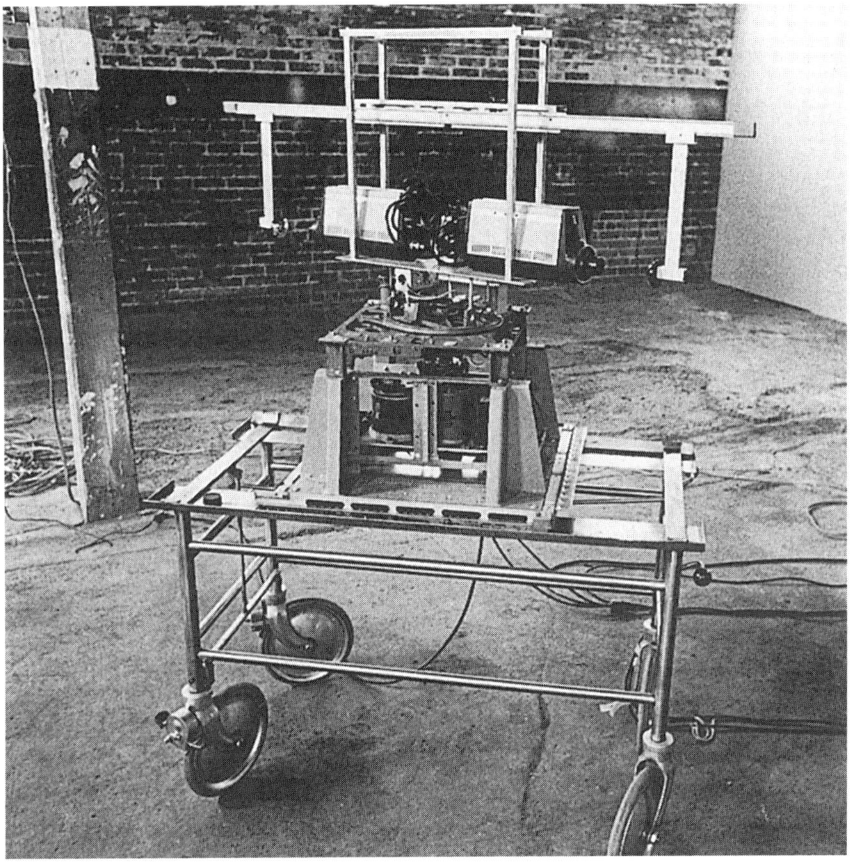

Figure 6.1
Steina, *Machine Vision,* 1976, series of video installations. In the first variation of this series, two video cameras on a motorized turntable were combined with two mirrored half-spheres and two monitors, emphasizing a camera view that moves beyond the restrictions of the human eye—all-seeing, all-encompassing vision. Photo by Kevin Noble.

These automatic motions simulated all possible camera movements without making the camera and its operator the center of the universe. Time and motion became the universe, with endless repetitive cycles and orbits.

I was a latecomer to this infatuation with machines, but I vividly remember, after moving to New York, going to Canal Street and looking at gears and motors as kinds of miracles that resembled life itself—mechanistic replications of the biological mystery. I love gizmos such as the ones I find in surplus yards that can be refitted to serve my purposes. If I had a lot of money, I would spend it on optical gadgets, mechanical toys, and state-of-the-art electronics. I would make gigantic environments such as monitors embedded in the floor from wall to wall, all showing imagery moving in either the same or a contrapuntal direction. Or I would build a four-sided corridor in which viewers could look down a long lane of images that keep moving toward and past them. In reality, though, I am very flexible about the size of the display, since, to me, the size of an installation is not determined by the number of monitors but rather by the complexity of the composition. I therefore often improvise how to conduct an installation based on what is available at an exhibition site. For example, for my favorite installation of *Geomania* (1986) (figure 6.2), I stack the monitors into a pyramid.

I always intend that these environments will be experienced in a dark place. A museum is potentially a good exhibition venue, but museum people always seem interested in placing video installations in maximally visible locations. They tell me triumphantly, "We are going to give you the lobby." It is always assumed that video ought to be loud and public, but I really want it to be quiet and private—a thousand monitors and one viewer, and not the other way around. I want the viewers to be so absorbed by the work that they experience another level of mind. I expect them to share the kind of strong feeling I have for the material, and to my amazement, they sometimes do. An old man who had watched *Tokyo Four* (1991) over and over once explained to me that this installation was all about death. At that moment I knew that he had really seen it—even though it is not all about death.

Borealis (1993) uses two video projectors that project through a split-beam mirror onto four translucent screens (*translucent* meaning that the

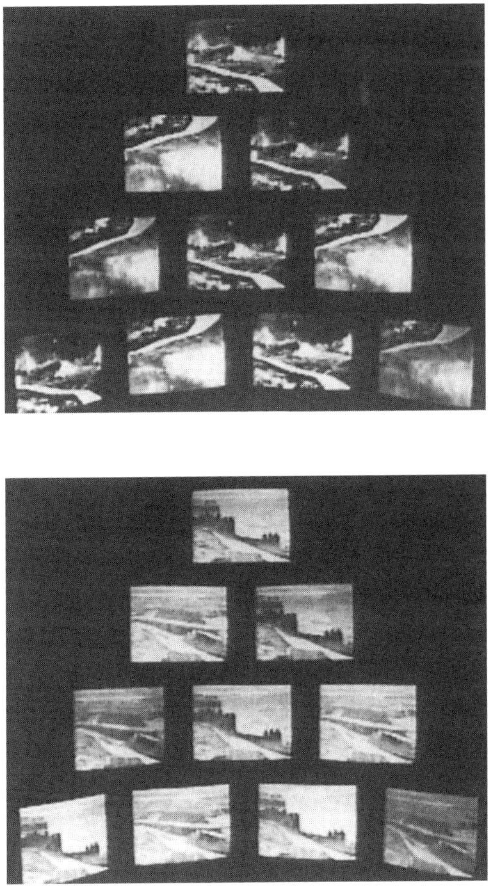

Figure 6.2
Steina, *Geomania,* 1986, two-channel sound and video environment. This work, presented on a pyramid of monitors, combines site-recorded imagery and sound in a layering of natural landscapes and electronically generated color and texture.

image appears with equal intensity on both sides of the screen). On entering the room, the viewer can watch the work from far away and see all four screens at once or walk directly up to and around one screen—a much more intense experience. The images are mostly of rivers, oceans, steams, and sprays.

SUBJECTS

The aspect of the process of creation I like most is the initial recording. Sleet or snow or howling rain, I love that part, especially if I am alone out in nature. In New Mexico, where I live, my images are rivers, mountains, and arroyos, but when I found myself in the metropolis of Tokyo, my material became the people. The Japanese have a social protocol that, for them, is a daily routine, but to a Westerner looks like a fabulous theater—the way they bow, the way they make certain signals. For example, when the Japanese want to cut through a crowd in a hurry, they put their hands forward in a chopping gesture and a magical corridor appears in the ocean of humanity. They have hand signals for *yes* and *maybe* (although *maybe* usually means the unutterable *no*). They seem to wear an invisible armor, a "no man's land" around their bodies. Elevator girls in a perpetual state of performance, train conductors, taxi drivers in their white gloves, and Shinto priests ritualistically pruning their arenas are all elements of *Tokyo Four.* Between taping and editing, there is usually an intermediary step during which I alter and mix the images, change color, or run things upside down or backward. This is where the particular uniqueness of working with the electronic image comes into play. It is somewhat akin to photographic darkroom techniques, but it really reminds me of playing an instrument. I change style, timbre, dynamics, and key in an improvisational and spontaneous way.

MUSICAL TECHNIQUES

In my multichannel video compositions, I often make a ground image of a certain duration, which I then duplicate as tape 2, tape 3, and so on. I then drop different but complementary images into the copies, and a phenomenon similar to musical composition starts occurring. Starting with a melody or theme, I add harmonic lines and discover that the melody is far less interesting than the counterpoint. Sometimes there is an emergent melodic structure that interweaves through the instruments or (in my case) the video screens. Late twentieth-century art is fast—too fast for me. But I realize that I am out of sync with the mainstream, which wants things fast. Multichannel compositions liberate me from this concern with speed, since they rely on different time principles and are therefore more like music.

TEACHING

I do not like teaching, just as I did not like going to school. It is an absurd theater, the teacher supposedly all-knowing and the students posing as eager minds waiting for illumination. So when I do teach, I go through the theory and the techniques—video is rather complex technically—and explain about frequencies, voltages, and the timing structure of the signal. I go into history, show a lot of tapes—mine and those of colleagues—and discuss them with the students. Then I ask them if they believe in UFOs (unidentified flying objects), at which point the whole class gets very uneasy. Half of them say they do, half say they do not. The class sessions that the students seem to appreciate most are the ones in which I present "the world according to Steina." We discuss the way galleries sew up the art scene and make the artists kiss ass. I always tell that them that they do not have to kiss ass. And they seem greatly relieved, almost as though they did not know this. I remember once overhearing a student say, "But we have to do this kind of conceptual/intellectual work because this is that kind of school." I turned around and said, "NO, YOU DON'T." And the whole class laughed because they realized that they really do not. I tell them that it is every artist's duty to be disobedient. We discuss what it means to be a mainstream person and have a comfortable life and how deciding to be an artist basically means deciding to live a materially uneasy but more rewarding life. They discuss this for a while—not that they have not already thought about it a lot, but they get lonely and confused. So I reassure them that there is no grander life than the creative, artistic life. It is the unknown, the exploration, the fact of being your own person on your own time.

After I ask my students about UFOs (and some of them say they do believe and others say they do not), I tell them we are not going to talk about UFOs anyway but how we must stick to our beliefs. I tell them that, if they believe in UFOs, they should raise their hands whether or not the other half of the class is going to sneer. The discussion turns to intimidation and lying about one's beliefs just to get along. It is emotionally stressful to admit to having an independent mind. One does not have to be an artist to experience this dilemma, but I believe it is the artist's duty to stay on the fringe.

CREATIVITY AND COMMUNICATION

The creative process, for me, is a tremendous pleasure, even when it is painful, such as when I feel inadequate to the task. People perceive this pleasure in my work and often object, "But you are just playing!" This comment gives me tremendous pleasure!

The motivation to make art seems to come from a deep desire to communicate; for some artists, it comes from a desire to communicate on a massive scale—something that does not particularly interest me. I see no qualitative difference in more people versus one person if I am communicating. Our whole existence seems to be about communication. It cuts through cultures, languages, and continents. It also cuts through time. We spend so much time with people we have never met—often, with people who are long dead. But the primary motivation for all art is the desire to communicate with oneself. This is a spiritual idea. It has been the sad lot of many artists to communicate only to future audiences, but, through lucky coincidences, artists and their audiences have sometimes found each other in the same place at the same time. Paris in the 1920s was like that. New York in the 1960s was like that for us. It was a luxury.

REFERENCES

Boyle, Deidre. *Video Classics: A Guide to Video Art and Documentary Tapes.* Phoenix, AZ: Oryz Press, 1986.

Cathcart, Linda, ed. *Vasulka, Steina: Machine Vision/Woody: Descriptions.* Buffalo, NY: Albright-Knox Gallery, 1978.

Clancy, Patrick, ed. *Video as Attitude.* (Albuquerque, NM: University Art Museum, 1983.

Dietrich, Frank. "The Computer: A Tool for Thought-Experiments." *Leonardo,* 20, no. 4 (1987):315–325.

Furlong, Lucinda. "Notes toward a History of Image-Processed Video." *Art Journal* (Fall 1985). The article is available at ⟨http://www.artscilab.org⟩.

Gill, Johanna. *Video: State of the Art.* New York: Rockefeller Foundation, 1976.

Huffman, Kathy Rae, ed. *Video: A Retrospective.* Long Beach, CA: Long Beach Museum of Art, 1984.

Pritikin, Renny. *Video Installed.* San Francisco: New Langton Arts, 1986.

Schneider, Ira, and Beryl Korot, eds. *Video Art: An Anthology.* New York: Harcourt Brace Jovanovich, 1976.

7

Transmission

Joan Jonas

My work consists of fragments and chance as much as materials and technology. In the late 1960s, after studying art history and sculpture, I became inspired by the idea of performance and began to work with time as material, transferring my concerns with drawing and the object into movement. At the time, "I didn't see a major difference between a poem, a sculpture, a film, or a dance."[1] Now, in 1998, working in video, performance, installation, sculpture, and drawing, I experience the forms as overlapping, not totally separate.

While I was studying art history, I looked carefully at the space of painting, films, and sculpture—at how illusions are created within a frame. From this, I learned how to deal with depth and distance. When I switched to performance, I went directly to real space. I looked at it, and I would imagine how it would look to an audience. I would imagine what they would be looking at, how they would perceive the ambiguities and illusions of the space. An idea would come from just looking until my vision blurred.[2]

At this time, in 1966, I visited Crete to research the Minoans. (I was interested in the imagery of early art forms—like the Cretan mother goddess.) I went to a wedding ceremony in the mountains that lasted for three days. The men sang songs to each other as guests arrived.[3] I was always interested in folk culture—the dance, the music, the objects—because it is a part of everyday life. I was especially interested in this particular wedding ritual because performance is not a space separate from ongoing activities of daily life. My own performance came from trying to communicate this experience to my audience—my community. That intent, and the community itself, would change over the years, but that's where I started.

At that time, I also traveled to the Southwest to see the Hopi snake dance. My reaction was complicated. I remember now the profound effect this dance—a ritual with live snakes—had on me, as well as the architecture of the pueblos and the amazing desert landscape. At the same time, I remember noticing that the audience of mainly white tourists wore huge squash blossom necklaces they had purchased at the pawn shops. I couldn't avoid the nonchalant display of these displaced symbols. Somewhat naively, I understood the reality of loss.

Were we an intrusion? Of course. The event was changed by our presence. Not long after that, outsiders were not allowed to witness the snake dance ceremonies. I was lucky to have been permitted to see these amazing events carried forward from another time in which people directly related with and communicated to the land, the environment, and the elements.

In a second ceremony at Ancoma, costumed figures were far away, in the desert, and then suddenly they were close up, in the plaza, dancing. What was striking to me was how these images from afar could be brought back home. What became apparent and of interest was how to think about one place and be in another. Was it possible to cross-reference rather than categorize? Was it possible to translate such concepts into one's own intuitive language, using technology as a tool of transformation and transmission?

Other references for me were the circus and magic shows that I saw as a child and the idea of alchemy or transformation of material and psyche. I especially liked sleight of hand—visual tricks that could be special effects. Perhaps I always like to have a reason in relation to structure and content—to know that something made it happen even if we don't know and can't see what it was. On the other hand, I'm interested in the obvious. In works of mine such as *Vertical Roll* (1972),[4] I reveal the mechanics of the illusion. I like to juxtapose high tech with the original gesture. In that way the touch, the body, and the machine are put into play.

Performance as a medium exists somewhere between "conceptual art" and "theater." For performance, a genre of multiple media, the critical material is time. This is said in the context of the visual arts—in my context. The artist builds a performance by designing and composing all aspects of the work—conceives, constructs, draws, and choreographs; makes the music or chooses it or selects a composer to work with; performs, produces, and directs film and video; often does camera or directs the camera work; and edits. The work is based on visual and aural concerns rather than text, although text can be used as material, and it can be written or chosen by the artist. Beyond this, there is also close collaboration with other performers and artists, filmmakers, editors, and producers.

The history of performance can be said to begin with prehistoric cave rituals and to extend through dada, multimedia events at Black Mountain in the late 1940s and 1950s (as well as Europe, Japan, and Central and South America in the same period), happenings in the New York art world of the 1960s, including the Judson Church group of dancers and artists working together, and the multimedia performance and installation work in the 1970s, 1980s, and 1990s. I wanted to look at sources from outside the art world. I wanted something that was not dance, not sculpture, not theater.

My work is often considered personal or private, perhaps because of the presence of the author as performer. Friends have told me that they feel they are looking into a private world. I do try to bring the audience into my space. There is an intimacy.

Finally, the attraction for me in performance is the immediate pleasure of working for a live audience. I am totally in a concentrated present. There is an unspoken communication and feedback that constantly changes. In 1968, when I first presented my work publicly in New York, most artists lived near one another downtown—that is, sculptors, composers, dancers, painters, musicians, performers, video artists, filmmakers, theater people. The geography of New York condensed things—we were friends, we attended each other's shows, we critiqued, supported, watched—and in this way, forms and boundaries were erased. There was also the desire to work outside the conventional spaces of museums, galleries, and theaters. The point of view of the audience was questioned. I step in and out of my work to direct the perception.

1968 TRANSMISSION: THE MIRROR

Inspired by the short stories of Jorge Luis Borges, I chose as my first technological tool the mirror, a device that transmits light. First, I made a long black costume for myself with mirrors pasted on it. I moved stiffly, parallel to the audience, quoting all references to mirrors in the short stories of Borges's *Labyrinths.* The piece was called *Oad Lau* (1968)[5] ("watering place," after a trip to Morocco; this work also related to the Greek wedding). Later, similar moving figures—a man and a women appeared in *Wind* (1968)[6]—my first film.

From the beginning, the mirror provided me with a metaphor for my reflective investigation. It also provided a device to alter space and to fragment it. By reflecting it, I could break it up. I could mix reflections of performers and audience, thereby bringing all of them into the same time and space of the performance. In addition to creating space, a mirror also disturbs space, suggesting another reality through the looking glass—

to see the reflection of Narcissus, to be a voyeur, to see one's self as the other. In this piece, *Oad Lau,* the reality was also to see oneself among and as one with others.

Then I did a series of works in which performers—about fifteen of them—carrying 5-foot-by-18-inch glass mirrors and glass moved slowly in choreographed sequences and patterns, reflecting the audience, themselves, and the space, fragmenting it, and yet always flattening it. The mirrors face front. The glass is heavy. The performers move slowly—in lines (*Mirror Piece, I & II,* 1969 and 1970).[7]

In another part of the piece, bodies were treated as material. They were carried stiffly—horizontally by feet and neck—like boards or glass. In another sequence, transparent glass panels are used. Two women roll across the floor with a 5-foot-by-18-inch sheet of glass between them, avoiding breakage. The panel is the same size as the mirrors used previously; here, though, at the same time, two men work with a larger piece of glass (four feet by five feet), turning it, shifting it. The audience, included by reflection, is part of a moving picture.

The mirrors and clear sheets of glass could break or shatter at a wrong move. We were barefoot. I was interested in this tension and that the onlookers might feel uneasy.

Narcissism provoked by mirrors is also disturbing. For *Mirror Check* (1970)[8] I stood naked, inspecting all parts of my body with a small round hand mirror. Using a slow circular movement, I began with my face and finished with the bottoms of my feet. The audience watches me checking myself. Vicariously, however, as they can't see what I see, despite the fact that they see more of me. The duration of the performance was about ten minutes.

TRANSMISSION: DEEP LANDSCAPE, THE DISTANT IMAGE

In 1970, I went to Japan and saw the Noh and Kabuki theater for the first time. This theater's highly developed visual vocabulary gave me new inspirations. I was aware of the attraction that Yeats and Fenellosa had for Eastern poetic forms. I later learned that Artaud had been inspired by Mexican rituals and Eastern theater, for similar reasons. I attended Noh as often as possible. This experience informed the work. I translated into my own language the familiar slow pace, the sound and use of wood, the masks, the costumes, and the idea of dance or formalized movement. After this trip to Japan, I began working in the medium of deep landscape space—again interested in altering what is perceived as reality in image and sound.

Beginning at Jones Beach, I worked with the transmission of the signal—distance flattens circles into lines, erases detail, delays sound. The mirror reflects light over distance. Working with the flat expanse of distant space, I was trying to work with the absence of depth over distance—in a sense, to displace the idea of the space or what happened in the space, to bring that forward to the audience. This is explored in two beach pieces—one in New York (*Jones Beach Piece,* 1970, figure 7.1),[9] one in Nova Scotia (*Beach Dance,* 1971),[10] and one at New York's Hudson River (*Delay Delay,* 1972).[11]

In the mud flats at Jones Beach, the audience is situated a quarter of a mile away from the performance, and in Nova Scotia the audience is on a cliff overlooking a beach. In *Delay Delay,* in lower Manhattan, the view was from the roof of a loft building overlooking the empty lots and distant docks of the Lower West Side. In Rome in 1972, the audience viewed a version of *Delay Delay* from across the Tiber River.[12]

The new element for the outdoor works was the sound delay. Performers clapped blocks of wood together at different distances from the audience. One saw the gesture of clapping in wide overhead arcs before hearing the sound, the lag depending on the distances and the atmosphere. This separation of action and sound, of sight and hearing, isolated for the audience the relativity of perception. The clapping gesture marked the perimeters of the space, but the sound transmission, the desynchronized delay, was its measure.

Being far away from the audience gave me freedom to move in strange or comic ways. Out on the mud flats at Jones Beach, I felt comfortable dressed in a long black skirt, head scarf, and heavy welding shoes, running with a shovel and a red bag of shells or sitting precariously on the top of the ladder in the distance and wearing a plastic hockey mask. I wore a blue dress with a long train, which was wet and blowing in the wind. The weight of the cloth caused the ladder to tip. I was holding a 5-foot-by-18-inch mirror and using it partly to balance myself while flashing reflections of the sun into the eyes of an audience away in the distance. Between my position and the audience, seven women dressed in black

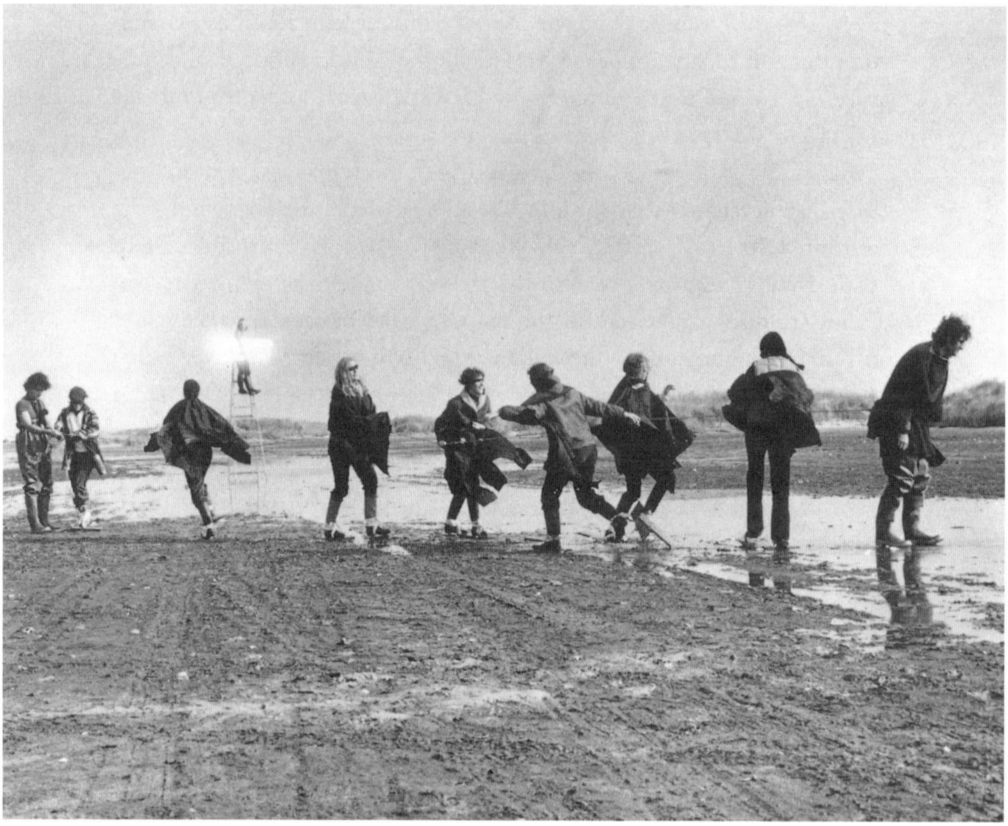

Figure 7.1
Joan Jonas, *Jones Beach Piece,* 1970, Jones Beach, Long Island, New York. Performed by Barbara Dilly, John Eldman, Carol Gooden, Tannis Hugill, Joan Jonas, Epp Kotkas, Kate Parker, Linda Patton, Susan Rothenberg, Gwen Thomas, George Trakas. Photo by Richard Landry.

capes, blindfolded, with blocks of wood tied to their feet, ran back and forth along a rope stretched between two men that was diagonal to the audience's view. It appeared to be parallel to the audience. Details of costume were not visible but affected the performers' movements. All movements were made to be seen in the distance. All was flat without color.

The structure of these pieces was simple—one thing after another like beads on a string.

In speaking of the movement of dance, I have to say that in the 1960s in New York the Judson Church project opened a way for visual artists like me to go into performance. In the works of dancers Yvonne Rainer, Deborah Hay, Steve Paxton, Trisha Brown, and Simone Forti, in particular, was an exploration of natural, everyday movement. Actions like walking across the stage to sit in a chair or performing a routine, simple task expanded the definition of dance. I began my work, first simply in relation to the job of moving or being moved by props. Slowly over the years I developed more complicated moves with music, sound, mask, object. And then I learned how to move in relation to the video camera—both as operator and as subject.

TRANSMISSION: MOVING IMAGES IN FILM, ELECTRONIC SIGNALS IN VIDEO

Wind (1968) and *Songdelay* (1973)[13] translated my live performances into the medium of film. In *Wind,* an indoor work—*Oad Lau*—was taken outdoors to a beach on Long Island's north shore. It was winter. The element of wind became the central force as mirrored figures slowly moved in a snowy landscape. We played with the wind, taking our coats on and off, again and again, with some effort, while moving along the water's edge in the strong wind.

In *Songdelay,* by using different lenses, a wide angle and telephoto, I translated the outdoor performance *Delay Delay* into film. This was the final development of the series of outdoor works that began at Jones Beach. I wanted to save my performances in a form that interested me, and since I consciously used film as a reference at times during the performances, film was appropriate to the task.

I was particularly drawn to early filmmakers such as Vertov, Vigo, Franju, Eisenstein, and Ozu. And the fragmentation of sequences in my performances comes partly from ideas that are based on film techniques such as the cut and the idea of montage. I felt the freedom to move from one element to another, cutting from one scene to the next like cut and paste.

In 1970, in Japan, I bought my first Portapak and began to work in video. The Portapak (a big heavy camera and reel-to-reel deck) was not often used for art making at the time. Some artists had begun to use it in the last few years of the 1960s, and artists such as Nam June Paik had worked with broadcast television in the early 1960s. It was definitely outside the mainstream commercial art world and television industry. The Sony Portapak was an appropriate tool for artists, who usually worked alone in their studios. It could be hand held. The technology was simple, and it did not require a crew. It was black and white.

The video camera did not have a history for me to refer to. In fact, history for me was film, a reference against which the new video possibilities became clear. I was aware of the work of independent filmmakers like Jack Smith, Kenneth Anger, and Stan Brackage (and in 1976 came to know the work of Maya Deren). What video offered was the opportunity to work live, to make a continuous series of images explicitly for the camera during live performance, which allowed me an added nonnarrative layer in a kind of condensed poetic structure that I had earlier found in the writings of the American imagists (including H.D., William Carlos Williams, Ezra Pound, and Emily Dickinson) and in Japanese haiku. I was also interested in how myth was used in the work of James Joyce, for instance. These forms were also models for work in time.

Video allowed for the immediacy and the continuity of television's live broadcast, while also allowing real-time, ongoing viewing via a monitor. It was simultaneously a recording medium. Video offered a continuous present—showing real-time actions, and incorporated a potential future, re-viewing and reusing actions thus recorded.

The monitor, at that time a critical factor of video, is an ongoing mirror. I explored image making with myself as subject: I said "this is my right side, this is my left side," and the monitor shows a reversal. I made a tape about the difference between the mirror and the monitor.[14] I worked with the qualities peculiar to video—the flat, grainy, black and white space, the moving bar of the vertical roll and the circle of circuitry formed by the Portapak, monitor/projector, and artist.

In the first tape that turned into the first performance, I imagined myself making a film. I sat on a white wicker chair facing the camera and monitor, and using props, objects, and sound, I improvised for the camera.

Organic Honey's Visual Telepathy (1972)[15] evolved as I found myself continually investigating my own image in the monitor of my video machine. Wearing the mask of a doll's face transformed me into an erotic electronic seductress. I named this TV persona "Organic Honey." (I stayed up all night wondering what to call my persona and then saw on the table a jar labeled "organic honey": it seemed perfect.) From a book on magic came the phrase "visual telepathy."

In translating this initial experiment into performance, I thought of my stage as a film set within my loft. I added a 4-foot-by-8-foot piece of plywood on sawhorses—a table for my objects. Among them were a big glass jar filled with water and a small shot glass, mirrors, silver spoon, old doll, silver purse, stone. On the wall, I tacked a drawing of my dog with one blue eye and one brown eye, doubled. I also used a tall, antique, wood accounting chair. Inside this set, I put the camera on a tripod. For some sequences, the camera would also be hand held by the camera woman. I showed the audience the video images in two ways—one on a small monitor, the other in a large projection on the wall of the set. I also placed a small monitor inside the set for me. All of my moves were for the monitor, which I monitored, keeping my eye on the screen as I worked.

The camera woman, holding the camera or placing it on the tripod, operated inside the set with me. She followed my rehearsed movements in close-up. This system—the set for *Organic Honey's Visual Telepathy* (figure 7.2) and *Organic Honey's Vertical Roll*,[16] the live performance and its related tapes[17]—was the model for all my subsequent black and white video works.

Video performance offered the possibility of multiple simultaneous points of view. Performer and audience were both inside and outside. Perception was relative. No one had all the information. I thought I had, but it was an illusion.

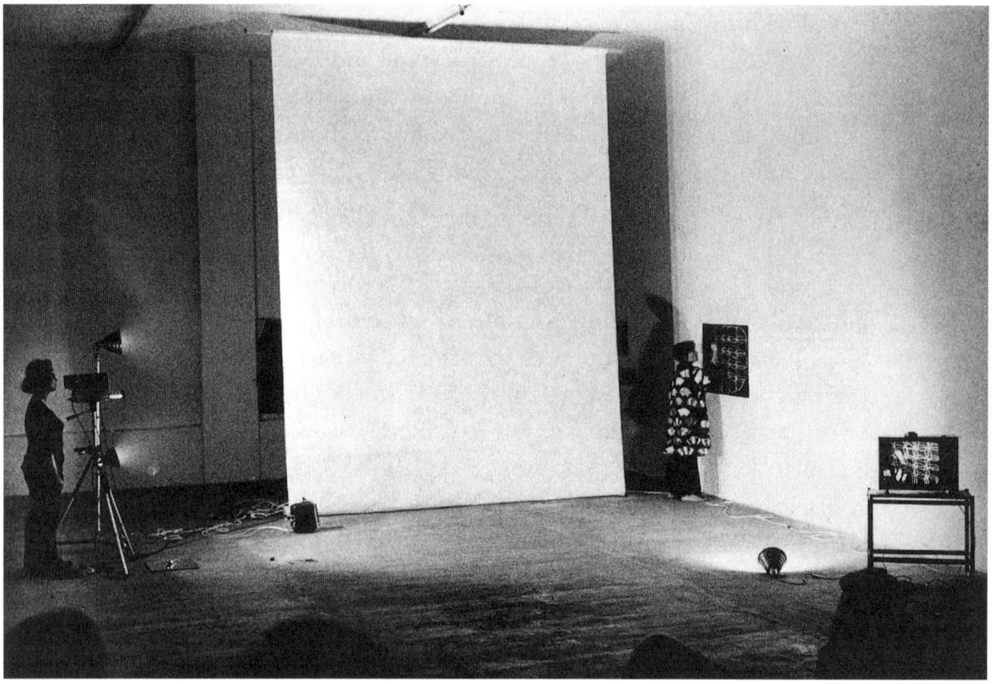

Figure 7.2
Joan Jonas, *Organic Honey's Vertical Role,* 1973, Gallenia Toselli, Milan. Performed by Joan Jonas. Camera by Barbette Mangotte. Photo by Giorgio Columbo.

The audience sees Organic Honey in her green chiffon dress. She kneels on the floor over a piece of white paper under the overhead camera. She slips back her mask to wear it like a hat, then draws a dog's head, the top half on the bottom of the paper, the bottom half on top. On the monitor, as the vertical roll bar rolls, the two halves of the drawing come together in proper position. As she draws, Honey's hands and arms become visible. Occasionally her mask looks up at the camera. I draw watching the monitor. These (and later performed drawings) are drawings for the monitor. The camera (woman) frames the drawing and the mask—and this detail is what is seen on the video monitor (and/or projector) in the live closed-circuit system. Then Honey removes the drawing, revealing a mirror also on the floor, and with a large silver spoon she bangs it methodically. The video sees and shows: a silver spoon hitting the mirror, with afterimages. The camera zooms into the spoon, the sound of metal hitting glass echoes against the walls and ceiling—hitting the mirror with a spoon, a tapping signal that loudly resonates as silver on glass. This began as anger. I was interested in translating emotions.

The audience sees, in fact, the process of image making in a performance and simultaneously with a live detail. I was interested in the discrepancies between the performed activity and the constant duplicating, changing, and altering of information in the video. The whole is a sequence of missing links as each witness experiences a different series by glancing from monitor to projection to live action. Perception was relative. There was a range of choices. Time and space in these performances were like Borges's *Garden of the Forking Paths.* Here were parallel worlds. I could inhabit, simultaneously, different fields of view, different channels.

Several parts of the Organic Honey performances were prerecorded tapes that could exist on their own as well as be part of the performances.[18] I worked back and forth between tape and video performance, translating ideas from performance into tapes, transferring elements from tapes back into performance. The word *tape* itself covers multiple types and uses: continuous tape, tape prerecorded to be included in a performance, or tape recorded to stand on its own. (Performances that were documented on tape to me are only documents, not artworks.) A performance recorded for a single-channel video work to be shown publicly would be altered—through special effects, change in camera angle, or working with and cutting back and forth with two cameras, inserting new material, cutting out parts, and so on.

The structure of videotape was not divided into visible frames, as in films. One could record for longer periods of time without breaks. Artist filmmakers recorded on ten-minute reels at that time. At first, the duration of videotape seemed to slow time. Time could be delayed. On the other hand, editing was fast—just pressing buttons.

Later performances included *Funnel* (1974),[19] with the camera person outside the set looking in. I used one monitor facing the audience and none for me. The set is made of paper in the shape of a cone. *Twilight* (1975)[20] employed two simultaneous closed-circuit systems of cameras with monitors. There were five performers. For *Twilight* and *Mirage,*[21] I designed the pieces for the theater space of Anthology Film Archives in New York. Because I had seen so many films in this space, I was inspired. I made 16 millimeter films for projection in both performances. *Mirage* was made after a trip to India (an underlying influence) and was the last of the black and white series. There was no live video, only prerecorded tape and 16 mm film. The film of drawing and erasing on the blackboard was a series of images from past and present pieces—a heart that looked like a bug, signs for a storm, a rainbow, and a mirror reflection, all to be read as a kind of sentence with no fixed meaning. There was also a five-minute documentary loop of volcanoes erupting and a film shot of the monitor of a television turned on its side with the vertical roll bar switching from right to left. My action of repeatedly stepping through a small wooden hoop was broken by the vertical bar. Rhythms were syncopated.

Around 1970, when I began the Organic Honey pieces, issues of feminism were important for all of us. In general, video became a vehicle for women's voices. It was unexplored territory. I was interested in the condition that both video and performance were unexplored.

As soon as I began to work with video, my focus shifted. I began to move away from a kind of minimalism to represent my concerns. I wanted more complicated layers of form and content. We could speak to the camera, record our movements, communicate our desires. I explored the possibilities of female imagery, questions of whether there is a female psyche, and representations of emotions. But anger was one of the main ones for many of us, expressed indirectly through a visual and aural form. Video as we used it was personal, and the personal was political.

To be without expression was the style of performance art. I used the mask as a way of exploring female identity. This instantly took away facial expression and my identity. Masking both concealed and revealed possibilities of representation that may not otherwise have been possible for me. Hidden, I was in a private world that seemed open and magical.

The particular mask of Organic Honey created a persona that seemed to be distinctly someone else. A mask here altered body language: I could add an erotic tone. I imagined playing roles like an electronic sorceress or a dog. I howled. I sang. I danced. I explored the place of women in history as outsiders—healers, witches, storytellers.

The video monitor's screen or the projected image was another mask for the construction and deconstruction of persona. Here there was also distance—even in the close-up. I did not act, I behaved. I performed activities. In a belly-dancer's costume, I jumped in and out of the bar of the vertical roll like frames in a film going by. This out-of-sync dysfunction of the television—the rolling pictures—presented on the screen parts of the body, never a whole. I had begun to dance with the TV.

TRANSMISSION: RETELLING STORIES—AN ORAL TRADITION, AN AURAL MESSAGE

In 1976, I wanted to change what seemed to be habitual dependence on the medium of film and TV—or the camera and the monitor. I was commissioned to do a performance for children and so decided the performance vehicle could be the fairy tale. I wanted to remain within the popular form of folk tradition, and so with the help of the son of a friend chose *The Juniper Tree* (1976),[22] a story told again and again mostly by women and then written down or recorded by the Brothers Grimm. Here is the technology of the human voice box handed down.

With the fairy tale, I was once again looking to see what roles women play and how they are represented. Again it is an exploration of the self. The story becomes the mirror of my projections. I look for how the stories reflect basic human psychology and behavior, while laying bare the hidden taboos.

I recorded the text on audio tape with music and sound effects, and this became a background for my image making—again, another kind of circuit. These images are in a way as imaginary and nonrealistic as the ones in the earlier video work, but they come directly from the story and how I experience it. I chose to represent the story again and again, first in a very theatrical collaboration in the style of Chinese opera[23] and then in a final version, more subjectively as a solo but going inside and allowing the demons to escape.

In the second version, I began with the stage set. It is a loose wood and string structure that represented the house that finally collapses at the end of the story. I also included a ladder—a shamanic representation of a magic tree or the ladders to the sky. This is now my sculpture.

Color enters the story—"as red as blood and as white as snow." On the wall behind the set are red and white drawings painted on red and white silk. These drawings are made during the performance, and the best are chosen for the set. They represent the boy and the girl of the story.

I play all the different female roles—the untouched daughter, the good mother, and the main character, the wicked stepmother. In one performance, the dancer Simone Forti played the bird that the boy becomes, and she sang the song "My mother she killed me, my father he ate me, and my little sister gathered up all my bones and laid them 'neath the tree, what a beautiful bird me."[24] Some parts were danced, some spoken, some drawn on the same mirror that showed up in early mirror works.

As I researched and analyzed the fragmented story, I found references to ancient Egyptian myths of Isis and Osiris, for instance, and other shamanistic practices from northern Asian and European migratory groups. I became interested in the literally devouring female in the story—or frightened by this representation. And this interested me, as it suggests our own hidden selves.

TRANSMISSION: MIRRORING LANDSCAPE, EVER-GREEN MYTHOLOGIES

I see *Volcano Saga* (1985–1989)[25] as the beginning of my synthesizing the development of female character, the story as mirror, and the volcanic landscapes as representation of narrative. Here in Iceland was the connection for the psyche to the elements. As in *Wind,* the elements become character. I was also interested in the fact that at the time of the sagas in Europe, Icelandic women were the most independent.

I chose as my story the *Laxdaela Saga,* which was written down anonymously in the thirteenth century but from stories that had been told and retold for centuries. In Iceland, storytellers went from house to house, and the stories all seemed to begin with real people and true events. This story is about a woman who married four times. The book begins with an historical account of the settlement of Iceland, tracing the character's ancestry, and then continues with the woman, Gudrun, who tells four dreams to a seer who then interprets the dreams. The second part of the saga involves the actual marriages and the carrying out of the prophecy.

I went to Iceland and explored the place, the landscape, and like many others, I became enchanted by the closeness of beliefs in the time of sagas to the present-day and the importance of the sagas in the culture.

A film crew recorded the landscapes on video. I photographed it for slide projection, and on returning to New York, I developed a solo performance with video projections and slide projections of the different mostly volcanic landscapes that for me represented parts of the story—the four dreams.[26] In this way, I worked by performing. I finally turned the live performance into a thirty-minute narrative for television broadcast, with Tilda Swinton playing the woman, Gudrun, and Ron Vawter playing the seer, Gest.

I developed a way of telling the story in video with the foreground shot in the studio against the Icelandic landscapes as backdrops. Tilda Swinton tells her dreams to Ron Vawter as they sit together in the hot springs—a very beautiful blue lagoon with wind, mist, and black volcanic rock. The tradition wasn't exactly this way in the book, and in the studio, Swinton and Vawter are shot in front of a blue screen with the lagoon later keyed in. Sitting in the steamy blue made the relationship of the characters in the story, who are actually old friends, erotic. I liked this added level of closeness in relation to our own ideas about how and when we tell our dreams to others and how they are interpreted.

To frame the story in the present, I began the tape by telling my story of an accident that actually happened in which my car was blown off the road by the wind. Otherwise I played a small part. The tape ended with an old couple talking about how fishing was invented—by a woman, probably, they say.

For a recent retrospective of my work art at the Stedelijk Museum in Amsterdam in 1994, I turned earlier performances into five installations—the outdoor and mirror works in one room, and *Organic Honey, The Juniper Tree,* and *Volcano Saga* each in their own rooms. I made a new installation—*Sweeney Astray* or "*Revolted by the thought of known places. . . .*"—that turned into a theater piece that was performed in collaboration with a theater group (the Toneel Groep, Amsterdam).[27] This was based on a Seamus Heaney translation of a twelfth-century Irish poem about a king who goes mad after battle. One of the props was a 6-foot-by-6-foot-by-7-foot glass table on which the actor (Sweeney) stood. A camera view from an underneath angle made it appear that he was floating. I like this way of presenting my work. My presence is not absolutely necessary.

And so I continue to work mostly in video and installation. What interests me now is making objects for video—in other words, sculptures—and to continue to relate to the culture of my immediate surroundings. In the summer of 1998, this was rural Canada.

My New Theater (1997), part of a series, is an object for video. It is a wooden cone, squared—a tubelike structure about 7 feet long. Inside is a miniature theater. One looks into the theater from the large open end of the cone to see a stage area and some miniature props.[28] Behind the stage is a back-projection screen. Behind that, a small projector, hidden, projects a video loop of an older man, a Cape Breton step dancer, step dancing on a board in front of a waterfall. This dream landscape is intercut with a closeup of a young Cape Breton girl's dancing feet. The sound of the feet reverberates in the wooden structure, a funnel-like form I've used several times before (in 1974, 1975, and 1976). It is an enlarged, yet low-tech amplifier, a variation on the hand-held megaphone. While video technology continues to be a vehicle for the transmission of stories, my theater can now travel without my body.

I am, however, still performing. I am now collaborating with Paul Miller, a.k.a. DJ Spooky. I am the VJ, mixing my tapes live to his live music mix. In front of a projected image, two performers speak and move. This narrative mix hovers between theater, music, and dance.[29]

I am also working with a very small digital camera. It has its own internal monitor. It weighs hardly anything, and it fits into the palm of a hand. The size and weight allow me to move very differently compared to how I moved with the high-8 cameras of a few years ago or the Portapak of thirty years ago. The fact is, the videos still dance and make music.

NOTES

1. Joan Jonas, *Scripts and Descriptions, 1968–1982,* ed. Douglas Crimp (Berkeley, CA: University Art Museum; Eindhoven: Stedelijk Van Abbemuseum, 1983), 137 p.

2. Ibid.

3. Joan Jonas, *Works, 1968–1994,* ed. Dorine Mignot (Amsterdam: Stedelijk Museum, 1994).

4. *Vertical Roll,* 1972, black and white video, 20 minutes, sound, camera by Robert Neiman.

5. *Oad Lau,* 1968, performance.

6. *Wind,* 1968, 16 mm, black and white, 7 minutes, silent, camera and coediting by Peter Campus.

7. *Mirror Piece I,* 1969, *Mirror Piece II,* 1970, performances, partial list of performers: Francis Barth, Eve Corey, Susan Feldman, Pam Goden, Carol Gooden, Deborah Hollingworth, Keith Hollingworth, Barbara Jarvis, Joan Jonas, Julie Judd, Jane Lahr, Lucille Lareau, Jean Lawless, Susan Marshall, Rosemary Martin, Tom Meyers, Judy Padow, Linda Patton, Corky Poling, Peter Poole, Susan Rothenberg, Andy Salazar, Lincoln Scott, Michael Singer, George Trakas, Pam Vihel.

8. *Mirror Check,* 1970–1974, solo performance.

9. *Jones Beach Piece,* 1970, performance, Jones Beach, New York.

10. *Nova Scotia Beach Dance,* 1971, performance, Inverness, Nova Scotia.

11. *Delay Delay,* 1972, performance, Manhattan Festival of Music and Dance, Tiber River, Rome, Documenta 5, Kassel, Germany.

12. See above.

13. *Songdelay,* 1973, 16 mm, black and white, 18 minutes, sound, camera and coediting by Robert Fiore; sound by sound technician Kurt Munkacsi, with Ariel Bach, Marion Cajori, James Cobb, Carol Gooden, Randy Hardy, Michael Harvey, Glenda Hydler, Joan Jonas, Epp Kotkas, Gordon Matta Clark, Michael Oliva, Steve Paxton, Penelope, James Reineking, Robin Winters.

14. *Left Side Right Side,* 1972, black and white video, 7 minutes, sound, camera and performance by Joan Jonas, produced by Carlotta Schoolman.

15. *Organic Honey's Visual Telepathy,* 1972, black and white video, 23 minutes, sound, camera and performance by Joan Jonas.

16. *Organic Honey's Visual Telepathy,* 1972, performance, Joan Jonas with Suzanne Harris, Kate Parker, Linda Patton; *Organic Honey's Vertical Roll,* 1973, 1974, and 1980, performance, camera by Robert Neiman, performed by Joan Jonas with Anne Thornycroft, Margaret Wilson, and Freuda; *Organic Honey's Vertical Roll,* 1973, 1974, and 1980, performance, camera by Barbara Mangolte and Joan Jonas.

17. Tapes that were made in relation to the Organic Honey series: *Organic Honey's Visual Telepathy, Vertical Roll, Duet,* 1972, black and white, 4 minutes, sound and camera by Joan Jonas; *Left Side Right Side, Two Women,* 1973, black and white, 20 minutes, silent, camera by Joan Jonas, with Christine Kozlov, Penelope.

18. Tapes that were used in the performance: *Anxious Automation,* 1972, by Richard Serra, sound by Phillip Glass, performed by Joan Jonas; *Duet;* and *Two Women.*

19. *Funnel,* 1974 and 1980, performance, performed by Joan Jonas, camera in 1974 by Babette Mangolte, in 1980 by William Farley, assisted in 1974 by Robin Winters and Christine Patoski.

20. *Twilight,* 1975 and 1976, performance, performed by Joan Jonas with Ariel Bach, Karen Helmerson, Chris Jonic, Paula Longendyke, Robin Winters, camera in New York by Andy Mann, in San Francisco by Pat Goudvis, film by Lizzie Bordon and Joan Jonas.

21. *Mirage,* 1976, 1977, and 1980, performance, performed by Joan Jonas, assisted in different locations by Tabea Blumenshein, Jane Crawford, Rosella Or, Chistime Patoski, Elsie Ritchie, Jane Savitt, film by Babette Mangolte and Joan Jonas.

22. *The Juniper Tree,* 1978 and 1979, performance, solo version with Joan Jonas, assisted in various locations by Pamela Raffaelli, Thomas Meyer, Pat Murphey.

23. *The Juniper Tree,* 1976 and 1977, collaborative performance, with Tim Burns, John Erdman, Simone Forti, Joan Jonas, Pooh Kaye, Shiela McLaughlin, Lindzee Smith, Linda Zadekian.

24. Song from *The Juniper Tree,* in *Grimms' Tales for Young and Old,* trans. Ralph Mannheim (New York: Doubleday, 1977).

25. *Volcano Saga,* 1989, color video, 30 minutes, with Joan Jonas, Tilda Swinton, Ron Vawter, music by Alvin Lucier, camera by Toon Illgems and Jules Backus, edited by Kathy High and Joan Jonas, online Robert Burden.

26. *Volcano Saga,* 1985, 1986, and 1987, performance, performed by Joan Jonas, narrated by Lindzee Smith.

27. *Joan Jonas Works 1968–1994,* 1994, Stedelijk Museum Amsterdam, curated by Dorine Mignot.

28. *My New Theater* (1997), *My New Theater II* (1998).

29. *Narcissus and Echo,* 1997, a video performance by VJ Joan Jonas, music by Paul Miller a.k.a. DJ Spooky, performed by Tomoko Take, Henk Visch.

8

The Individual Voice as a Political Voice: Critiquing and Challenging the Authority of Media

Dara Birnbaum

Hans Ulrich Obrist: . . . you went into single-channel video work with Wonder Woman and Kojak—you changed speed, you accelerated, you slowed down. Was this about creating other narratives?

Dara Birnbaum: Never. Not at the beginning. I took exactly what was there.[1]

AT THE BEGINNING: APPROPRIATIONS FROM MASS MEDIA, 1977 TO 1982

In 1977, the Nielson ratings stated that the average American family was watching television some seven hours and twenty minutes per day. This reality formed my landscape. It painted a picture of a society that was seemingly more capable of relating through mediated communication than through direct contact—a society of stereotyped relationships as portrayed through mass media rather than a society capable of relating through interpersonal contact and social engagement. I reacted first by appropriating images from that landscape, images that blazed across my TV screen, like Wonder Woman. The star of TV's *Wonder Woman* transformed through a burst of light while spinning in space three times. In that burst of light she changed from an ordinary woman into a superwoman who was capable of saving mankind. In 1977, this was a special effect. My reaction was simple: "This is not acceptable." Wonder Woman typified the type of image forged for television by multiconglomerates and corporations. In addition, this image was being transmitted only one way—at me. I felt that it was necessary to find a way to "talk back to the media."[2] I decided to paint my own version of this media-induced landscape with my first single-channel video work, *Technology/Transformation: Wonder Woman* (1978–1979) (figure 8.1).

HUO: . . . Lucy Lippard once said that the goal of feminism is to change the character of art and directly attack the infrastructures of the art world. Was it also, at the beginning, this idea of attacking the infrastructures of the art world? Or at least undermining. . . .

DB: . . . I don't think I made my work so much directly to undermine. I had such a nonbelief in that structure that it didn't matter to me. My belief—and I want this still to be true—was in the message. . . . I just knew something had to be said regarding mass-media imagery. And I did perhaps see art—for me very idealistically—as an activist position. It was

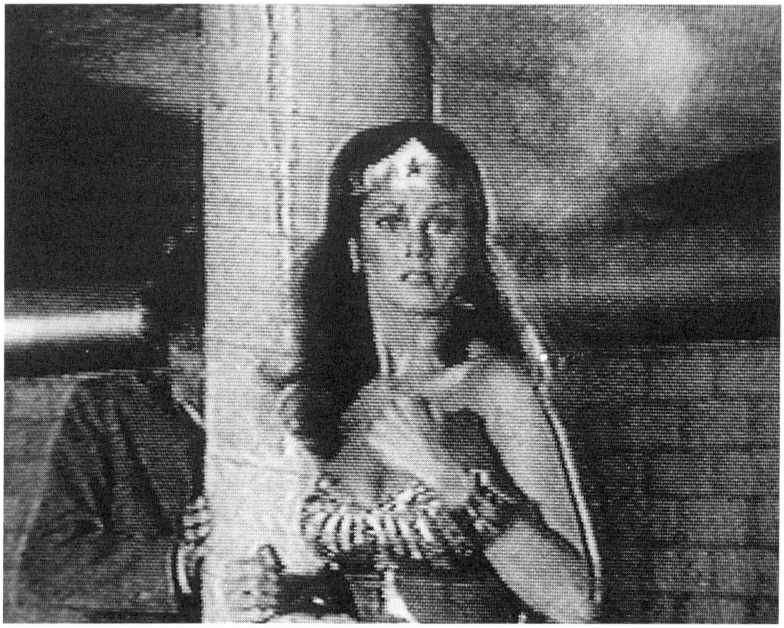

Figure 8.1
Dara Birnbaum, *Technology/Transformation: Wonder Woman,* 1978–1979, still from the videotape, color, stereo, 5:50 minutes.

one of the few positions that could be held in a society that was activist. Then, the more I learned, the more I saw what was controlled. How do you become the one who penetrates—who hacks through this system? Do you want to be the hacker? I don't know.
HUO: You always saw video as a vehicle.
DB: Maybe I saw art as the vehicle.[3]

However, I was also in a dilemma: I took images that I felt belonged to me but that were technically owned by corporations. Working from outside of their originating system, I "pirated" images—taking them, without permission, from their original context and from their original narrative flow. By subjecting these now displaced images—floating signifiers—to formal devices, such as repetition, these images performed a new abbreviated narrative: one that seemed capable of deconstructing the original narrative while revealing its hidden agendas. I then placed this new narrative construction back out into the culture through a variety of public venues. *Wonder Woman* was once again transformed, this time by placing her newly forged image inside of the storefront window of H-Hair Salon de Coiffure, Inc., SoHo (1980). This videowork became a replacement for the only videotape the store owned at that time—the Italian version of *Woodstock.* To have the tape shown, I needed to approach the store's proprietress:

DB: Please, can you show my tape in your window, especially on the weekends?
H-Hair: What is your tape about?
DB: Wonder Woman.
H-Hair: Oh, I love Wonder Woman. I've been told I look like her.
DB: (She did.)[4]

Technology/Transformation: Wonder Woman also screened on Manhattan Cable TV opposite the network version of *Wonder Woman,* which was broadcast during prime time on CBS-TV. I also produced a 16 millimeter kinescope of the tape and placed this film version into an avant-garde film festival organized by the Kitchen in 1980. It became a type of barroom B-film within that context. I then moved her to a popular club—The Mudd Club—for its series called *Early Evenings at the Mudd* in 1980. I lent her for an evening of *Guerrilla Girls at the Palladium* in New York City, where by utilizing the club's two analog-based video-wall systems (among the first video walls in the United States), where she broke through the isolation of her singular TV frame. On the club's two video walls, she burst through the seams of twenty televisions per wall unit, as these large walls floated on hydraulic pumps above the main dance floor. The imagery beamed down at the dancers—as if its electronic radiation could stimulate, emancipate, or irradiate the people below. It was as if these images could squash them to the ground, as either an aggressive wallpaper or as a type of video raid.

WORK UTILIZING MASS-MEDIA IMAGERY WITHIN DIVERSE ENVIRONMENTS AND PUBLIC SPACE, 1989 TO 1992

A History of Television as a Public Site

I had previously thought of television, due to its dominant position within American society, as a public space—a space of flow. Of course, tightly controlled and dominated by the industry, television mainly consists of programs that are formulated by adhering to the lowest common denominators between people to reach as large an audience as possible.

In the late 1970s and throughout the 1980s, the possibilities of a more democratic voice within television seemed strengthened by cable TV. Whereas the real contract, regarding cable TV, remained relatively undisclosed, for a moment it was as if one could project a utopian vision onto cable TV—community and public access through newly made available studios, channels that could specialize in content, and no commercials. In other words, an American apple pie from which everyone could take a bite and then even express their points of view.

Public Space: A Redefinition

Public space, historically, has been thought of as a physical site—a place for people to engage in social and productive activities. However, media has now redefined public space not as a physical and particular place but as a network of points. Within this newly formed domain, we are confronted not by spatial boundaries but by an organization of information (spatial data management). These boundaries are no longer fixed but shift and move through a territory of receivership. Through a variety of media frameworks, complex systems of data, image, and sound map out a newly engaged public space where inclusion and circulation are the most significant features. Within the reality of electronic networks, it is no longer necessary to engage bodies directly to engage in public events. Instead, there is the provision for an expanded public flow, as electronic pulses enter our bodies and act on our senses. We are no longer eyewitnesses to experience or to place but are observers to its mediation. Our previous notions of social and spatial experience have been challenged. The result is that our previous pathways, which forged readable distinctions between public and private space, are no longer operational. Instead, a contemporary definition of public space consists of a fluid architecture.

INSTALLATIONS: 1989 TO THE PRESENT

Rio Videowall (1989)

Awarded commission for the public plaza of the Rio Entertainment and Shopping Complex, Atlanta, Georgia.

Technical Description of the Work

Twenty-five color monitors (27-inch diagonal each), arranged in a five-by-five grid (Delcom Videowall System). Programming software: C-Through. Two live closed-circuit cameras with special-effects generator and preset video luma-key; custom-made enlarged light box (approximately 8 feet by 12 feet), vertically mounted on a wall of a main entryway to the complex (for use with the luma-key); and satellite receiver.

Housing: structural steel and black spandrel glass. The glass used is highly reflective, similar to architectural details found in the surrounding storefronts. Black spandrel glass has the ability to reflect the surrounding areas adjacent to the video wall (a darkened mirror view of the surroundings, the activity, and inhabited public social space.)

Intention of the Work

As the winner of an international competition, *Rio Videowall* was actualized as an electronic public artwork to be sited within the newly developed Rio Entertainment and Shopping Complex, Atlanta, Georgia. The project conjoins the three elementary approaches inherent to the structure of video as a medium: prerecorded imagery, broadcast TV, and interactivity. The artwork utilizes these three significant features simultaneously, constructing a continuously shifting, fluid montage.

Footage of the site, an urban park before construction of the Rio Complex, forms a tableau that is disrupted by a continuous feed stream of news from the Cable News Network (CNN), whose home base is Atlanta.

Two interactive cameras, located at strategic points within the Complex, register images of passers-by. The form of each passer-by appears as a silhouette within the prerecorded landscape footage, through the use of a special-effects generator. In addition, a switcher and a preset luma-key then allow these moving silhouettes to be filled with images directly from a satellite feed of the news of the moment. Thus, the passer-by's physical presence, within the architectural space of the mall, disrupts a historical "electronic memory" with an influx of temporal information.

Further Discussion

Rio allows the viewer a controlled feeling of interactivity and engagement. However, the viewer's effective interaction produces a sequence of events that are virtually uncontrollable as the viewer's "body" delivers the "news of the moment," unceasingly and relentlessly.

In *Rio Videowall,* the television box is now simply one of twenty-five modular units—a metaphoric pixel, a microcosm to the macrocosm of the total system. With the onset of digital video-wall systems, the television image seems to break through its frame for the first time in history. The television image, previously contained within an individuated box, now pushes through this box onto a multiframed matrix, stretching past its previous boundaries.

Tiananmen Square: Break-In Transmission (1990)

Technical Description of the Work

Four LCD color video monitors (4-inch diagonal); one large-screen color video monitor (27-inch diagonal); four video laser disc players; eight stereo speakers (self-amplified); one surveillance switcher; and one custom-designed, ceiling-hung support system.

Intention of the Work

The main thematic issue of *Tiananmen Square: Break-In Transmission* is the examination of fragmentation and disunity between the viewer and televised news broadcasts. Four microsized LCD video monitors are strategically located within the exhibition space, each portraying different aspects of demonstrations during the student occupation of Tiananmen Square in the summer of 1989. The custom-designed structural units that support the individual monitors, disc players, and audio speakers are created from industrial materials that can be easily altered to reflect the architecture of any given exhibition space. Each unit displays an isolated event gathered from the networks' news coverage—the first moment of violence; the cessation of satellite coverage by CNN and CBS networks; a student-composed song, *The Wound of History* (which directly expresses the Chinese students' feelings to gain sympathy for the position of their democracy movement); and the use of readily accessible devices (such as fax machines), which the students used, in the absence of satellite transmission, to transmit updated information to the outside world. These displayed images are meant to function as totems, portraying a historical note taking of the mediation and receivership of these events as culled from numerous television sources. A fifth, large-scale, TV monitor is connected to an interactive surveillance switcher that randomly selects sequences from each of the other channels. Thus, as you view one of the four LCD monitors, the image that you are seeing may be taken away from you (by means of the surveillance switcher) and suddenly reappear in the large monitor as TV. This exemplified disunity of narrative is meant to counter the simplified and singular news broadcast that is made available to most viewers. Furthermore, it questions TV's claim to render the viewer an omnipresent vision of eyewitness news,[5] since the installation provides no single vantage point from which to regard all the isolated and miniaturized video elements. Instead, the viewer is asked to maneuver within the installation and choose a plan of interaction in the reconstruction of meaning.

Further Discussion

Tiananmen Square was, for me, a very large image. It was CNN bringing images, around the clock, that I felt I had no way of really absorbing. Therefore, I elected to form a landscape of LCD images, each of which could only be seen up-close and frontally. If you move to the side of this type of LCD monitor, the image ghosts out, due to the mechanics of its technology. At a distance, the LCD monitors seem as if they are utilitarian minilights hanging from the ceiling. When you move closer, you can see images. So given the necessitated proximity of the viewer to the image, the viewer must travel through the entire space to see all of the presented views. The ceiling becomes the basic support structure, as if all information is coming down at you. It was important to avoid previously used (historical) support structures, such as the convention of placing video monitors on floor-based support pedestals, which had tended to reduce video artworks to a traditional form of sculpture. The attempt to hang from the ceiling, reversing this previous position, is a symbolic attempt to forge more of an alliance with broadcast and satellite transmissions: "It's coming from out there." For me, all news transmissions can bring only a mediated portion of the news. The Chinese government chose to shut down all satellite transmissions of news shortly before they began a violent attack against the protesters, who were mainly students. The major networks sensed that the demonstrations would become violent; they were looking for the break—the rupture—when events would turn. In one LCD display, Dan Rather is seen with headphones on and a satellite dish in the background. What he's hearing in his headphones is: "They just broke off the transmission of CNN." He's hearing from CNN's field reporters: "We're getting the first images of violence, of the police cracking down on people." By comparison, the re-presented footage from CNN's news headquarters in Beijing presents a different posture to these events. When the government comes in to terminate CNN's transmission, the news team tries to push the government representatives back out. They try physically to say: "No. What are you doing? You can't stop us!" Whereas, Dan Rather reacts diplomatically at the moment of cessation of transmission. Rather stalls, he says to the government representatives: "I'm sorry, I don't understand. Oh, you need to shut us down." What he's doing is buying time. He bought approximately one-sixteenth of a second. By buying just enough time, CBS was able to transmit the first images of violence to a worldwide audience.

Transmission Tower: Sentinel (1992)

Technical Description of the Work

Eight color monitors (20-inch diagonal with built-in stereo speakers); nine video laser disc players; nine stereo channels of sound (total eighteen channels of sound); custom-designed computer hardware and software system; custom-designed housing for hardware; external amplifier with four speakers; three prebuilt sections of a Rohm Sentinel steel transmission tower (10-foot sections), used as the main support structure; a custom-designed base, cap, and supports for suspended monitors and video laser disc players.

Intention of the Work

Originally commissioned by Documenta IX in Kassel, Germany, *Transmission Tower: Sentinel* (1992) is composed of a singular column of eight video monitors, custom fitted to tilted sections of a Rohm transmission tower (figure 8.2). As a significant feature within the contemporary land-

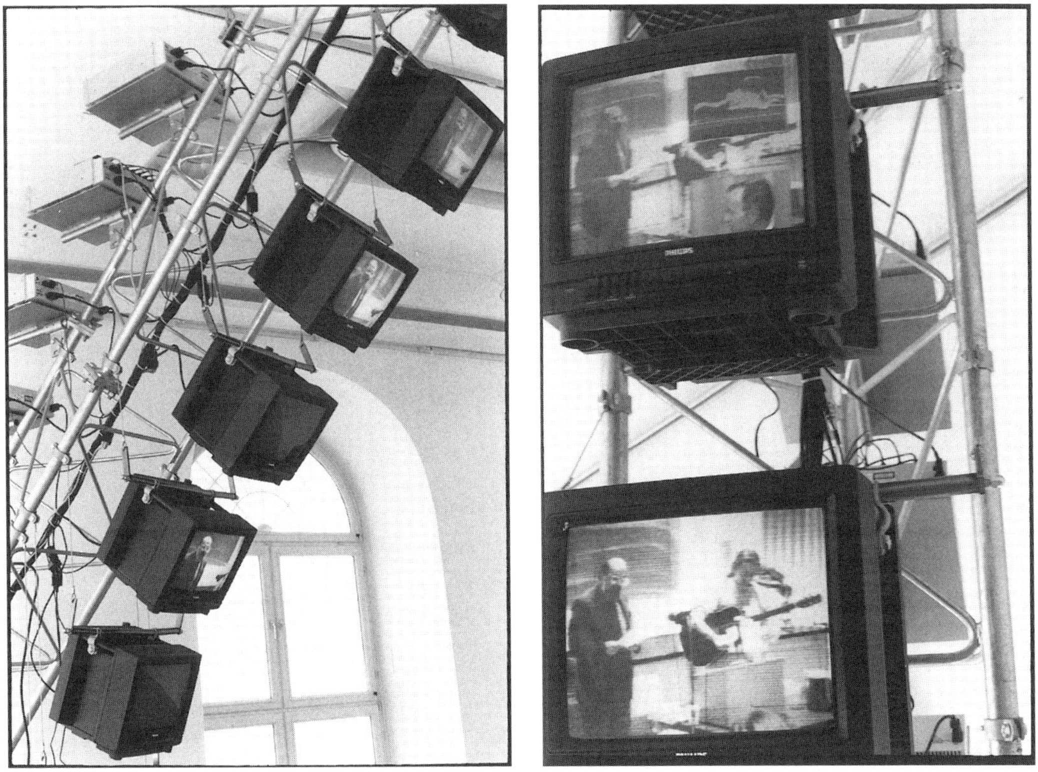

Figure 8.2

Dara Birnbaum, *Transmission Tower: Sentinel,* 1992, Documenta IX, Kassel, Germany. *Left:* Partial installation view. *Right:* Installation detail: video-still sequence of two of eight monitors, forming a linear video wall.

scape, the transmission tower is capable of subverting its architectural form by denying the importance of its physical site. The tower's spatial form is more accurately delineated through a network of transmission points, which suggest a larger, more encompassing, more important, and invisible architecture. A linear video wall, suspended from the angled tower (by custom-designed support elements), simultaneously plays a montage of low-end documentary footage and broadcast TV imagery. The work is meant to represent events that were obscured at their original spatial and temporal points of transmission. This strategy exhibits—and examines—these images within a newly forged relationship, as differentiated from their more sanctioned expression in mass-media news reportage.

Further Discussion

The linear video wall, angled and suspended by airplane cable off the transmission tower, forms a graphic depiction of a typical bomb drop by the United States during the Gulf War. Three distinct sets of images are simultaneously presented. Crawling down, slipping across each of the eight monitors in a small inserted box, are selections from televised imagery of George Bush, reciting his vision of global politics as delivered during his acceptance speech for the Republican Party's nomination as candidate for president in 1988. In a chanting, rhythmic pattern, images of Allen Ginsberg reciting his antimilitaristic poem *Hum Bom!* descend. Finally, as a major third element, footage from the National Student Convention at Rutgers University in 1988 is used to disrupt these downward patterned flows. This imagery of the students ascends in a smooth wave, counter to the other two downward elements. This cutting between contrasting and incommensurable narratives is a reference to the exclusion of independently expressed dissent from mass and industrial communication networks.

Four Gates for St. Pölten (1996–2002), Commission (Unrealized)

Technical Description of the Work

Equipment list not finalized.

Intention of the Work

Four Gates for St. Pölten is an awarded commission for a site-specific public work within the newly developed Government Quarter of Lower Austria,

St. Pölten. The work is composed of four distinct and yet interrelated, interactive elements, each of which is to form a gateway between adjoining government administrative buildings.

Because the original city walls of St. Pölten contained four gates, this project symbolically reestablishes a reference to the original gates within the context of the new Government Office Complex. Thus, a passage of transition is to be established between the original historical elements of the city and the architectural elements of the newly declared capital.

Four critical periods of Austrian history have been chosen to proclaim these new electronic entry points—the Noricum (Roman times), the Habsburg empire, the Austrian-Hungarian empire, and the present Federal Republic of Austria. The largest expanse of geographical boundaries for each historical period are to be etched into tempered glass as a base map for each of the display units. Behind each of these four glass elements exists a large-scale (approximately 5-foot diagonal) video display, which presents a constant flux of the continuously changing boundaries of each era. When a proximity detector is activated by a viewer for any of the gates, the constant flow of "territorial changes" is immediately montaged—on-screen—with "military portraits" of soldiers in uniform that portray the same time period and territory.

Each pedestrian who transits the administrative spaces of the Government Office Complex is therefore exposed to a sequential set of historical motifs representing St. Pölten's place within Austria's history. *Four Gates* is meant to allow for a procession and a declared entry, which is sequenced from Roman times to the establishment of this city as the new capital for Lower Austria.

Further Discussion

The architecture of a city offers glimpses into an evolving history under erasure and revision. Similarly, a map offers a momentary registration of an evolved history of national identity through recorded traces of economic and political histories. The historical transformation of a country's shield and military display offers similar systematic access to the social and economic conditions of a given historical moment. In a sense then, these military objects yield yet another sense of the architectural order, as each portrays extended "mappings" of historical cartography.

However, despite the empirical promise of any map, the exact boundaries of identity are difficult to locate. *Four Gates* represents four gateways that enable people to be informed of the long lineage of the city's history.

The occupant of this new seat of government is to be made all the more aware of St. Pölten's contemporary presence through exposure to its past identities. In fact, the numerous signs of a national identity in flux might be considered to be the symbolic gates of access to a nation's collective memory.

CONCLUSION

A presiding and connecting link between the projects that have been presented here in this chapter is the necessity to forge work that ultimately questions the control that is available to, or possible for, the individual within a developed technocratic society. How can a person remain in control of identity, individual voice, and form of representation within the scope of vast technological development?

For example, in *Kiss the Girls and Make Them Cry* (1979), women are seen presenting themselves within a structured grid, a tic-tac-toe board of boxes. I originally appropriated these images of women from the TV game show *Hollywood Squares,* stereotyping them even further. The clichés embedded within the original footage—a blonde, a brunette, a young girl, a redhead—are brought into an even more precise focus. Each woman is seen to make a different gesture. You can see this very clearly because each representation is dislocated from its original source and then repeated. Yet you can also see how each woman is fighting to find an identity that is capable of overriding her stereotype.

How do you introduce yourself to an audience of millions? What does your identity become? Is it in the smallest nuance of gesture or form? With the World Wide Web, we have an even larger television audience. You're introducing yourself; you're going on a bulletin board. Which bulletin board? As Nam June Paik said: "Video—very good—no gravity." Now you have no identity. You can take on alternative identities. You can live out your fantasies through easily adopted identities. The first legal case against a man stalking a woman on the Internet was in 1995. Can this stalker also be persecuted? Can he be sued? With what form of legal indictment? Can you rape through the system? Can you rape identity?

We have evolved into a society that is in a crisis of identity. It seems increasingly difficult to act as an individual within highly developed technocratic societies. Beginning with *Technology/Transformation: Wonder*

Woman (1978–1979), I sensed that this form of representation expressed a technocratic view of woman: you either heroicize her, or you underrate her as a secretary. What place did you ever create for me within this representation? Where I am is in between. However, in advanced technocratic societies, there is no space in between. The burst of light says that I'm a secretary—I'm a Wonder Woman—I'm a secretary—I'm a Wonder Woman. Nothing exists in between—and the in-between is the reality we live in.

ACKNOWLEDGMENTS

The author wishes to thank both Rick Pirro and Scott Lyall for the diligent and creative assistance that they provided on numerous writing projects. Their assistance and coauthorship are reflected here in the descriptions of my installation works.

NOTES

1. From the transcription of an interview with Hans Ulrich Obrist, September 1995, Vienna. Total length of interview: 85 minutes, videotape. Originally for *KünstlerInnen: 50 Positionen—Zeitgenössischer internationaler kunst, Videoportraits und Werke,* an ongoing project of the museum and of Peter Kogler for the Kunsthaus Bregenz, 1997.

2. In November 1985, a citywide art event entitled *Talking Back to the Media* was held in Amsterdam. The event was, according to two of its organizers, formulated with my work in mind. This early work, which directly appropriated images from television, had recently been presented within a solo exhibition of my work at the Stedelijk Museum, Amsterdam.

3. Interview with Obrist, September 1995, Vienna. Total length of interview: 85 minutes, videotape. Originally for *KünstlerInnen: 50 Positionen—Zeitgenössischer internationaler kunst, Videoportraits und Werke,* a project of the museum in progress and Peter Kogler for the Kunsthaus Bregenz, 1997.

4. An accurate reconstruction of a conversation with the proprietress of H-Hair, Salon de Coiffure, Inc.

5. "Eye Witness News" is the phrase used by the American Broadcast Company (ABC television network) to define its approach to local and international news reportage. This slogan is used promotionally and as part of the opening introduction to its news programs.

9

Small Leaps to Ascend the Apple Tree

Jo Hanson

A technologically challenging early work, *Crab Orchard Cemetery,* opens this discussion, followed by an innovative piece seeking an ecological peace with snails in the garden. Urban trash becomes the source of my increasing focus on environmental and ecological concerns. And nature makes a dramatic entry.

Understandably, for me and other ecoartists, such a focus acts like a zoom lens enlarging the view until existence on the planet becomes the subject matter—how we got where we are, where we are now, and where we are going, with technology almost a touchstone through all of it. I think of internal as well as external technology, and of the Earth's technology as well as human-devised technology. In this, I have an assist from quantum theory and DNA research.

Finally, I ask questions about the role of technology in a world facing (or failing to face) ecological crisis, which has the potential to be reflected in economic and political crises. I have tried to create work that is practical and visionary, and I offer certain practical and visionary propositions here.

CRAB ORCHARD CEMETERY (1974)

Crab Orchard Cemetery re-created in gallery terms an actual rural cemetery in my native southern Illinois (Carbondale). The preparatory work spanned several years, but the exhibition opened at the Corcoran Gallery in Washington, D.C., at the end of 1974. It toured for three years and was revived in part by Michael Schwager at the Richmond (California) Art Center in 1989.

In its time, the exhibition was regarded as a technical tour-de-force. When it opened at the San Francisco Museum of Modern Art in 1976, two discussants were overheard to say, "It would have to be a sculptor who did that; a photographer would know better." They referred to a scenic surround of registered positive images on 12-foot-high film panels that hung around the gallery walls to recreate the foliage of the original cemetery (figure 9.1). In scale and ambition, that work is still a point of reference, and its conceptions and production are described below.

Other elements of the installation also were innovative at that time. I recorded the ambient sounds of the cemetery through the day and night and edited them into a sequential raga of three hours that repeated continuously in the exhibition. These were typical rural sounds, with stellar performances by birds, night insects, and frogs. Soon after the exhibition

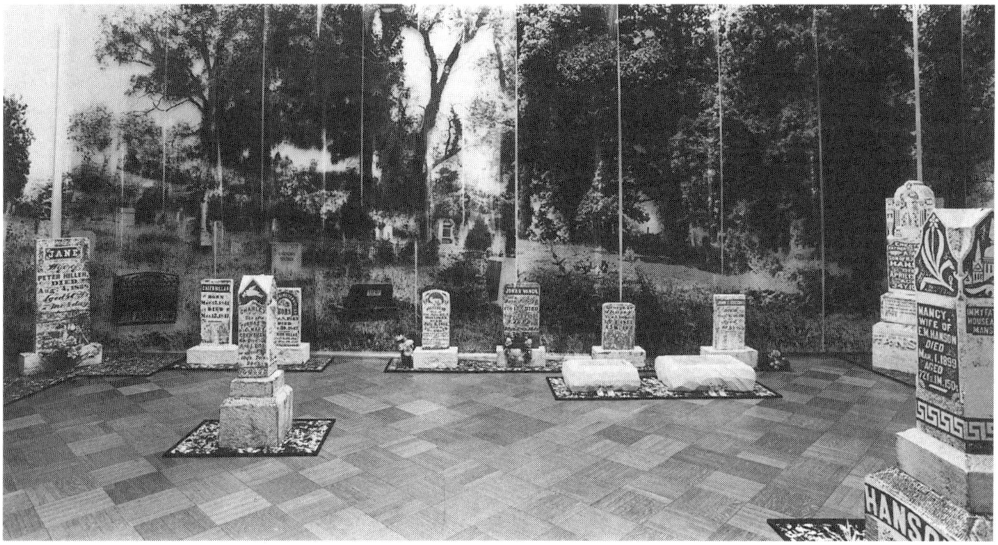

Figure 9.1
Jo Hanson, *Crab Orchard Cemetery,* 1976, installation, San Francisco Museum of Modern Art. A scenic surround of registered positive images on 12-foot-high film panels hung around the gallery walls to recreate the foliage of the original cemetery.

opened in San Francisco, similar bird sounds were heard playing on a department store elevator. Occasionally, I heard a brief excerpt of a nighttime dog bark and howl on the *Hearts of Space* radio program and recognized it from my tape.

The processes that I developed for faithful re-creation of tombstones involved (1) tombstone rubbings made with photogenic black etching ink on pH neutral mulberry paper, (2) positive and negative 4-inch-by-5-inch "negatives" of the rubbings that were used to (3) blow up and produce positive and negative life-size photo silkscreen stencils. On slabs of Styrofoam precut to size and shape, the positive image was printed in white or gray to etch in the texture of stone. The negative image was screened in raw umber. It etched the intaglio of weathering and carving. The solvent of the ink did the etching when it was placed in sunlight while the ink was wet. I registered each image through the screen. Conventional registration techniques were impossible. The completed object usually was mistaken for stone, though analytic examination revealed that it could not be so. Rubbings were used instead of photos to convert to silkscreen stencils because of the literalness of rubbings. The camera interprets what it sees.

Regarding the reference above to "precut" Styrofoam slabs, my cutting procedure used a hot wire stretched across a table at the height of the desired thickness of the slab. As the Styrofoam on the supporting table is pushed through the wire, the heat cuts and seals the cut surface. My jerry-built jig accomplished the same process as commercial cutting. *Hot wire* means a resistance wire that heats as current acts against resistance and flows through the wire. It is the same principle as the coil of an electric stove. Yes, toxic fumes are produced in heat cutting.

The scenic 12-foot high blow-ups began with my taking a Photography 1A class at San Francisco State University and learning what makes a good photo, how to use a view camera, and how to print a good image. I was motivated to learn a great deal quickly.

I set up the camera in the middle of the cemetery and photographed sequential images, 360 degrees around, three times (figure 9.2). I believe "three times" has magical implications and is an underrecognized area of technology. I relied on getting a useable sequence in three tries, and I did. With the foregrounds cropped, the 4 × 5-inch negatives were enlarged to print 12-foot high images. They held detail remarkably well.

The film was kodaline, which was like kodalith but slower. The slowness avoided the filling in that characterizes kodalith. It permitted large

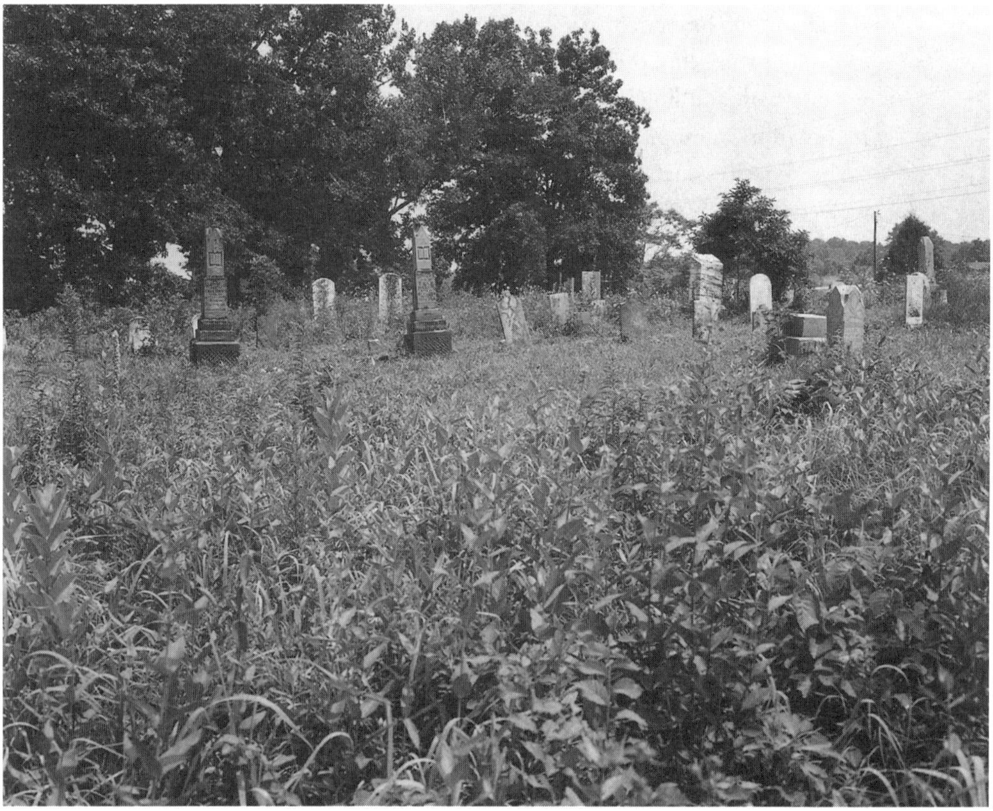

Figure 9.2
Jo Hanson, *Crab Orchard Cemetery,* 1976. The camera was set up in the middle of the cemetery, and sequential images 360 degrees around were photographed three times.

film panels to be removed from the developer and placed into the stop without losing detail. Diluting the developer yielded a beautiful gray range instead of the film's intended all-or-nothing printing.

My technology employed standard procedures that were manipulated for the logistics of coordinating materials, chemistry, and physical management. Management involved a jig for cutting 13-foot film lengths accurately by safelight; hanging the 42-inch-wide film length edge to edge on a 12-foot-by-16-foot exposure board; testing for the long exposures that the Bessler enlarger needed for horizontal projection over a distance of 25 feet to enlarge the image to 12 feet high; managing the film panels through the chemistry of developer, stop, and fixative followed by washing and drying; and finally preparing a means of suspending the panels a few inches in front of the gallery wall, which had to be white. The film had to be handled as rolls for most operations. There was trial and error, problem solving and innovation, ingenuity and amazing generosity and help from other people, who responded to the excitement of the challenge.

A photomural group offered the use of their jerry-built but essential facilities during the night hours. In the same building, an antique dealer allowed use of his warehouse space for setting up exposure facilities. Friends came in to help. The crucial help was a friend, Jim Weeks—the photographer, not the painter, with the Bessler enlarger and with so much creative photographic expertise that he always had three answers to the things other people said were impossible. Also, Jim (not I) could stand on a high ladder in the dark and suspend film panels straight for me to fasten down with edges butting.

The decision to use film to establish the scenic surround came after the failure of other concepts, including applied emulsion on transparent fabric. As I was about to give up, grieving that apparently I would have to use conventional photomurals, I realized that I had been holding the answer in my hand, literally, all the time, thinking of it as a transitional means. The answer lay in flexibility of perception.

Positive image on film, in its own terms, matches the beauty of foliage in nature. The suspended panels reflect light, make shadows, and stir with every opening and closing of a door anywhere in the building.

Crab Orchard Cemetery steers a hazardous course between profundity and the sentimental. "A masterpiece of somehowness" was Alfred Frankenstein's characterization in a San Francisco newspaper review.[1] None of the individual parts stood alone as compelling artwork, but the totality integrated into a moving experience of thought and physical beauty.

Eleanor Dickinson spoke of this in the Corcoran Gallery catalog:[2] "*Crab Orchard Cemetery* . . . may initially be seen as a visually powerful re-creation of a rural cemetery in Illinois. On reflection it will lead the viewer/participant past its reportive quality to a vision of American history and life. It is this unusual combination of the trained artist and passionate historian that in this exhibition works a transcendence of documentation to a universality we experience as art. It is the super-reality of the tombstones, the huge floating film panels, the opposition of film to photographic ground that redeems *Cemetery* from any hint of sentimentality and transforms it to the realm of art."

Asked often why I would "make a cemetery," I explained that in the same catalog: "My involvement with *Crab Orchard Cemetery* began with the sensory seductiveness of image and sound there: the weathered stones in their richly textured setting of trees, weeds, and wild flowers, all shimmering in summer sun and the constantly modulating southern Illinois light; the milieu of bird songs and insects flecked with barking dogs, cows, little airplanes, and a continuo of mowers and other farm machinery."

My next level of involvement derived from a passion for history as expressed in the work of art—the socioanthropology of art history, the way the commemoration of death reflects life in middle America. *Crab Orchard Cemetery* presents a continuous evolution from frontier times to the present. The changes in materials, styles, sizes, and messages (inscriptions) presented information that made it necessary for me to restudy American history and art. It has been a profound experience and has changed my concepts dramatically.

I looked for roots, personally and culturally, before Alex Haley's published search[3] and at a time when death-and-dying courses were edging into college curricula in response to the work of Elizabeth Kubler-Ross.[4] This was the Vietnam War period, an anguished time when television held up a mirror of the suffering and devastation that were exploding from the American role in Vietnam. The information that I received from the cemetery spoke in part of a century and a half of declining value systems, of declining quality of feeling, of the commercializing of death and commemoration, and of a culture groping with insufficient guidelines and seeking security though an enlarging *I*. My restudy of American history turned me to the conclusion that our war in Vietnam was not an aberration, as I had regarded it earlier. It was consistent with the pattern of American development.

THE MONTH OF THE SNAIL (1981)

I am a problem solver, and ideas and innovation attract me more than mastery within a given medium. A sequel to *Crab Orchard Cemetery* reflected this attitude also.

Crawling on my belly around the garden, I photographed snails at their eye level and so close up that sometimes the camera bumped into the snail. These images were often of exquisite beauty. I projected the images at floor to ceiling size, an unusual dimension at that time. A narrative discussed the life and history of the snail and ecological solutions to the problem of snails in our garden. The scale of image to viewer and the equality of eye level of snail and viewer gave the snails a stronger voice and presence than they had in the garden.

The snail work began at The Farm in 1981 in a residency called *The Month of the Snail—"Don't Tread on Me."* The projection show was called *Super-Snail—The Featured Creature.* It went through many manifestations, the last one called *Good Eats* because it was set up like a restaurant in a storefront exhibition space. Every Friday, we had a Happy Hour and served escargot. A big sign that read "Good Eats" hung across the store front.

Both these pieces, regardless of their innovative aspects, were strongly humanistic. They functioned in my development of an alternative awareness of technology that evolved in stages and has sharpened with the passage of time.

Stage one had to do with the gradual realization of the materials waste and environmental offense of everything I used and did. My photographic chemicals were dumped into the sewer system. The only salvage was of silver from the fixer, and that occurred only in the commercial facilities that I used—not in the individual and school darkrooms. Most of the things I used in photography were pollutants in their manufacture and use.

Equally, silkscreen processes involved harmful processes and chemicals that most commonly were dumped into the environment in one way or another. When venting systems existed, they removed harm from the workplace to dump it into the general environment.

Even at that time Styrofoam was known to be harmful in its manufacture and continuous out-gassing. Recycling or re-using were not even considered. I used discards thrown away by a packing plant, astounded at the quantity and size of their throw-aways. The situation focused my thoughts on the relationship of materials cost and labor cost. Materials were conserved as long as they cost more than labor. When the labor cost of careful use of materials exceeded the cost of materials saved, industry responded by throwing away materials. (That also explains in part the large quantities of good materials thrown away at construction sites.) In my travels, I observed that materials in poor countries were carefully conserved. Discards of any kind disappeared quickly into somebody's work or living situation.

All of my work and projects have pushed me to think about the unsustainability of our attitudes and processes in producing, persuading, consuming.

ART THAT'S SWEEPING THE CITY

My next stage began with active involvement in environmental issues. Against the background of the kind of awareness I have described, it is not surprising that the next phase of my work dealt directly with the urban waste that surrounds me and most other urbanites. Prompted by the astounding volume of discards on my own windy San Francisco street, I have cleaned the street since 1970 and documented its contents and their changes through dated samples preserved in plastic covers in binders.

My work led to relationships with city agencies, especially the street-cleaning department, and with my community. An intention of my work was to promote citizen cooperation with street-cleaning services. This activity provided the content of a major exhibition in San Francisco's City Hall and the San Francisco Museum of Modern Art sponsored by a city agency and the museum. I organized a dada-style bus tour of illegal dumping and curated an exhibition contest sponsored by a city agency and the school district, with entries from 1,700 school children on themes of litter and recycling. I served six years as a San Francisco arts commissioner and subsequently applied the unusual expertise gained during that experience to propose and advise an artist-in-residence program in the disposal company in San Francisco. It has remarkable outreach into the community and the schools, promoting awareness and education in environmental issues.

GAIA DOES THE LAUNDRY (1994)

In mid-1994, my urban experience became supplemented by part-time living in California's Russian River area, where ecosystem restoration, floods, storms, and their attendant liabilities of slides, erosion, silting, and washed-out roads became part of my experience and therefore part of my work.

In this context, I began thinking about how Earth's technology responds to violations of Earth's large systems and tries to heal the wounds. Violations of large systems include such things as factory emissions in Ohio or Michigan that cause the death of trees and lakes in upper New York state or, in my case, logging and construction that impairs waterways and destroys habitats far from the logging site.

Watching "my" flood (intently, for at that moment my deck was an island), I became aware that any given action creates and destroys in the same act, just as we saw demonstrated in recent newspaper photos of stars being born. I saw the Earth intent on preserving plant diversity throughout the flood plain—which was expanded due to erosion and silting from logging and construction. I saw the creek destroying its banks because of damage to the channel.

These are heady experiences, provocative of intense thought. The piece that I did to discuss the phenomena (and my own crash course in flood) was called *Gaia Does the Laundry*. A subsequent piece dealt with watershed.

LEAPS OF IMAGINATION

Leaps of imagination and conceptualizing are the artist's working mode. Involvement with nature at any point carries all of nature's systems with it so that all that is becomes the context of one's work. A popular John Muir quote says, "When we try to pick out anything by itself, we find it hitched to everything else in the universe."[5]

My leaps of conceptualizing, which follow, are based on simple propositions:

1. The way Western cultures live on the earth is unsustainable—meaning that we will exhaust the life support system on Earth, especially as populous developing countries rush to emulate American consumerism.
2. Defining and addressing the problem requires
    - Understanding the historical base of the problems (just as I don't understand ivy eradication from my redwood ecosystem without comprehending the ivy's root system,

- Having the will to change, which involves knowledge, intelligence, and morality—a morality that serves common well-being, so that rich corporations will pay taxes, conserve resources, and would not even dream of promoting excessive consumption or buying the votes of lawmakers to permit harmful practices, and
- Accepting that the enormous mystery of existence and consciousness far exceeds our present knowledge, exceeds the reach of Western science, and requires us to be open to natural process and learning.

I observed that the technology that I used in my own work was harmful. Looking at the evolution of industrial technology and especially now of communications technology, I've seen a staggeringly destructive course. In the work of artists and others, I have also seen transformation and transcendence.

My awareness now includes both internal and external technology. It is not really controversial anymore that some people have telepathic skills, clairvoyance, the ability to alter physical forms, or controls over automatic body systems. The probability that most people are not ready to accept that information into their belief systems does not negate the information. It is a common phenomenon that information exists before it is generally accepted. I call these skills *internal technology.* Despite the choice that actually was made, the human potential demonstrated by a limited number of people holds out such exciting dimensions that one could regret leaving such inner resources undeveloped. Millennia of effort have gone into our present external technology. Had we committed as much time and effort to internal technology, who knows what grand achievements might have been possible.

Quantum theory and quantum physics have jolted previous understandings of reality and matter.[6] I apply them to the interconnected webs of energy systems that link the webs of the most minute life forms into larger and larger systems to form the totality system of all that is—a constantly modulating totalitity.[7] But quantum theory also tells us to keep our minds open because it demonstrates that many of the favorite concepts of classical physics are wrong and that much more is to be learned.

In *The Cosmic Serpent: DNA and the Origins of Knowledge,* the hypothesis that DNA energy emissions may be the means for worldwide communication networks is developed along a path of scientific investigation and leaps of revelation by author and anthropologist Jeremy Narby. His supporting data are quite conventional, but the totality of his conclusions is mind bending. I draw two points from the book. The first is speculation that circuits in human DNA could be waiting for us to realize more of our potential and our evolutionary next step. The second, more speculative, point is that the circuits might be present because they were once in use in an unknown past of our organism. Equally, they might be in use now without our understanding.

The question that hangs in my mind is whether technology, especially communication technology, in its present reality and its future will somehow become the salvation of existence on Earth or part of the ultimate guarantee of disintegration. Can it create community appropriate to the oneness of all existence? Or is it more likely a guarantee of isolation? Can it emerge in any way to mitigate environmental harm, or will it continue as a major environmental liability? Can it in some way become an instrument of spiritual evolution appropriate to the reality of interconnected existence? And a final question: Will we use DNA resources to continue human evolution toward dimensions of consciousness that some few people seem already able to achieve? I don't know the answers. I would really like to hear from readers who have thoughts about the questions.

NOTES

1. Alfred Frankenstein, "A Masterpiece of Somehowness," *San Francisco Chronicle,* 20 May 1976

2. Eleanor Dickinson, *Crab Orchard Cemetary—Jo Hanson,* 1974, Corcoran Gallery of Art, Washington, DC.

3. Alex Haley, *Roots* (Garden City, NY: Doubleday, 1976).

4. Elizabeth Kubler-Ross, *On Death and Dying* (New York: McMillan, 1969)

5. John Muir, *My First Summer in the Sierra* (Boston: Houghton Mifflin, 1911), 11.

6. Jo Hanson, "The Healing Role of Eco Art," *EcoPsychology Newsletter* (Fall 1996): 5.

7. Jeremy Narby, *The Cosmic Serpent: DNA and the Origins of Knowledge* (New York: Tarcher/Putnam, 1998).

10

Shifting Positions toward the Earth: Art and Environmental Awareness

Helen Mayer Harrison and Newton Harrison

Our work begins when we perceive an anomaly in the environment that is the result of opposing beliefs or contradictory metaphors. Moments when reality no longer appears seamless and the cost of belief has become outrageous offer the opportunity to create new spaces—first in the mind and thereafter in everyday life.

We understand the universe as a giant conversation taking place simultaneously in trillions of voices and billions of languages, most of which we could not conceive of even if we knew that they existed. Of those voices whose existence has impinged on our own to the degree that we can become aware of them, we realize that our awareness is imperfect at best. Therefore, it seems to us that the casual and wanton destruction and disruption of living systems of whose relationships we know so little requires extraordinary hubris.

For us, everything started with a decision made in the late 1960s to deal exclusively with issues of survival as best we could perceive them. Each body of work sought a larger or more comprehensive framing or understanding of what such a notion might mean and how we, as artists, might express it. For example, in *The Seventh Lagoon* of *The Lagoon Cycle,* we came up with the statement, "But that would require reorienting consciousness around a different database."

We are now exploring what such a statement might mean—unpacking our intuitive ideas. Our most recent work opens up the idea of setting up an eco-security system, a safety net for the ecology not unlike a social-security system. However, there are issues such as the population explosion that need a separate and comprehensive address, for just as prairie grass would displace everything that is not itself, so would any expanding population. The notion that ingenious technology will resolve population pressures on the one hand and generate infinitely expanding markets on

---

This paper, originally published in *Leonardo* 26, no. 5 (1993), documents some of Helen Meyer Harrison and Newton Harrison's seminal projects using art to expand environmental awareness. Since this paper was written, their new projects have included *Peninsula Europe: The High Ground* (2000 ongoing to 2005), *The Green Heart of Holland* (1994–1997), and *The Endangered Meadows of Europe* (1994–1997).

the other is simply an illusion. It is too easy to forget that every entrepreneurial act, even recycling, is itself a tax on the ecosystem.

The works that we describe and present here are excerpts from narratives that have engendered projects or stories that may engender projects in the future. *The Lagoon Cycle* is its own story, presenting a series of our adventures and thereby telling of expanding ecological awareness. *Breathing Space for the Sava River* was on its way to happening when the war in the former Yugoslavia began. *Tibet Is the High Ground* is a project we are working to make happen. *The Sacramento Meditations,* originally exhibited in 1977, is still only in its beginnings. The latter three projects required, or would require, the cooperation of many others in our dialogs—ecologists, landscape architects, engineers, and politicians.

THE LAGOON CYCLE (1973–1985)

This work is, in part, a mural 360 feet long, averaging 8 feet tall, in sixty parts (figure 10.1). It was completed over the period 1973 to 1985. It is portable, done on photomural paper mounted on heavy cotton duck. The materials are photography, oil, graphite, crayon, and ink. It was first exhibited in complete form at the Johnson Gallery at Cornell University 1985 and then later at the Los Angeles County Museum of Art.[1]

The Lagoon Cycle can be read as a story in seven parts; each part, as in a picaresque novel, is its own story. It can be read as an array of storyboards for a very unusual movie. As artists, we see it as an environmental narrative, one of whose properties is to envelop the viewer with its form and subject matter. For us, this work relates to other twentieth-century environmental works as well as to the myriad mural programs of the past.

The Lagoon Cycle unfolds as a discourse between two characters who discuss the ways in which the metaphors we live by affect what we do to each other and to the environment. It casts light on how we create our worldview and are in turn created by it. *The Lagoon Cycle* is named for the estuarial lagoons that are endangered everywhere; the lagoons are used as a metaphor for culture and even for life itself.

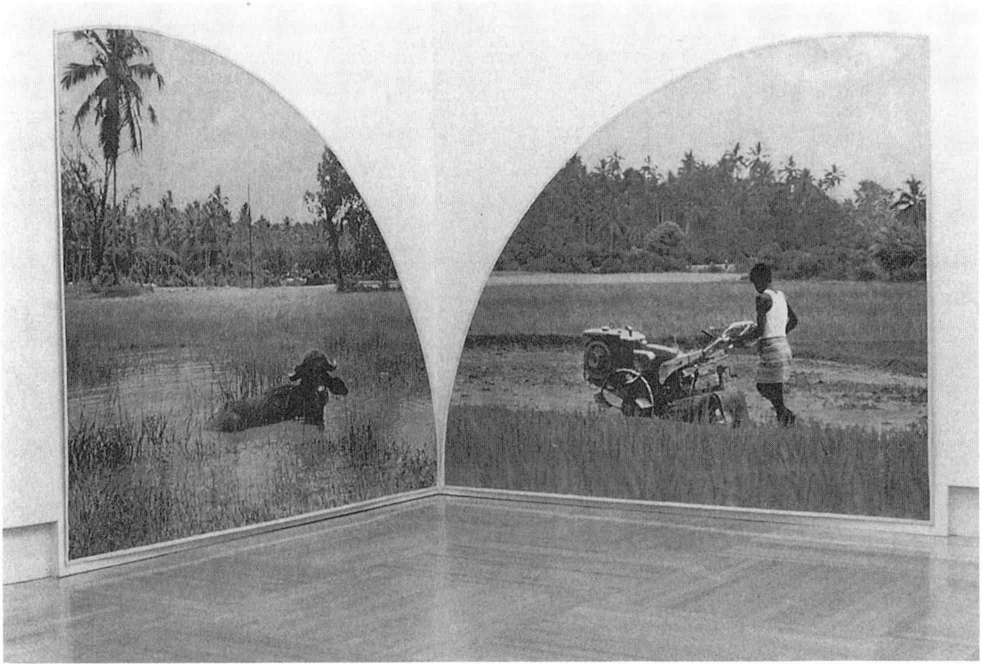

Figure 10.1

Helen Mayer Harrison and Newton Harrison, *The Seventh Lagoon of The Lagoon Cycle,* 1988, installation, Panels 3 and 4, 8 by 9½ feet each. Part of the artists' twelve-year work including text, photos, drawngs, maps, site-specific artworks, and performance. Courtesty of Los Angeles County Museum of Art. Photo by Barbara Lyter and Steve Oliver.

The story concerns two characters who begin a search for a "hardy creature who can live under museum conditions" and who are transformed by this search. The characters define themselves in *The First Lagoon* by the differences in their values and perceptions, with one naming himself Lagoon Maker and the other naming herself Witness. Both proceed to live up to their names, although they finally surrender them as circumstances push the two characters into constructing ever-larger frames for their discourse.

The Sixth Lagoon: On Metaphor and Discourse

The Fifth Lagoon deals with the Salton Sea, which was formed by flood flow released by human error from the canals along the Colorado River. *The Sixth Lagoon* treats the entire Colorado River basin. Lagoon Maker and Witness reflect on the insights they have gained through observing aquatic systems. They expand the scale of their thinking from the Salton Sea to the Colorado River watershed, which has been changed by lifestyles that demand vast amounts of electricity and irrigation. The exploding megatechnology of the twentieth century has shocked the environment and does not have time to "niche itself in." Witness sees all nature as a discourse between the elements, and both characters urge, "Pay attention to the discourse between belief systems and environmental systems":

  Pay attention to the flow of waters
Pay attention to the integrity of the waters
flowing
  Pay attention to where the waters are flowing
Pay attention to where the waters desire
to flow
  Pay attention to where the waters are
  willed to flow
Pay attention to the flow of waters and
the mixing of salts
  Pay attention to the flow of waters and
  the mixing with earth
Attend to the integrity of the discourse
between earth and water   the watershed
is an outcome

Pay attention to the discourse between
earth water and men interruption
is an outcome

Pay attention to the meaning of the nature
of such discourse and the nature of the
meaning of interruption   After all
a discourse is a fragile transitory form
an improvisation of sorts

And anyone may divert a discourse of any
kind into another direction if they do not
value its present state

Pay attention to changes of state

For instance
If
the flow of the waters of the Colorado River system has
been interrupted by dam after dam which demand of
the river that it generate electricity and serve as a source
of potable water for cities outside the watershed while
requiring the river to act as a sewer for agricultural wastes
then
the state of the river has been changed and that change
must reverberate back through the system

If
the flow of waters has been made to behave like rain to
irrigate millions of acres of arid land
then
the state of the land has been changed to give advantage
to that which would not naturally grow and
disadvantage to that which did

If
the state of the water has been changed to disadvantage that
which evolved in the flowing waters and give advantage
to recent forms such as urbanization and industrialized
agriculture
then

since a steady output of industrialized energy must be spent
for the maintenance of the system
and the energy is electrical and of high voltage
and the energy is hydraulic and of high cost
concomitant beliefs have evolved that these
recent forms are valuable and self-justifying
and necessary
for the maintenance of life and well-being

Pay attention to the cost of belief

For instance
If
for millennia the fresh waters of the Colorado
and the salt waters of the gulf
have been improvising the form
and disputing the southern boundary
of the delta the silt-laden Colorado
dropping its load and extending the delta
the gulf with its thirty- to fifty-foot tidal bores
washing the delta away
but slowly retreating before the massive silt load
and if this dialogue has been interrupted
to divert the waters of the Colorado elsewhere
then
advantage has been given to the tides of the gulf
which will once more slowly seep up the delta
to again reclaim
the Colorado-irrigated farmlands
of Mexico
and the Imperial and Coachella valleys
and
disadvantage has been given
to that vegetation which had adapted
to the inconstant flow of waters
and to all parts of the life web
dependent upon those parts
already disadvantaged

and advantage has been given to Phoenix
Arizona the Denver plains
the Platte River the cities of Los Angeles
Long Beach San Diego
and the counties of Riverside Imperial
Orange Los Angeles and San Diego
and all of their populations and industries

Pay attention to the cost of the giving
of advantage and disadvantage

For instance
If the flow of waters within the Sri Lanka tank system
has been operating for more than 2,500 years
utilizing depressions in the ground for off-stream storage
and gently sloping canals to bring water to smaller tanks
and smaller tanks again
through the villages to the paddies
And if
the water in the canals serves for daily bathing
and washing
and as potable water
and as a sanctuary for fish and animals
and bird and plant life
Then
over time
the tank and the canal system has niched itself into the ecology
and the state of the land has been minimally changed
giving advantage to such historic forms as villages and farms
while not disadvantaging the rivers and riverine life
Thus
although a steady output of manual energy must be spent
in order to sustain the system
the energy spent is collaborative and of low force
the energy is harmonic and of low cost
Therefore

concomitant beliefs have not evolved
that high-force high-energy systems
are valuable and useful
and necessary
for the maintenance of life and well-being

Pay attention to the cost of belief

Pay attention to the state of belief
Pay attention to the belief stated
Pay attention to the flow of belief and the willing of desire
Pay attention to the flow of belief
and the enacting of desire
Pay attention to the system upon which desire is enacted
and the system that generates desire
Attend to the discourse between belief systems and
environmental systems
Pay attention to the meaning of the nature
of such discourse and the nature of
the meaning of redirection   After all
a discourse is a fragile transitory form an improvisation
of sorts
and anyone may divert a discourse of any kind into
another probable direction
if they do not value its present direction
Pay attention to the choosing of probable
directions   and the authorship of
changes of state
Pay attention to changes of state

The Seventh Lagoon: The Ring of Fire; The Ring of Water

The Lagoon Maker and the Witness deal with enlarged perceptions of time and space, shape and size. For example, they envision the Pacific Ocean as an estuarial lagoon with "rivers feeding it like tiny streams." They perceive that "the business of the universe" is conducted in an odd kind of dialog but that the business of technology is conducted as a monologue that has become more seductive and "does not like that which is not itself." Lagoon Maker and Witness begin a search for new guiding metaphors to replace those of force and fire, as they perceive the accelerating greenhouse effect as nature's response to the millennia of force and fire. Finally, they muse, the oceans will rise gracefully, but will people withdraw with equal grace?

Now I see the Ring of Fire as
the wave front of an ocean of fire
beneath an ocean of water
mostly separated by rock

Of course a more literal mind
could see the mountains as froth
on top of a wave of fire
moving at the speed of
one to ten centimeters a year

And in less than a second
I can visualize any section of the Ring of Fire
the Kuril Trench   for instance
with the Pacific plate subducting
uplifting the Kuril Islands thereby

And in less than a second I can shrink the Pacific
by orders of magnitude   and make its size
no more than that of an estuarial lagoon

And in less than a second I can imagine
a corresponding simplification
of biocultural complexities
That would require reorienting consciousness
around a different data base

Sometimes I dream of the water buffalo
in its wallow in Sri Lanka
the one that ran afoul of the gasoline engine
and is being replaced by the tractor
Now that tractor does not replicate itself freely
nor provide milk nor utilize weeds as fuel
nor produce fertilizer and fuel with its dung

Yet the tractor maker would say that
the tractor is a bold invention
an improvisation that will change the state of farming
It is more efficient
it can cover more ground in a day
It is modern and cheap
and helps bring people into the technological domain

Yes
yet in some places the buffalo and its wallow still continue
their several-thousand-year-old discourse
their collaboration
and one of the consequences of redirecting their discourse
into the technological monologue will be a peculiar
subtraction of possibilities   for gone will be the fish
that eats the larvae of the malaria mosquito
while itself serving as a source of protein
and gone will be the vermin-eating snake
that breeds in the wallow's surrounds

while fertilizers will be added
and insecticides   and herbicides

And the refugia disappears
though the tractor is not graceful on the land
and the buffalo will yield to that tractor
although the buffalo
finally
is more efficient
and its dialogue with the land
more lucid

Clearly there is something about
technology that does not like that
which is not itself

Yet this is not
a necessary condition
this unfriendliness
to the land

SACRAMENTO MEDITATIONS (1876–1977)

(MEDITATION ON THE CONDITION OF THE SACRAMENTO RIVER, THE DELTA, AND THE BAYS AT SAN FRANCISCO)

I WHO DAMMED ALL THE RIVERS

DYKING   CHANNELING   PUMPING THE WA-
TERS DIVERTING THE FLOW OF THE SAN
JOAQUIN AT FRIANT   AND THE SACRAMENTO
AT THE DELTA   LIMITING THE FLUSHING
OF THE DELTA AND THE BAYS   CRISSCROSSING
THE CENTRAL VALLEY WITH DITCHES AND
CANALS

WHO DAMMED ALL THE RIVERS AND MOST OF
THE CREEKS THAT FLOW INTO THE DELTA AND
THE BAYS

WHO DAMMED THE SACRAMENTO
THE TRINITY   THE MCCLOUD   THE PITT
FALL CREEK   HAT CREEK   COW CREEK
STONY CREEK   BATTLE CREEK   PUTAH
CREEK   BUTT CREEK

WHO DAMMED THE FEATHER ON THE NORTH
FORK   THE SOUTH FORK THE WEST BRANCH
AND ALL THE BRANCHES OF THE YUBA AND THE
BEAR

WHO DAMMED OREGON CREEK   CANYON
CREEK   FRENCH DRY CREEK   THE SOUTH
FORK AND THE MIDDLE FORK OF THE AMERICAN

WHO DAMMED THE RUBICON   BRUSH
CREEK   SILVER CREEK   TELLS
CREEK   GERLE CREEK   DRY CREEK
AND MOKELUMNE   THE STANISLAUS
THE TUOLUMNE   ANGEL CREEK CHERRY CREEK
SULLIVAN CREEK   DAMMING AND REDAMMING

THE MERCED THE SAN JOAQUIN THE
KINGS THE KAWEAH THE KERN

II FROM THE SATELLITE THE CENTRAL VALLEY IS ON FARM

WERE THEY "VISIONARY" PLANNERS INGENIOUSLY USING MODERN TECHNOLOGIES TO SECURE THE INHABITANTS OF CALIFORNIA FROM FLOOD AND DRAUGHT HAVE CONTROLLED THE FLOW OF WATER IN THE CENTRAL VALLEY DEVELOPING A COMPREHENSIVE INTERCONNECTED ARRAY OF RESERVOIRS DAMS POWER STATIONS PUMPING STATIONS DITCHES AND CANALS TO IRRIGATE THE CENTRAL VALLEY AND TO SEND WATER OVER THE TEHACHAPPI MOUNTAINS TO THE METROPOLITAN WATER DISTRICT IN THE SOUTH CREATING THE LARGEST IRRIGATION SYSTEM IN HISTORY GENERATING AN EIGHT BILLION DOLLAR INDUSTRY THAT SUPPLIES FOOD AND FIBER TO THE STATE THE NATION AND THE WORLD

AN IMPROBABLE PROFITABLE EXPANDABLE SYSTEM

WERE THEY "TECHNOCRATIC" PLANNERS SUBSIDIZED BY THE TAXPAYERS OF THE NATION AND IN HIDDEN INTEREST GIFTS BY THE STATE AT THE EXPENSE OF NONIRRIGATED FARMING ELSEWHERE PRIMARILY FOR THE PROFIT OF A FEW LARGE LANDHOLDERS AND AGRIBUSINESS HAVE TURNED THE ENTIRE WATERSHED OF THE CENTRAL VALLEY INTO ONE LARGE IRRIGATION SYSTEM SERVING OVER SIX AND ONE HALF MILLION ACRES COMPOSED OF DAMS THAT BECOME USELESS THROUGH SILTING A PUMPING SYSTEM THAT WILL USE MORE ENERGY THAN IT CREATES AND A DYKING SYSTEM THAT REQUIRES ONGOING

REPAIR THAT IN CONCERT REDUCE THE QUALITY AND LONG TERM PRODUCTIVITY OF BOTH THE LAND AND THE WATER THROUGH PROGRESSIVE SALINIZATION

AN ENERGY-EXPENSIVE SELF-CANCELLING SYSTEM

—TEXT I AND TEXT II OF *SACRAMENTO MEDITATIONS (MEDITATIONS ON THE CONDITION OF THE SACRAMENTO RIVER, THE DELTA, AND THE BAYS AT SAN FRANCISCO)* (FIGURE 10.2)

The nine texts of the *Sacramento Meditations* comprise the revised version of the work that was commissioned by the Floating Museum, San Francisco (director, Lynn Hershman), in 1976 and was hung in the San Francisco Museum of Art in 1977. The text and slides were slightly revised for an exhibition at the Ronald Feldman Fine Arts gallery in New York that then traveled to the Chicago Museum of Contemporary Art (1980). The nine mappings of the state of California are a drawing, a satellite photo, a political boundary map, and maps of state and federal water projects, irrigable and potentially irrigable land and topology in duplicate.

The poster sketch contains eleven posters, each beginning with "For instance . . . if" and ending with "What if all that irrigated farming wasn't necessary?" While *Sacramento Meditations* was hanging in the San Francisco Museum of Modern Art in 1977, we put these posters up on street corners throughout San Francisco with the help of San Francisco Art Institute students and others; we also erected two huge brown and blue billboards that said "Water." We wrote street graffiti on sidewalks in chalk, such as, "Somebody's crazy, they're draining the swamps and growing rice on the desert" and "Let every community empty its wastes upstream from where it takes its drinking water." Performances at the Art Institute and the San Francisco Museum of Modern Art included the simultaneous but slightly offbeat reading of a four-page water resources bibliography as a work scored for two voices. We showed a videotape of a Bakersfield (California) Water Board meeting on local TV, obtained several spots on local radio, performed readings of the work in situ, and published a series of bits of "advice" to important water personnel in the personal columns of a local paper. Each bit of "advice" was

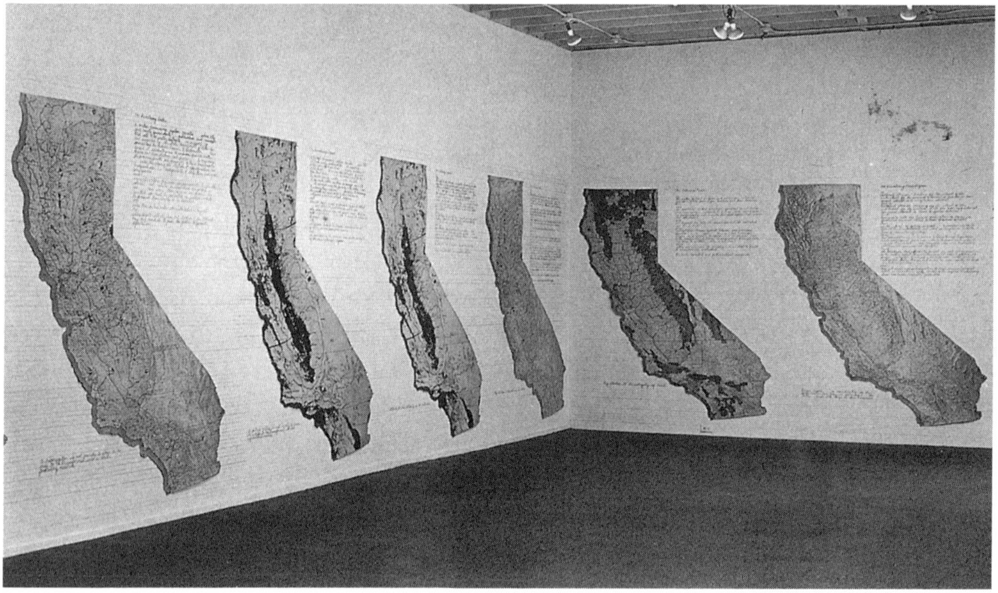

Figure 10.2
Helen Mayer Harrison and Newton Harrison, *Meditations on the Condition of the Sacramento River, the Delta, and the Bays at San Francisco,* 1977, installation, Maps and text, with site-specific works of street graffiti, posters, billboard, personal newspaper advertisements and video. Los Angeles Institute of Contemporary Art.

derived from Buddha's eightfold path and applied to water. For instance, "Dear Ron Robie: Right water thought," "Dear Jerry Brown: Right water action."

The work owed its existence to six months of research at the Berkeley (California) water resources library, which displayed a set of the original posters for several years. All the information it presents that was not public knowledge when the work was first done has since become readily available. However, although 650,000 of the 6,500,000 acres under irrigation became too salty to farm long ago, the wetlands at the end of the reversed flow of the San Joaquin River have become deadly, and although several severe draughts in California have made people begin to question water priorities, subsidized intensive irrigated farming continues as usual through good times and bad.

THE SAVA RIVER PROJECT (*ATEMPAUSE FÜR DEN SAVE FLUSS*) (1988)

In 1988, while on a Deutsche Akademische Austauch Dienst (DAAD) fellowship in Berlin, we were introduced to Yugoslavia by Hartmut Ern of the Berlin Botanical Gardens. He asked if we could not help with the formation of a nature reserve in the area that had once been a no-man's land situated at the border between the former Austro-Hungarian and Ottoman empires. After spending time in Yugoslavia and viewing the site, we felt that the idea of a nature reserve, although important, could not succeed in the long run because a reserve would suffer from all the problems of an island ecology and its uniqueness would be under attack from the surrounding industrial farming. Therefore, we proposed a nature corridor that would run the length of the Sava River from its twin beginnings above Ljubljana to its ending in Beograd where it joins the Danube River and supplies the lower Danube with one-third of its water.

Our work took the form of an exhibition—*Atempause für den Save-Fluss: Die Summe Sener Geshichte, Beginn Einer Neuen Geshichte*—funded by the Neuer Berliner Kunstverein in cooperation with the Berliner Kunstlerprogram das DAAD and the Moderna Galerija in Ljubljana.

But the Sava project was more than an exhibition; it was a proposal for change.

This is a work that wanted to happen. Support immediately appeared from many sources. All agreed that we should begin an address to this marvelous but obscure place—really a 340 square kilometer place, an island of biophysical providence and diversity, an exquisite cameo of cultural and ecological history existing within a 7,000 square kilometer factory farm.

It was obvious to us, as it is to all who pay attention to such things, that this island was under assault from petrochemical invasion and factory-farm wastes, that even the Sava River that flowed through it was under assault, and that the assault was and is both overt and covert. The defenders were few and mainly passive, and it seemed to us that the strategy of its defense was a guarantee of defeat. It had all the limitations of the Maginot Line. Therefore, we began to invent a work building support for the allocation of territory as a nature reserve but recontextualizing its strategy, generating a new set of tactics for the nature corridor. The manifestation of the story of this place was larger than the nature reserve itself.[2]

Since we first made the Sava proposal, the aftermath of Chernobyl led the Slovenian and Serbian peoples to consider shutting down a relatively safe American-made nuclear energy plant and trying to construct a series of ten to fifteen dams on the Sava to replace nuclear power with hydroelectric power. This would turn the Sava into a series of lakes and effectively kill the river. Our work served as a rallying point and a counterproposal for the opponents of this idea. This work was supported by the Croatian Water Department and the Department of Nature Protection. The nature reserve was declared, and the corridor was being considered. Until the tragic invasion of Croatia, the even more tragic ethnic cleansing attack on Bosnia-Herzegovina, and the ensuing ongoing disaster, the World Bank was going to support the clean-up of the Sava River. Since then we have heard that the invasion has severely damaged, if not destroyed, the nature reserve, and Yugoslavia has remained a series of separate warring states.[3]

TIBET

For two decades, we have pursued the art of conversation. Over the years we have created photographs, maps, drawings, poems, and performance scores that document a diverse array of conversations, most of which take up the issue of global survival. The specific subject matter ranges across a wide spectrum of eco-political issues, often generating new metaphors that in turn are the basis for proposals and projects.

What follows is a proposal for the early stages of a conversation about one million square miles of emerging man-made desert in what used to be the rainforests of Tibet. We are exploring the potential of pulling together a team of experts to create an "analog" forest—a proposal that seems to run parallel to the Dalai Lama's vision of a "world peace park." As envisioned, the analog forest would be only 20 percent as complex as the ecosystem it replaces but would behave in much the same way. Its future survival would be aided by the presence of species most useful to man if harvested selectively.

Tibet is the High Ground

FOR THIS PROJECT, WE INTEND

To create a very large-scale model of Asia, where Tibet as the High Ground is manifest and the Seven Principal Rivers of Southeast Asia—the Indus, Yellow, Yangtze, Salween, Mekong, Brahma Putra, and Ganges Rivers—are clearly stated. This model would make obvious the implications of the deforestation of Tibet for the many nations of the continent of Asia.

To come to a clarified statement of the research on the outcome to the rivers, riverine life, and riverine surrounds, including the dams and other river projects, in other countries as well as Tibet, of the deforestation of Tibet.

To develop a conceptual design for the regeneration of Tibetan forest, farm, and meadowland. We would hold that a conceptual design for this kind of transformation carries a moral force in advance of criticism of any kind.

To develop a conceptual model that envisions a probable life Web for the deforested area of Tibet. This ecological model, working with the idea of an analog forest or a simplified woodland-rainforest ecosystem, would also have yields of recognizable value to those who inhabit the surrounds. The process would suggest an analog for the forest that once existed, less complex than the original but nonetheless with overstory, understory, canopy, and appropriate niches. This forest would also serve to conserve existing or replaced topsoil and protect riverine and wetland ecologies—first in Tibet proper and then in the countries below.

More is at stake here than the simple regeneration of a new kind of forest. The forest knits Tibet, China, Nepal, Kashmir, Bhutan and all the bordering countries together from an ecological point of view. The rivers that spring from Tibet are also waters that bind these countries. We would hold that a conceptual design expressing these understandings can be the basis for generating conversation based on common interest as opposed to conflicting interest.

We hold that putting in place a working model of this kind is a necessary addition to the present discourse on saving the Tibetan culture, peoples, and children and is necessary to preserve the terrain from emerging abuses such as atomic waste dumping.

As we stated earlier, we believe that the universe is a giant conversation and that any introduction of new ideas, new metaphors, or new possibilities can change that conversation. Although we have built works, we think that changes in the conversation that lead to attitudinal and behavioral changes are as significant as any "built" work.

NOTES

1. This work was commissioned by the Metromedia Company and is presently in its collection.

2. A catalog of this work with full-color illustrations exists in German, English, and Slovene, with translations into Serbo-Croatian, French, and Japanese. The text, with several accompanying black-and-white images, has been reprinted in the *Journal of International Synergy.* The work has been exhibited in Berlin at the Neuer Berliner Kunstverein, in Zagreb (Croatia), in Ljubljana (Slovania), at the Centre International d'Art Contemporain de Montréal, at the Palmer Museum at Penn State University, at Ronald Feldman Fine Arts in New York, at Washington University in Saint Louis (Missouri), and at the Nagoya Biennale in Japan in 1991, where it won second prize. The work was part of a traveling exhibition entitled *Fragile Ecologies,* which toured the United States for three years, beginning at the Queens Museum in New York.

3. We wish to offer our thanks to Hartmut Ern and Martin Schneider-Jacoby of the Siftung Europaiches Naturerbe (European Natural Heritage Fund) for their generous ongoing assistance in the Sava River Project. In addition, our work has been made possible by the support of the Water Department of Croatia, the Nature Protection Agency of Croatia, the Croatian Zoological Society, the Natural History Museum of Zagreb, and various other official and semiofficial organizations.

11

Process(ing) Interactive Art: Using People as Paint, Computer as Brush, and Installation Site as Canvas

Sonya Rapoport

My work describes people's responses to queries in an installation environment. The questions have concerned their emotional condition[1] and their attitudes about shoes,[2] death,[3] and sexual jealousy.[4] The interaction with the artwork extends to participants making choices among images to be placed in word themes,[5] selecting words to be arranged into poetic phrases,[6] and placing their shoes on coded floor tiles.[7] Responses to the above comprise the media—the "paints"—that are manipulated into an art form. Each installation—the "canvas"—where these inquiries take place contains elements of another time, place, or culture. Ancient Egypt,[8] Southern India,[9] the early American Indian Southwest,[10] and mythical systems[11] such as alchemy have been introduced along with the "here and now." This chapter describes specifically how the diverse elements of the process—starting with a pictorial calendar diary that focuses on biorhythmic cycles and ends with a philosophy of life by Rabindranath Tagore[12]—are integrated into an interactive installation.

THREE PHASES TO *DIGITAL MUDRA*

Phase 1: *Biorhythm Calendar*

In 1980, I kept a pictorially descriptive diary on a large calendar. Each month was illustrated on a 45-inch-by-30-inch format of vellum. As I collaged (figure 11.1) information on the square of each day, I also notated for that day an evaluation of my three biorhythm conditions—emotional, intellectual, and physical.[13] At the end of each month, I compared my own assessments with printouts of a computerized biorhythm program that predicted what my biorhythm cycles would be.

The following year, 1982, I showed the biorhythm material as a computer art exhibition.[14] The twelve monthly calendars, the printouts of my projected biorhythmic state, and various plots were included. The

---

This classic *Leonardo* paper (from *Leonardo*, 24, no. 3 [1991]) documents some of Sonya Rapoport's seminal work in interactive art. Since this paper was written, she has produced many other works including *Redeeming the Gene, Molding the Golem, Folding the Protein* (2001); *The Transgenic Bagel* (1994–1996); and *Smell Your Destiny* (1995), all available on the World Wide Web at ⟨http://www.sonyarapoport.net⟩.

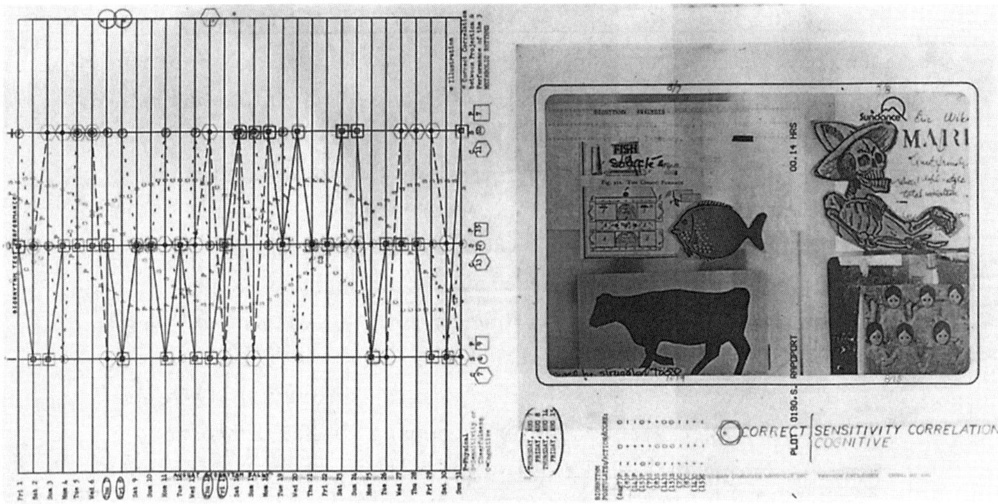

Figure 11.1

Sonya Rapoport, *Biorhythm: Postulate and Performance,* 1981, computer plots on vellum, color Xerox transparencies, color pencil, 10 × 10-inch format artist's book, 28 pages. *Left:* A plot describing the success between the computer prediction of biorhythm cycles and the author's own assessment. The helix in the center is the computer biorhythm diagram for August. The horizontal lines mark the three levels of success — emotional, intellectual and physical — for each cycle. The top borderline indicates agreement in all three cycles for those days. *Right:* A detail of August 8 from the pictorial calendar diary. Photo by Robert Rapoport.

success of each month's correlations between my personal assessment and the computer calculation based on my date of birth were described by using a calcomp plotter to lay out x and y coordinates on vellum (see figure 11.1). This output extended 26 feet. Each month's plot was illustrated by a color transparency detail of one calendar day of that month. The visual presentation culminated in three spiral calcomp plots on transparent vellum, one for each cycle, superimposed to reveal the agreement between the technical and person evaluations of the biorhythm cycles. Only thirteen days had identical predictions and personal evaluation rhythms for all three metabolic cycles.[15]

Phase 2: *Biorhythm Participation Performance*

In May 1983, at Works Gallery in San Jose, California, I invited viewer-participants to compare their own assessments of "how they felt" with computer-derived analyses of their biorhythm conditions for that day (see figure 11.1). How participants felt determined which color of hospital wristband they were given: blue for bad, pink for good, or white for intermediate. The hospital bands served the purpose of reminding the participants of their own assessments and commitments for the duration of the performance. Some, however, did return for a different wristband. The participants were then asked to have their hands express those same feelings while being photographed. During this sitting each subject wore a dentist's bib to serve as a background for the hand gestures. I was nearby to give directions to each participant for producing manual and verbal expressions of his or her emotional state. The verbal exclamation was to be declared at the same time as the gesturing. The comments were recorded on an audio cassette recorder. The participants then each received a biorhythm reading of their emotional state from the computer of their choice. At the site were three computers—a teletype, a portable calculator, and a personal computer (PC). Those who had made assessments that agreed with the computerized analyses of their biorhythm cycles were "winners" and received blue ribbons. Of the total number who participated, 33 percent were winners. To enliven the activity with a point of view that was different from that of high technology, a palmist gave readings of the thumbs of the participants. Each reading, according to cryptic symbological analysis, reveals the immediate emotional state of the thumb's owner. Finally, each participant signed a chart recording the three evaluations: personal, computer, and palmist.

Phase 3: *"The Computer Says I Feel . . ."*

In 1984, a San Francisco artist[16] conceived of an exhibition that addressed the futuristic theme *SF/SF: San Francisco/Science Fiction.* The opportunity arrived to use the material from my book *Biorhythm: Postulate and Performance* (see figure 11.1) in creating a fictional future for the computer's biorhythm program. *"The Computer Says I Feel . . . "* addressed a future fantasy world in which we must consult a computer to find out "how we feel." The visual portion of the work consisted of the chart filled in by the participants in phase 2 and an accompanying panel that displayed the black-and-white photographs of them gesticulating. These panels were made of plotting vellum, each 11 feet by 30 inches. The photographs of the participants on the one panel were aligned with the statements by same participants on the other panel. The wristbands in full view indicated the personal assessments of their own emotional conditions by the letters *G* (good), *B* (bad), or *I* (intermediate). In words beneath the rows of 8-inch-by-10-inch photographs were the dictates of the computer analysis of each condition. Nearby on the wall were earphones from which could be heard the audio tape of the previously recorded phrases about "how they felt."[17] These phrases were interrupted by an officious British-accented voice insisting on how the participants felt. As mentioned earlier, 33 percent of the pronouncements by this voice matched the participants' phrases on the audiotape.

DIGITAL MUDRA: AN INTERACTIVE INSTALLATION

The impetus for creating *Digital Mudra* was my desire to utilize in a visual context the "paint" material—the photographs of the hand gestures from *Biorhythm Participation Performance.* When I decided to have photographs taken of the participants gesticulating about how they were feeling that evening, I had no idea about how the material would be incorporated into future artworks, though I knew that I would enjoy the challenge of eventually using the information. In this latter endeavor, one obvious direction for me was to investigate American Sign Language (ASL). I had lived near a school for the deaf for many years. However, the use of the fingers in ASL is too detailed for association with the hand gestures used in my biorhythm photographs. So, influenced by my earlier studies of cross-cultural correlations between the Anasazis and my own environment,[18] I returned to the subject of early American Indian culture to try to find a gesture language that correlated somewhat with the gestures used in the biorhythm participation event. I came across some relevant material among both Plains and Athapaskan tribes but not enough on which to build an artwork. Perhaps subconsciously the word *Indian* took me eastward to India, where I had enjoyed many exciting experiences. I was somewhat familiar with its art forms. Among the hundreds of mudra gestures in the southern Indian dance called Kathakali, thirty-four are similar to the gestures in the photographs from the biorhythm performance event.[19] Furthermore, the duality of the mudra gesture verbal translations and the participants' verbal expressions for the similar gestures was provocative.[20] The similarities and dissimilarities of transcultural meanings of the gestures were the decisive factors for proceeding within this context. The thirty-four mudras similar to the gestures in the photographs became the core of the interactive installation *Digital Mudra*.[21] The title reflects the use of the computer with the ancient tools of Kathakali drama and dance.

The structure for the piece became a loop from gesture to word, word to gesture, and back to word. In implementing this pattern, the two cultures—Western as experienced in northern California and Eastern as experienced in southern India—are woven in and out of the framework.

I started with the photographs of people gesticulating in *Biorhythm Participation Performance.* Then, moving on to the word, I sent a list of the mudra word meanings[22] (no drawings of the gestures) to art-world friends, asking them to create poems using any number of the thirtyfour words and phrases on the mudra word list. Other words could be included with the mudra selections. One poem included all thirtyfour words and was used as a reference source for the *Digital Mudra* installation.[23] Each gesture, accompanied by its gesture meaning, was painted on the cover of a transparent 8-inch-by-10-inch Plexiglas box. Visible through this image on the inside back of the box was a photograph of the related gesture with a caption of the verbal expression from the *Biorhythm Participation*

Performance event. The boxes were hung on the wall in the sequence of the written poem that used the complete mudra word list.

A videotape was made of Kathakali dancer Kunhiraman transforming six of the Western poems that had been responses to the mudra word list.[24] While Kunhiramans wife read the poetry aloud, he danced the corresponding mudra gestures and innovated a transition to the next mudra. This videotape was included in the installation. The poetic phrases created by the Western participants had become integrated with the ritualized dance movements of an Eastern culture.

A slide show projected the same mudra word images as drawn on the boxes. Following each mudra slide was a slide of a political figure using this same gesture. Occasionally, the sequence continued to a third slide, a comic strip character also using the same gesture.[25] All captions were included to emphasize cultural congruence and incongruence.

After the viewers had become acquainted with the visual aspects of the installation, they were familiar enough with the elements to participate in an interaction with them. The interactive part of the installation began when participants chose three Mudra word cards from among an array of complete sets of thirty-four cards that were displayed on nearby tables. White cards, 3¼ inches by 2½ inches, had been printed in black ink with the same mudra word images as seen on the slides and the source boxes. With these three words and any other words they wished to include, the participants were asked to compose brief poetic phrases. Pencils and paper were provided for this activity.

Each participant's creative effort was typed into the computer (figure 11.2), which up to this point had been continuously displaying the entire cycle of mudra gestures on the screen.[26] The mudra words selected by the participant became a gesture-dance sequence on the monitor and were then printed and given to the participant.

After the poem was printed, the participant took it to the "philosopher" stationed at another computer. He read the poem and asked the participant which word was the key mudra word in the poem and which word was its support. The information was then entered into the "crystal ball" computer,[27] and a relevant philosophical commentary by Rabindranath Tagore was printed.

As a final step, the participants "engraved" their poems on the wall of "Temple Writings" (see figure 11.2). This was achieved by choosing red-

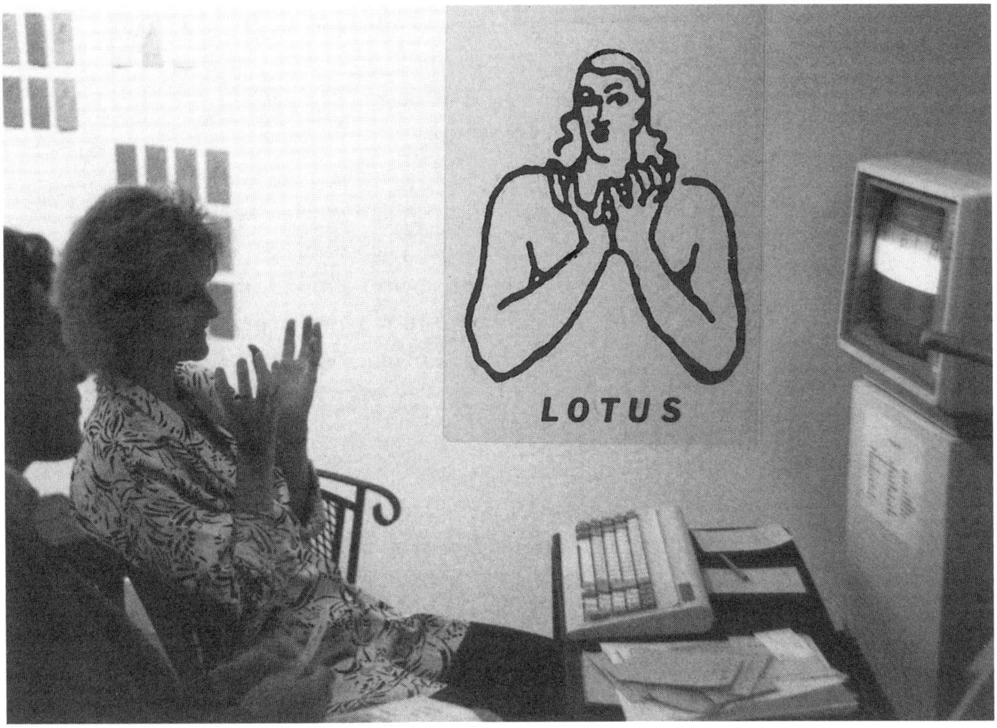

Figure 11.2
Sonya Rapoport, *Digital Mudra,* 1987, installation detail. A participant has entered her poetic phrase into the computer by keyboard and views her poetic phrase *lotus* on screen. She imitates a gesture. Behind her is the Mudra Wall of Temple Writings—red-imprinted gestures on gold-colored tags.

imprinted, gold-tinted metal tag duplicates of their mudra word-card choices and hanging them on "golden" nails.

CONCLUSION

The above phases comprise my work as process art wherein the execution of each step produced material for the next. The various responses of the participants created the media—the "paint" to be mixed, manipulated, and applied with the use of the computer "brush"—which developed into a grand finale of an integrated interactive artwork. My artistic concern is that those experiencing the installation, the "canvas," add new dimensions to their traditional aesthetic concepts.

Digital Mudra thus far completes the development of my process interactive art. However, I may add one more set of correlations to the original gestures—expressions of newborn infants. Who knows? The project may never end.

NOTES

1. *Biorhythm,* 1983, exhibition at Works Gallery, San Jose, CA.

2. *A Shoe-In,* 1982, exhibition at Berkeley Computer Systems, Berkeley, CA.

3. *The Animated Soul: Gateway to Your Ka,* 1991, interactive installation at Ghia Gallery, San Francisco.

4. *Coping with Sexual Jealousy,* 1984, interactive performance at Pauley Ballroom, University of California, Berkeley, CA.

5. *Objects on My Dresser,* 1984, interactive audience participation event at New School for Social Research, Sarah Lawrence College, Bronxville, NY; at University of California, Berkeley, CA, 1983.

6. *Digital Mudra,* 1988, installation at Kala Institute, Berkeley, CA; at Saint Mary's College, Moraga, CA, 1988.

7. *Shoe-Field,* 1986, exhibition at MEDIA, San Francisco.

8. *The Animated Soul.*

9. *Digital Mudra.*

10. *An Aesthetic Response,* 1978, exhibition at Peabody Museum, Harvard University, Cambridge, MA.

11. *Interaction Art and Science,* 1979, exhibition at Truman Gallery, New York City.

12. Rabindranath Tagore, poet and philosopher, was born in Calcutta, India, in 1861. He received the Nobel Prize for literature in 1913.

13. "Our body produces, stores, and releases energy in regular cycles. These cycles begin at birth and continue. . . . There are three biorhythm cycles—the physical, the emotional, and the mental—with recurring periods of 23 days, 28 days, and 33 days respectively." Lee Martin, *The Biorhythm Handbook, Workbook, and Forecaster* (New York: Harper & Row, 1978), 4. Particular days of caution are chartable in advance by arithmetic calculation, computer programs, and biorhythm calculators. Wilhelm Fliess of Berlin is responsible for the pioneer work done on biorhythms.

14. *Biorhythm,* 1982, exhibition at New York University Graduate School of Business Administration, New York City.

15. The correlations were as follows: 117 days had identical physical predictions and personal evaluation rhythms, 96 days had identical cognitive predictions and personal evaluation, and 127 days had identical sensitivity predictions and personal evaluation rhythms.

16. *SF/SF San Francisco/Science Fiction* (1984–1985) was an exhibition conceived by Paul Hasagawa-Overacker and organized with Jo Babcock, Nat Dean, and Robert Atkins. It traveled from the San Francisco Arts Commission Gallery to the Clocktower in New York City and to the Otis Art Institute of the Parsons School of Design in Los Angeles.

17. Susan Stone, a San Francisco sound artist, reassembled the audio material.

18. *Overview,* 1978, exhibition at Union Gallery, San Jose State University, San Jose, CA. This installation included work that interpreted archaeological research computer printouts by drawing visually related material (of early American Indian artifacts, the Navajo language, and general information about traditional American Indian territory) over the computer information.

19. Mohan Khokar, "The Tradition" and "A Dictionary of Mudras," *Marg Magazine* (Bombay, India) 11, no. 1 (1957): 7, 18. Kathakali originated in Kerala, India. The meanings of the gesture language—of both the gesture and the word spoken aloud to prompt the dancer performing the gesture—were conceived in Sanskrit and expounded in Malayalam. Later, the Sanskrit was replaced by the Malayalam language.

20. In one photograph, a woman raises her hands above her head. The caption of what she exclaimed reads, "I feel right up there!" In the dictionary of mudras, an almost identical mudra gesture means "wings of bird."

21. *Digital Mudra.*

22. Examples of mudra words and phrases include *rice, greet with a bow, book, approval, greeting, skin, send away.*

23. Vance Martin created the poem using all thirty-four mudra meanings. I am grateful for this contribution to the installation.

24. The poems of Terry Ellis, Nancy Genn, Karen Hirchmiller, Judy Malloy, Vance Martin, and Tina Singer were "danced" and videotaped. My thanks to Tom Bates, who operated the video camera.

25. An example of a slide sequence in the slide show is President Mitterand of France smiling with President Mubarak of Egypt. The latter is raising his hands, obviously delighted by French support. The next slide is of the similar mudra gesture for *sun.* And the next slide is of a comic-strip character raising his hands in the same way and saying, "We're celebrating!" Many thanks to Alan Marshall for taking the slides.

26. The mudra images had been scanned and programmed by John Watkins, the software engineer for *Digital Mudra.* He used SHOW PARTNER (Brightbill-Roberts & Co.) for the basic software. The graphics equipment required for this program are an IBM AT-compatible computer with 640K memory, centronics-compatible printer interface card, enhanced graphics EGA adapter, EGA monitor, and EX 800 or 1,000 Epson printer. In the weave cycle, a picture forces the previous picture off the screen in two directions simultaneously, with even lines pushed to left and odd to the right. This gives the effect of the images weaving in and out of each other. The "dance" gestures are simply slide fade-ins and fade-outs. These fades overlap the second and third gesture-word images as they impose on and replace the previous ones in the sequence.

27. John Watkins created a database program using the C language. The selection of the two important words by the participant triggered an output that contained either these two words or words associated with them. There was also a question by the on-site philosopher, Jeffry Tibbets, that required a plus or minus answer. This influenced the "philosophical advice" selected from a database of negative and positive commentaries by Rabindranath Tagore.

12

Touch-Sensitivity and Other Forms of Subversion: Interactive Artwork

Lynn Hershman

In his seminal article "The World View of Subversive Cinema," critic Amos Vogel states that "at the core of modern art stand concepts that still strike scorn or fear into the hearts of many, as they are subversive to the very notion of conventional 'reality.'" *Dissolution, fragmentation, simultaneity, decomposition*—these words are in the service not of obfuscation but of clarification. They denote a more fundamental analysis by a dissection of its ever-growing complexity. The art of our century, says Katherine Kuh, has been characterized by "shattered surfaces, broken color, segmented composition, dissolving forms, and shredded images."[1]

The art that was apparent to Vogel and Kuh reflected a gestating anti-aesthetic. Work produced at the time of Vogel's article (1974) seemed to be based on destruction, decomposition, and deconstruction. This type of work functioned perversely and pervasively to erode aesthetic preconceptions, offering, instead, art that was random, fragmented, schizoid, nonlinear, disruptive, antinarrative, and interactive. The tension of alienation and disorientation became part of both form and content. This new art was, as might be predicted, a reflection of an emerging shift in perceptions of what reality meant.

Avant-garde art, particularly performance art, has historically been hostile toward the notion of audience as a collection of passive viewers. Jarry, Brecht, Artaud, and Apollinaire reconfigured environments in order to blur distinctions between stage and audience, art and life. John Cage sought to create an audience of active listeners, and La Monte Young searched for participatory socializing systems with his *Composition* (1960).[2] Even Allan Kaprow, once a student of Cage at the New School for Social Research, staged audience-dependent "happenings." Incorporated into the form and content also were ideas of rupture, participation, interactivity, and escalated danger. In events of the Fluxus movement, particularly, the audience and artists were often put at risk. This is especially true in pieces such as Nam

---

This paper, originally published in *Leonardo,* 26, no. 5 (1993) documents some of Lynn Hershman's seminal work in interactive art. Since this paper was written, she has produced many other works including the feature films *Conceiving Ada* (1993–1998) and *TeknoLust* (2002) and the interactive, telematic installation *Difference Engine No. 3* (1993–1998) that explored the physical and the ethical boundaries of virtual space.

June Paik's *étude for Pianoforte* (1963) and Yoko Ono's *Cut Piece* (1964). Eventually, artists such as Chris Burden, Karen Finley, and Peter Campus involved live audiences in vivid horrors of participatory encounters.

What was not immediately apparent was that the delivery systems through which this new art reached its audience would change. Ultimately, these new systems of communication have had a radical effect and have also proved wildly subversive—at least as subversive as the art itself.

This shift began to occur when form, content, and delivery methods commingled. The means by which the new art reached its audience depended on how it was transmitted. This caused a new audience to emerge. The new audience consisted of users of fax, computer, e-mail, and compression technology. Members of this audience comprised a privatized public linked by its varied uses of electronic networking systems. The interactions of this broad community of receivers and participants helped shape a revolution.

In 1876, the invention of the telephone allowed people to operate in two places simultaneously. This phenomenon created a model for disengaged intimacy that was experienced through a mechanical prosthetic. I use the term and idea of *prosthesis* to represent a method of extending one's body into space and time, a way of extending both the human reach and the human gaze. A few years after the invention of the telephone, the automobile was introduced. This provided another means by which to displace time, space, and distance. A revolution this grand in scale did not occur again until computers became accessible to a broad public, which again radically shifted conceptions of time, space, and distance. The electronic revolution changed things even more. Impulses of flowing current fostered cross-fertilizations of art and technology, present and past, logic and chaos, reality and illusion.

Computers, interactive discs, postsymbolic communication, virtual reality, and interactive television are but a few of the new information processes that defy linear structure. They require decisions from a participant and incorporate chance into their structures.

There is a relevant John Cage–Marcel Duchamp anecdote. After composing the music for Hans Richter's film *Dreams Money Can Buy* (1956), Cage mentioned to Marcel Duchamp that he had used Duchamp's ideas for scoring the film. Duchamp replied, "I must have been fifty years ahead of my time." In fact, Duchamp was right. His experiments with Rotodisks and chance operations were conceived nearly a century before the technology was invented that would allow his vision's potential to be fully manifested. Duchamp was not alone in his insight. New technologies and their interactive uses by artists now extend many previously conceived ideas such as the use of multiple perspective and simultaneous viewpoints as explored by the cubists; incorporation of randomness, everyday experience, and the audience as investigated by the surrealists; and the destruction of form as explored by the dadaists. The conceptual basis for the new reality of art technologies is firmly rooted in art history.

Interactive technologies encourage artists to engage in (at a minimum) two-way dialogs. They have created a process through which illusion and icon have become a simulacrum of origin and authenticity. The illusion is the substance of truth. Authenticity and originality have assumed a nostalgic veneer. One of the more subversive elements of art that uses interactive, community-based systems is a shift in the relationship of artist to audience.

Artists who work in this genre are not separate from society. When effective, they create nonhierarchical systems that address fundamental perceptions that require responses. The interaction itself therefore becomes a political act. For example, the recently popularized technology of the camcorder (a lightweight video camera with a built-in recording system) allowed a bystander to capture on video the infamous 1991 beating of Rodney King by Los Angeles police officers. Thus a personal point of view accounted for the magnitude of the public furor to the case. It has been said that if U.S. President John Kennedy had been assassinated in the era of camcorders, there would have been at least forty-five angles captured of the event rather than the lone, 8 millimeter Zapruder view. Such interactions can constitute a form of intervention, in that members of the audience can take their viewpoint of an event (literally, in the form of film) and distribute it to the public—democratizing media-based viewpoints.

Consider the following works, all produced since 1990. Daniel B. O'Sullivan, a student at New York University's Tisch School of the Arts, created an interactive cable-television show for Manhattan Cable, titled *Dan's Apartment.* In it, viewers could, at will, navigate through O'Sullivan's home by speaking into their telephones. David Rokeby's sound systems scan viewers' gestures and movements, transferring them into a computerized score that participants play through the motion of their own bodies. Jeffrey Shaw's *Legible Cities* (programmed by Gideon May) also uses the participant's body to pedal a bicycle that, in turn, directs the movement through a projected computerized city of words. Sara Robert's *Early Programming* involves interaction between a mother (who is a computer) and a child. Roberts is currently working on combining four computers into synchronized engines that drive the emotions of four corresponding actors.

Roberts and I collaborated on *A Room of One's Own,* an artwork that operates principally through the viewer's voyeuristic eye movement. In 1989, we created *Deep Contact,* an interactive videodisc that was activated when the viewer touched the screen. Both these works utilize the viewer-participant's body. Like the telephone, new interactive works extend the participant's reach and gaze into artificial, computerized space.

Large corporations are also experimenting with interactive techniques. In Montreal, Videotron allows home-television viewers to install (for a small fee) a small computer that tracks choices and then simulates personality-reflective programming based on these viewer preferences. For example, a white female in her forties would get different program choices (and commercials) than would a teenage African American male. In California, laboratories are working with the interface between human motion and actual immersion into computer space. For example, Electronic Arts has developed a system called 3DO that compresses video images, thus allowing interactive multimedia systems to be brought into the home market.

Recently, a psychiatrist disguised his identity through computerized enhancements so that he appeared as an eighteen-year-old female. This led to "software porn"—sexually explicit correspondence with many e-mail respondents who were eager for relationships. Respondents did not realize that they, like the postgendered cyborg described by author Donna Haraway,[3] were being electronically intimate with an artificially created persona.

Art that uses new technology generally has an architecture that is invisible. In that sense, it is ecologically compact. That is, this art occurs within a disc: it does not pollute the environment. Some artists feel that it is a way to digitally immortalize themselves.

Perhaps the most subversive element of new technologies is their ability to force "real life" to transgress space and enter artificially based environments. They thereby diabolically transfigure the essence and authenticity of the participant, who not only becomes artificial through the process but can be recognized only when electronically disguised.

INTERACTIVE PERFORMANCE

My path to interactive works did not begin with video but with performance. In 1971, I created an alternative identity that I named Roberta Breitmore. The character of Roberta belonged to the creation of a history and identity for women, as Roberta represented "allwomen." At the time I created her, many women were obsessed with defining themselves. Like Roberta, women found that one way to define themselves was through the cultural props around them. I consistently try to deal with how the media stereotype women and how the media create an identity to which women are supposed to aspire. There is something ambiguous, artificial, and self-defeating about women's participation in that process.

Roberta's manipulated reality became a model for a private system of interactive performances. Instead of being kept on a disc or hardware, documents on Roberta were stored as photographs and texts that could be viewed without predetermined sequences. This allowed viewers to become voyeurs of Roberta's history. Their interpretations shifted, depending on the perspective and order of the sequences.

INTERACTIVE VIDEO PROJECTS

Lorna (1980–1984)

Two years after Roberta's transformation, *Lorna,* my first interactive art videodisc, was completed.[4] Unlike Roberta, who has many adventures out in the world, Lorna, a middle-aged, fearful agoraphobe, never leaves her tiny apartment (figure 12.1). The premise is that the more she stays at home and watches television, the more fearful she becomes, primarily because she absorbs the frightening messages of advertising and news broadcasts. Because she never leaves home, the objects in her room take on magnificent proportions.

Every object in Lorna's room is numbered and becomes a chapter in her life that opens into branching sequences. The viewer-participant accesses information about her past, future, and personal conflicts via these objects. Many images on the screen are of the remote-control device that Lorna uses to change television channels. Because the viewer-participant uses a nearly identical unit to direct the disc action, a metaphoric link or point of identification is established between the viewer and Lorna. The

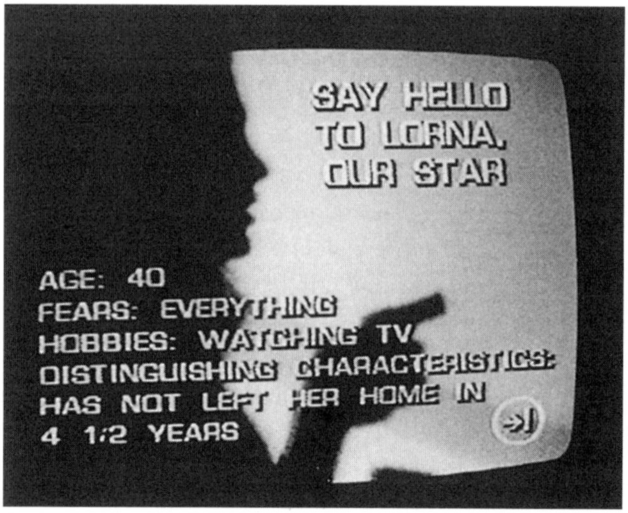

Figure 12.1
Lynn Hershman, *Lorna,* 1983, still image from the interactive videodisc.

viewer-participant activates the live action and makes surrogate decisions for Lorna. Lorna's choices are designed as a branching path.

Although there are only seventeen minutes of moving image on the disc, the thirty-six chapters can be sequenced differently and played over a period of several days. There are three separate endings to the disc, and the plot has multiple variations that include being caught in repeating dream sequences. Multiple soundtracks can be seen backward, forward, at increased or decreased speeds, and from several points of view.

I established no hierarchy in the ordering of decisions. This idea is not new: it was explored by such artists as Stéphane Mallarmé, Cage, and Duchamp, particularly by Duchamp in his music. These artists pioneered ideas about random adventures and chance operations fifty years before the invention of the computer technology that would have allowed them to exploit their concepts more fully.

Lorna's passivity (presumably caused by being controlled by the media) is in counterpoint to the direct action of the participants. As the branching path is deconstructed, players can become aware of the subtle yet powerful effects of fear caused by the media, and my hope is that they become more empowered (active) through this perception. By taking action on Lorna's behalf, viewer-participants travel through their own internal labyrinths to their innermost transgressions.

Despite theories to the contrary, the currently dominant Western cultural presumption is that art making is active and art viewing is passive. Radical developments in communication technology—such as the marriage of image, sound, text, and computers and the interactivity that these new technologies provide the "viewer"—challenge this assumption. Viewer-participants of *Lorna* have reported that they had the impression of being empowered because they held the option of manipulating Lorna's life. Rather than have Lorna be remotely controlled by the artist—as in conventional cinema, where the viewer passively watches the filmmaker's choices about character and narrative—a unit for making decisions was literally placed in the viewer's hands. Such artistic acts interrupt the flow of the media bath of transmitted, prestructured, and edited information in which our society is submerged. Alteration of the basis for exchange of information is subversive in that it encourages participation and therefore creates a more empowering audience dynamic.

Interactive systems require that viewers react. Choices are made by means of a keyboard, mouse, or touch-sensitive screen. As technology expands, more permutations will become available, not only between the viewer and the system but between elements within the system itself. Some people feel that computer systems will eventually reflect the personality and biases of their users. Such systems only appear to "talk back." That they are alive or independent is an illusion: they depend on the architectural strategy of the program.

There is, however, a space between the system and player in which a link—a fusion or transplant—occurs. Truth and fiction blur. Action becomes icon. According to Sigmund Freud, reality may be limited to perceptions that can be verified through words or visual codes. Therefore, perceptions are the drive to action that influence, if not control, real events. Perceptions become the key to reality.

Television is a medium that is by nature fragmentary and incomplete, distanced and unsatisfying, like platonic sex. A precondition of a video dialog is that it does not "talk back." Rather, it exists as a moving stasis, a one-sided discourse, a trick mirror that absorbs rather than reflects.

Lorna was developed as a research and development guide, but it is inaccessible to most people, as it was pressed in a limited edition of twenty, of which only fourteen now exist. *Lorna* is only occasionally installed in galleries or museums. Creating a truly interactive work demands that it be available and accessible on a mass scale.

Deep Contact (1985–1990)

Because interactive media technology is becoming increasingly visible in many areas of society (particularly outside the art world), the political impact is spectacular. Traditional narratives (those with beginnings, middles, and ends) are being restructured. At the same time, genetic engineering advances are reshaping the meaning of life. As a result, a growing number of people feel an increasing need to participate personally in the discovery of values that affect and order their lives, to dissolve the division that separates them from control (freedom), and to replace longing, nostalgia, and emptiness with identity, purpose, and hope.

This videodisc provides players with the ability to "travel" through the fifty-seven segments of *Deep Contact*[5] into its deepest part. The title refers to the players' use of their own intuition to determine the route to the center, while they simultaneously try to find and to feel the deepest, most essential parts of themselves. Viewers choreograph their own encounters with voyeurism that I have incorporated into *Deep Contact.*

This touch-sensitive, interactive videodisc installation compares intimacy with reproductive technology. It allows viewers to have adventures in which their sex, age, and personalities are changed. Participants are invited to follow their instincts as they are instructed to touch their "guide" Marion on any part of her body. This is done via a Microtouch monitor attached to a Macintosh IIcx personal computer. Different adventures emerge from touching different body parts.

When the video begins, Marion knocks on the projected video screen and asks to be touched (figure 12.2). She continues her pleas until the participant touches the parts of her body, which have been scanned and programmed to rotate onto the Microtouch screen.

For instance, if the viewer touches her head, a choice of TV channels is provided—some giving short, humorous analytic accounts of "reproductive technologies" and their effect on women's bodies and some showing how women see themselves. The protagonist also talks about "extensions" into the screen that she describes as similar to "phantom limbs"; the screen thus becomes an extension of the viewer-participant's hand, similar to a prosthesis. Touching the screen encourages the sprouting of phantom limbs that become virtual connections between the viewer and the image.

If the viewer touches the area below Marion's neck but above her legs, the video image shifts to a bar where the viewer can (1) select one of three characters (Marion, a Demon, or a Voyeur) to follow through interactive fiction that has a video component or (2) choose to have his or her own image replace the one on the screen via a surveillance camera.

If viewers touch Marion's legs, they enter a garden sequence in which they can follow Marion, a Zen Master, an Unknown Path or a Demon. Participants make selections via images that have been photographically scanned onto the touchscreen. In the garden, for example, the image on the touchscreen is a hand that jumps forward (depending on selections) and allows the viewer to follow the lines on the hands to different routes.

The viewer-participant usually follows a character or a segment to a fork in the road. At this point, the disc automatically stops, requiring the viewer to make a selection to go left, go right, return to the first segment of the disc, or repeat the segment just seen.

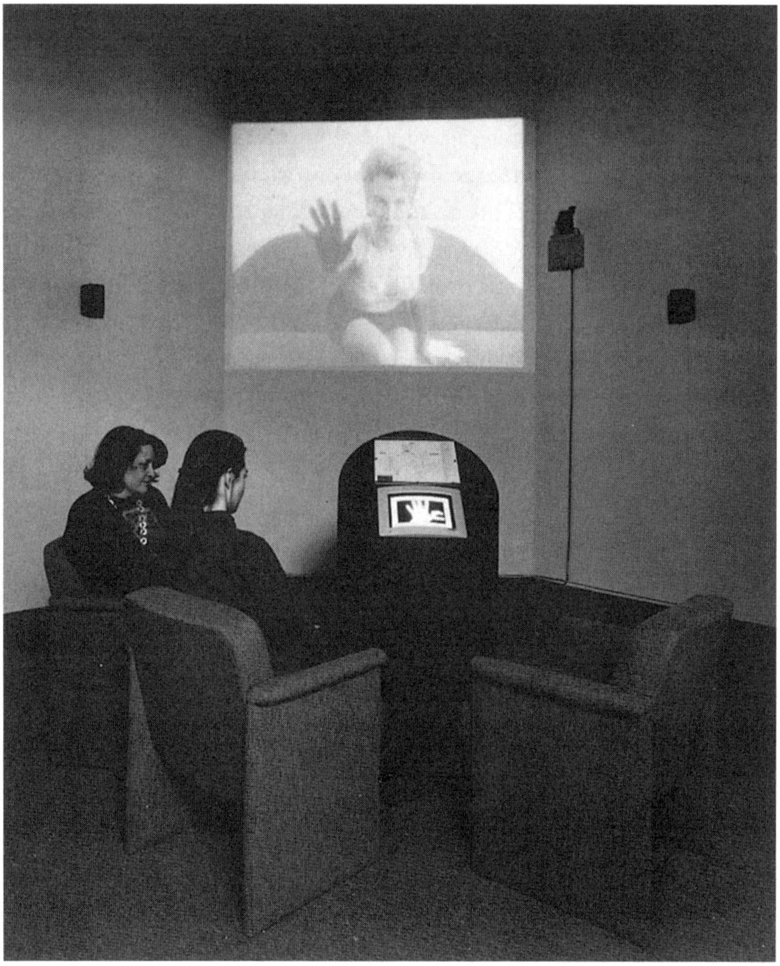

Figure 12.2
Lynn Hershman, *Deep Contact,* 1990, video installation, San Francisco Museum of Modern Art.

At certain points, viewers can see details of what they have just passed. For example, Marion runs past a bush that, examined closely, reveals a spider weaving a web. This allows new perceptions of the same scene, depending on the speed at which it is seen.

In some instances, words flash on the screen for just three frames, forcing the viewer to go back and view at a slower speed to see what has been written. At other points, the Zen Master speaks his lines backward, forcing the viewer to play the disc in reverse to understand what the master has said. The Demon and the Zen Master are played by the same actor, indicating different aspects of personality and suggesting that the same event can appear frightening or enlightening, depending on the context in which it is seen.

A surveillance camera is programmed to be switched on when a camera operator's shadow is seen. The viewer's image instantaneously appears on the screen, displacing and replacing the image. This suggests a "transgression of the screen"—being transported into "virtual reality."[6]

A Room of One's Own (1990–1993)

Constructed to reference early "peep shows," *A Room of One's Own* (with Sara Roberts) is a computer-operated installation that operates by triggering the viewer-voyeur's eye movements.[7] The viewer looks into a small, specially constructed bedroom scene. A tiny video camera digitizes the viewer's eye movements and sends the signal to the computer, which causes the videodisc to access particular segments.

In the tiny bedroom are several objects—bed, telephone, table, and television. The viewer-voyeur accesses modular video segments. *A Room of One's Own* also responds to the viewer's physical presence: there are audio sensors beneath the mat that the viewer stands on. When a viewer is on the mat, sounds and words are "spoken" to the viewer. Furthermore, the camera allows images of the viewer's eyes to be inserted onto the television and become part of the scene.

USING TECHNOLOGY TO CREATE COMMUNITY

Voyeurism is a strong element in my work. Information about how we define ourselves is really information about how we see ourselves; how we see ourselves depends on the point of view from which we are seen.

It is a cycle. I do not believe that other artists have dealt with this cycle of how external views of women (including media views) influence women's internalized views of themselves. In *A Room of One's Own,* I place the viewer as voyeur literally in the center of the picture.

This placing of the viewer at the center renders all other movement peripheral. Furthermore, the cycle can be seen from several perspectives, each with a subtle variance of a perception of truth based on that singular viewpoint. This combinination of shifts in perception allows meaningful overlaps to emerge and form a relevant overview.

In a world addicted to dreams, there is a danger of "interiority complexes." What is important is the implications that new technology offers. Mistakes will be made, and no doubt there will be chain reactions to the errors. Media have caused a confusion of truth and fiction as well as of good and evil. By communicating directly in an interactive community, we may be able to relearn the idea of trust, a term that has itself become suspect. We may discover ways in which we can share dreams and so create a community of enhanced values driven by a vision that though we may be terminal, we remain connected and capable of creating an enhanced and shared environment.

NOTES

1. Amos Vogel, *Film as a Subversive Art* (New York: Random House, 1974), 11.

2. Ann-Sargent Wooster, "Reach Out and Touch Someone: The Romance of Interactivity," *Illuminating Video* (New York: Aperture Press, 1991), 285.

3. Donna Haraway, "Manifesto for Cyborgs: Science, Technology, and Social Feminism in the 1980s," *Socialist Review* 80 (1985): 72.

4. *Lorna,* 1980–1984, interactive laser video art disc, edition of 20, interactive Level I videodisc produced in association with Texas Tech and Art Com, programmed by Ann Marie Garti, music by Terry Allen.

5. *Deep Contact,* 1985–1990, touch-sensitive interactive videodisc installation, pressed in glass, limited edition of 3, programmed by Sara Roberts.

6. The videodisc is an ODC glass DRAW disc composed of about sixty segments. It plays on a Pioneer LDP 6000 player that can be driven by the computer over a serial port.

The particular driver used in the stack is an XCMD external command written and compiled in LightSpeed C by Jim Crutchfield. The user interface was designed and programmed by Sara Roberts. The program exists as a single 300K stack in HyperCard in which each segment of the videodisc represents a card in the stack. Graphics were scanned using HyperScan and altered using SuperPaint, both Macintosh graphics programs. The HyperCard program runs on an Apple Macintosh IIcx with 2 megabytes of random access memory (RAM). The monitor is a 13-inch Microtouch Touchscreen. The HyperCard program works on most Macintosh computers and can be genlocked to a disc player or a compact-disc–video player, or it can be used alone. It can access moving or still images, has a wide range of sound capabilities, and is relatively inexpensive.

The process of making this piece spawned the collaboration of many people. John Di Stefano had the unwieldy task of composing music that would work in modulated segments, as well as backward, forward, and in slow motion. Jiri Vsneska contributed immeasurably with the shooting and scanning of photographic images, and Marion Grabinsky, the leather-clad protagonist, gave the piece the erotic appeal so necessary for sexual transgressions. Toyoji Tomita played the Zen Master and Demon with equal charm, while the crew of camera operators, editors, and production managers added tremendously to the success and joy of making this piece.

7. A *Room of One's Own,* 1990–1993, interactive videodisc installation, edition of 1, operates with a hypertrol, miniprojector, and Pioneer LDP 5000 videodisc player, programmed by Sara Roberts.

REFERENCES

Baldwin, James. *Notes on a Native Son.* New York: Bantam, 1955.

Benjamin, Walter. *Illuminations.* New York: Schocken Books, 1969.

Fast, Julius. *Body Language.* New York: Evans, 1970.

Freud, Sigmund. *Civilization and Its Discontents.* New York: Norton, 1961.

Lifton, Robert J. *History and Human Survival.* New York: Random House, 1961.

All the works of Emma Goldman.

All the works of Henry Miller.

13

Bicycle TV: Expo '92 Installation

Nancy Paterson

A North American obsession with simulation and pseudo-realities is evident in the increasing popularity of science fiction (literature and screenplays) and high-budget amusement theme parks. No longer satisfied with mere descriptions or depictions of utopias and alternate realities, North Americans apply computer technology to the creation of simulated environments in which we may physically participate. Video (games and television) was once considered the most powerful and compelling mass medium. The sociological effects of video have been the subject of numerous studies and remain a controversial topic of debate. Measuring the influence and effects of interactive virtual-reality environments on mass culture will doubtless prove to be an equally challenging and difficult task.

Despite its successful applications in science, medicine, and architecture, virtual reality will be judged a success or failure by popular culture based on its applications in the amusement and entertainment industries. Tourism and pornographic publishing are the first industries to adapt this new technology to their needs. I have developed a virtual-reality system that explores tourist applications of this technology and I hope will provoke serious thought regarding the current direction of commercial virtual-reality research and design.

Consumer applications of virtual-reality technology in amusement and entertainment industries must be critically evaluated. This is an important step toward understanding our hopes and expectations for this new technology. The sociological and psychological effects of simulated travel and experience (unlimited by three-dimensional space and linear time) must also be considered in any analysis of this revolutionary technology.

As a video-touring device, my interactive installation *Bicycle TV* demonstrates a new approach to the design of virtual environments. Selected for exhibition in the Canadian Pavilion at Expo '92 in Seville, Spain,[1] this installation simultaneously showcases creative applications of videodisc technol-

---

This classic paper, from *Leonardo* 25, no. 2 (1992), documents some of Nancy Paterson's seminal work in interactive art. Since this paper was written, she has produced many other works, including *Stock Market Skirt* (1998); *Coppelia* (2001); and *The Library* (2002), a project that transforms an existing interactive 3-D environment from a self-contained experience into a meta-data creative reference tool utilizing dynamic 3-D objects.

ogy and also features and highlights the small town of Bracebridge and the beautiful fall colors of the foliage in the Muskoka region of Ontario.[2]

The primary difference between my installation and many of the systems being designed by companies such as VPL Research Inc. or Fake Space Labs, California, is that I utilize real-time, full-motion, prerecorded video imagery to place the viewer in an authentic rather than digital landscape. My rejection of computer-generated graphics was deliberate; as an independent artist I did not pursue the state-of-the-art tools available to companies designing and marketing similar systems. In addition, my system was available for a reasonable price because it used available technology, such as the industrial videodisc player.[3]

As seen in amusement theme parks, virtual-reality systems offer predetermined and collective (although not yet interactive) experiences. The market for personal virtual-reality systems, meanwhile, is anxiously anticipating an improved level of image sophistication at a more moderate price. My video imagery, which employs natural landscapes, provides an alternative to the artificial environments presently offered. *Bicycle TV* also provides the viewer with an intimate experience of interactive technology without the isolation of a head-mounted display or specially designed clothing. By mounting and pedaling a stationary, 1950s-style bicycle, the participant activates video of winding, country roads displayed on a large 50-inch color monitor (or video wall) that faces the rider (figure 13.1). The speed of the video playback is controlled by the rear-wheel motion-detector assembly and is directly determined by the cyclist's pedaling speed. Use of a multispeed videodisc player allows video that was shot at approximately 17 miles per hour (mph) to be slowed by the rider to less than 10 mph or increased to 35 mph. The direction of the video is controlled by the turning of the handlebars at all intersections and crossroads where a choice of direction is indicated by arrows superimposed on the video. The front-wheel direction-sensing device sends data to the controller, which is programmed to select the appropriate video. The rear-wheel custom-designed optical sensor counts pulses via black markings on the rear wheel of the bike. I rejected the use of momentary contact closures for both the front-wheel direction-sensing device and rear-wheel speed detector in order to improve reliability in the field.

Both the front-wheel and rear-wheel sensors are linked to the videodisc player through a custom-designed control system. A 2-inch-by-3-inch cir-

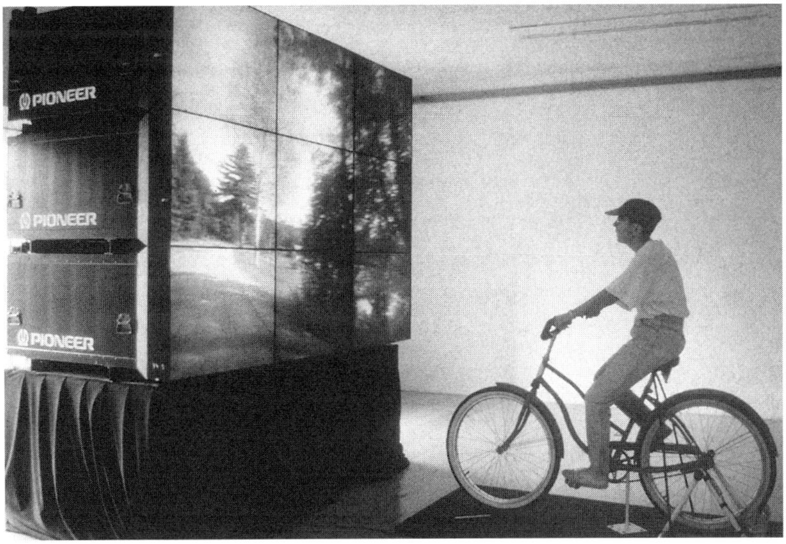

Figure 13.1

Nancy Paterson, *Bicycle TV,* 1991, interactive videodisc installation. The viewer is provided with an intimate experience of interactive technology without the isolation of a head-mounted display or specially designed clothing. By mounting and pedaling a stationary, 1950s-style bicycle, the participant activates video of winding, country roads displayed on a large 50-inch color monitor (or video wall) that faces the rider.

cuit board contains a Z8 microchip that reads both data and instructions from the 16 kilobytes of the Erasable Programmable Read Only Memory (EPROM), which is also on the board. The RS232 communication driver enables control of the videodisc player. For the purposes of the EXPO '92 installation in Seville, Spain, it may be necessary to install a joystick user interface to keep the system functioning under expected heavy use by a large number of viewers.

It is possible for the cyclist to explore many different roads and paths within the "world" offered by each videodisc that I have produced. Although there are a limited number of roads on each disc, the program is constructed so that the cyclist may continue to ride indefinitely. There are no beginnings or ends to the video programs, and loops have been constructed to prevent the rider from falling off the "edge" of each disc.

The scenery was selected for its beauty and pastoral effect. Two channels of Digital Video Effects (DVE) were used extensively in postproduction to produce special effects highlighting scenes of special interest via display of a "picture within a picture" in one or more corners of the video screen.

At the preproduction stage, I developed a branching system. First, road characteristics and features (gravel, pavement, houses, wooded lots on either side of the road, etc.) of the video footage were noted for possible match-ups. Once the required number of roads were scouted and logged, a Betacam SP master videotape was produced, utilizing numerous source tapes of roads and turns approached from all directions. Arrows were overlaid at the appropriate decision points, and special effects were incorporated. The Betacam SP master was transferred to a Constant Angular Velocity (CAV) level-one videodisc, and final planning for programming was done. I then gave the videodisc, a series of seemingly disjointed roads and turns, to a programmer, along with detailed notes outlining the necessary logic of the program to be written.

The interactive nature of my videodisc installation invites comparison with the construction of labyrinths and mazes. Through electronic technology, *Bicycle TV* updates a concept that has ancient and, indeed, primordial origins. Offering a multitude of choices and possible paths, *Bicycle TV* is most appropriately described as a maze. The labyrinth, by definition, allows only a single, uninterrupted trail traversing the entire interior of its space. Historically, however, *maze* and *labyrinth* have been used interchangeably, and *Bicycle TV* combines the challenges of both.

The maze has traditionally been associated with spiritual initiation and personal triumph. The process of solving a maze has been a vehicle for profound growth and transformation. *Bicycle TV*, however, holds no sacred beast of power such as the Minotaur housed in the legendary Cretan maze of Daedalus. No length of thread or trail of bread crumbs, for that matter, will lead participants out the same way from which they came. This maze has no center and therefore no possibility for a complete reversal of direction, which can symbolize rebirth.

The Roman ceremony that took place on horseback in a labyrinthine pattern at the founding of a city demonstrates the archetypal characteristics of the maze and the purpose that they serve. Taking into account the process of delineation, self-recognition, and self-definition, whether of an individual, a community, or a new technology, it is no accident that the pattern of the maze reappears in a virtual-reality environment such as *Bicycle TV*. The maze presented in *Bicycle TV* can serve as a metaphor for the challenges and opportunities offered by new electronic technology.

My experience of producing and exhibiting *Bicycle TV* has resulted in a greater awareness on my part of the issues and questions that will become increasingly relevant as more people are directly exposed to virtual-reality environments. In exploring the creative potential of virtual-reality systems, artists such as myself have the opportunity and obligation to address related issues such as physical control and psychological manipulation in the electronic age. If our greatest challenge thus far has been the development of the technology to make these systems possible, then our next challenge will relate to the use and commercial application of these systems.

NOTES

1. The dates of Expo '92 are 20 April to 20 October 1992.

2. Two videodiscs are currently available for use with this installation.

3. I selected the Pioneer LD-V8000 videodisc player for a number of unique features, especially its design for industrial use. A digital memory ensures that the video program is not interrupted during search periods—for the very short search periods (1/4 second maximum duration) that a digitized image is held on the screen. As previously described, the multispeed feature of this VDP is utilized in this installation. This player also has four available channels of audio.

14

Acoustic and Virtual Space as a Dynamic Element of Music

Pauline Oliveros

As a musician, I am interested in the sensual nature of sound, its power of release and change. In my performances throughout the world, I try to transmit to the audience the way I experience sound both when I hear it and when I play it. I call this way of experiencing sound "deep listening." Deep listening is listening in every possible way to everything possible: this means one hears all sounds, no matter what one is doing. Such intense listening includes hearing the sounds of daily life, of nature, and of one's own thoughts, as well as musical sounds. Deep listening is my life practice.

In this chapter, I attempt to relate my experiences as a performer and composer to the development of my interest in acoustics and technology. I have been involved with the use of technology for live performance since the late 1950s. I will review several of my pieces as they relate to the development of the Expanded Instrument System (EIS), described below, which I use in composing both my solo and ensemble performances with the Deep Listening Band. I will discuss my interest in acoustic and virtual space. Virtual acoustics—a perceptual phenomenon—is created with electronic processing within an actual physical space. Simulated walls or reflective surfaces may cause a listener to perceive differences in room size and the tone quality of a musical instrument.

DISCOVERING ACOUSTIC SPACE

Early in my student career as a performer, I noticed that I liked playing in some rooms or concert halls much better than in others. When I played my French horn in a dry hall (a room with little or no reverberation), the sound felt stuffy, and it seemed harder to play. I would wonder why my tone quality seemed to sound poor or very thin in such rooms. More

---

This paper, originally published in *Leonardo Music Journal* 5 (1995), documents some of composer/musician Pauline Oliveros's seminal work. Since this paper was written, her interest in live electro-acoustic improvisation and composition has continued, with Expanded Instrument System (EIS) as the primary tool. Her recent work has included *The Library of Maps An Opera in Many Parts* 2002, with text by Moira Roth—digital arts; *Red Shifts* (2000), for solo trombone, oscillators, and electronics, and *Primordial/Lift* (1998), for ensemble oscillator and electronics.

reverberant rooms always felt better—especially rooms with wooden floors and walls as the reflective surfaces. My tone was fuller, richer, and rounder, and it was easier to play in these rooms. I learned about resonance, reflections, reverberation, and how to play in a room through these experiences. This situation was similar to one I was to experience later as a composer when I worked with electronic sound. When I used dry sounds (sounds recorded directly from the source without a sense of room space), reverberation of some kind was needed to make the sounds seem more musical. In my first attempts at composing tape music in the late 1950s, I used the resonance and reflective surfaces of my bathtub as an approach to help solve this problem. I recorded small acoustic sound sources with the microphone in the (empty) bathtub.

Music, as I understand it, is played in acoustic spaces. Concert halls, theaters, and cathedrals all act as mechanical amplifiers, which, by their architectural design, capture the sounds of voices and instruments and impose resonances, reflections, and absorption that color the sounds. Instrumental and vocal sounds are enhanced or distorted by these mechanical amplifiers, depending on the nature of the sound and the purpose of the design. Resonance, reflection, and absorption are determined by the relationships and materials of the enclosure as well as by environmental factors such as air temperature and humidity.

Particular styles of music are often associated with preferred architectural designs. Music that is intended for reverberant cathedrals, such as Gregorian chants, may not sound well in dry halls, whereas contrapuntal music needs a dryer and smaller hall for clarity of all the contrapuntal lines. Good musicians adjust their performances to the nature of the hall as best they can. Seasoned audience members seek out the seats where the balance of direct sound and reflected sound is the most pleasing. Generally speaking, the architectural acoustic space (concert hall) is assumed to be fixed, with relatively unchangeable characteristics. Harmonies, melodies, rhythms, and timbres change in more or less intricate relationships, while the acoustic space does not change; it is the container of the music. As my experience of numerous performance spaces accumulated, I began to wish for the possibility of changing the acoustic space while performing. I also wished that I could hear as if I were in the audience while I was performing for it.

With the advent of signal processors and sophisticated sound systems, it is possible to tamper with the container of music in imaginative ways. The walls of an electronically created virtual acoustic space can expand or contract and assume new angles or virtual surfaces. The resulting resonances and reflections, which change continuously during the course of a performance, can be used to create spatial progressions, much as one creates chord progressions or timbre transformations by changing the tone quality of an instrument while performing a single pitch. The audience and performers can experience sensations of moving in space as well as perceiving sounds moving through space. They can also experience the relationship of moving in space in relation to sounds moving in the same space while the space itself is changing. Such audio illusions or virtual acoustics can function as a new parameter of music, much as timbre became new with the advent of *Klangfarben Melodie* (tone-color melody), in which the notes of a melody are distributed to different instruments successively (as in the music of both Arnold Schönberg, who coined the term, and Anton Webern).[1]

As I gradually became more and more sensitized to acoustic phenomena and its effects on my sound as a performer and composer, I began to listen carefully to each space I encountered. I noted that changes in the position, height, or direction of an instrument could affect the tone quality by enhancing it or detracting from it. I worked back and forth between acoustic and electronic sound.

TAPE DELAY

Beginning in the early 1960s, I strung tape from the supply reel of one machine to the take-up reel of another so that the tape passed the heads of two to three tape machines. This tape-delay process allowed me to return the signal from the playback head to the record head in a variety of configurations to create various accumulations of layering, echoes, and rhythms. I used these techniques to enhance electronic sounds and process sounds made with acoustic instruments.[2]

A seed idea appeared in my early work *The Bath* (1966) for Dancer's Workshop in San Francisco (Anna Halprin, director). I wanted to create the music for this dance out of the intentional and unintentional sounds made by the dancers during the course of the performance. I wanted to change the acoustic characteristics of the performance space as well. To accomplish this, I used several tape recorders to collect the sounds the dancers made during the first part of the piece without introducing any other sonic elements. There was no accompaniment other than the dancers' own sounds and random room sounds. For the second part of the piece, I used feedback from the playback head to the record head of the tape recorder to create a virtual reverberant space that seemed to grow gradually in size as I increased the amplitude of the feedback. At the same time, I continued to record the sounds of the dancers. For the third part of the piece, I played back earlier parts of the piece while selectively overlapping segments of the earlier parts of the piece onto this new version of the space. Repeated material was transformed by the virtual space and the layering. The density and textures grew increasingly complex as the reverberations colored the materials.

I worked extensively with tape delay and became quite involved with timbral transformation in real time with delay and mixing techniques. I discovered that repeated layering of a single tone contributed to a transformation of its quality. *I of IV* (1966) and *Bye Butterfly* (1965) are recorded examples of real-time electronic pieces from the 1960s that came out of this involvement.

THE EXPANDED INSTRUMENT SYSTEM

After using tape-delay systems for some years for live acoustic instrument performance, in 1983 I acquired a pair of pre-MIDI Lexicon PCM 42 digital-delay processors. These instruments gave me the opportunity to apply my knowledge of tape-delay techniques to my solo accordion performances, as well as the capacity to smoothly vary settings in real time with greater ease. The PCM 42 is a real-time performance instrument. Partly analog in its nature, it has foot pedals that the performer can use to change delay time, allowing the bending of delayed sounds—a phenomenon not easily possible to achieve with tape delay (figure 14.1). Other functions, including mix control, feedback, and capture, can be performed by use of the pedals as well. Modulation with sine and square waves can be handled with front panel knobs.

Unfortunately, the performance-oriented direction of this processor was discontinued by Lexicon. Although I still use it, the PCM 42 is no longer manufactured.

I used a PCM 42 delay processor for each hand while playing the accordion and quickly wanted to have multiple delays for each hand. Multiple delays can more nearly simulate the numerous reflections in an acoustically interesting space. I performed using a variety of configurations, sometimes with as many as four processors per hand. I liked the results

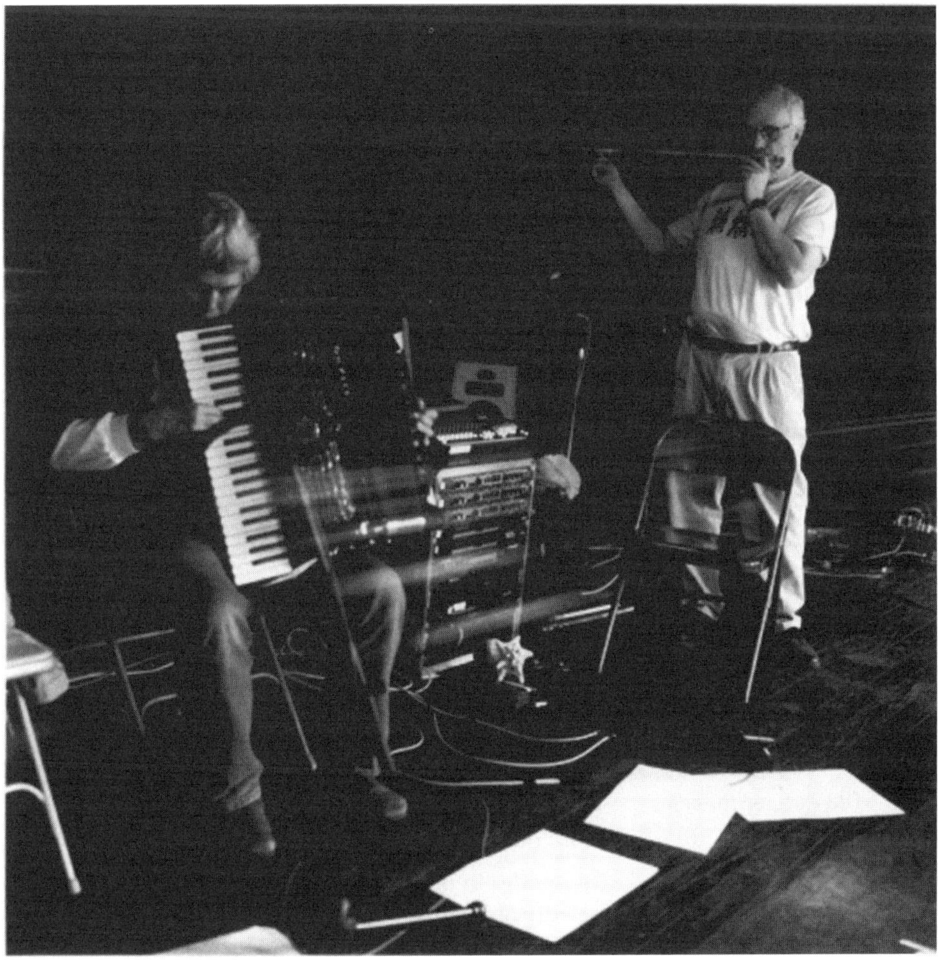

Figure 14.1
Pauline Oliveros and David Gamper performing with PCM 42s, Austin, Texas, 1994.

and called the accumulating collection of microphones, amplifiers and signal processors that I manipulated in complex networks through matrices and mixers the Expanded Instrument System (EIS) (figure 14.2).[3]

I wanted more. I wanted control over the apparent acoustic space represented by the multiple delays. I wanted my maneuvering of the delays of direct sound to be heard as new acoustic spaces. I wanted to stay and dwell in a selected space or change spaces as rapidly as the limits of the processors allowed, knowing that the changes would color the sound or transform the timbre of any acoustic instrument utilized. I was interested in hearing how a bat might perceive sound as it sends out signals on the fly or in listening to the nuances of whale song as it reverberates in great underwater locations. I loved the challenge of increasing the amount of information with which I could deal as a soloist. Whatever I played could come back as a delay and be layered with the present moment or be modified by pitch bending or modulation. At the same time, I had to anticipate the future of the sound that I was making in the present moment.

My exploration as a soloist trying to control the equipment while performing reached an early limit. It became necessary to engage others in the process of developing the EIS. I knew that I would need software and controllers to satisfy my imagination and advance the EIS. From 1987 to 1993, Panaiotis, a young composer and musical engineer, helped in the development of the EIS. My direct and processed sound was further processed and routed to multiple speakers by Panaiotis during performances. We explored endless equipment configurations: different configurations offered varieties of sonic possibilities. This work resulted in two compact discs: *The Roots of the Moment* (1987), issued by HatArt, and *Crone Music* (1988), issued by Lovely Music Ltd.[4]

During two development residencies, I focused on control possibilities for the EIS. In 1989, at the Exploratorium in San Francisco, Panaiotis and I experimented with a foot-controlled track ball to send control information to the PCM 42s. Larry Shaw of the Exploratorium designed an interface circuit between the track ball and the PCM 42s for direct control of the PCM 42s. The track ball worked in our demonstration, and we were able to send control information on the x and y axes so that more than one function would be available simultaneously. The track ball ultimately proved to be impractical. We hoped to have more functions available to the performer with a computer interface.

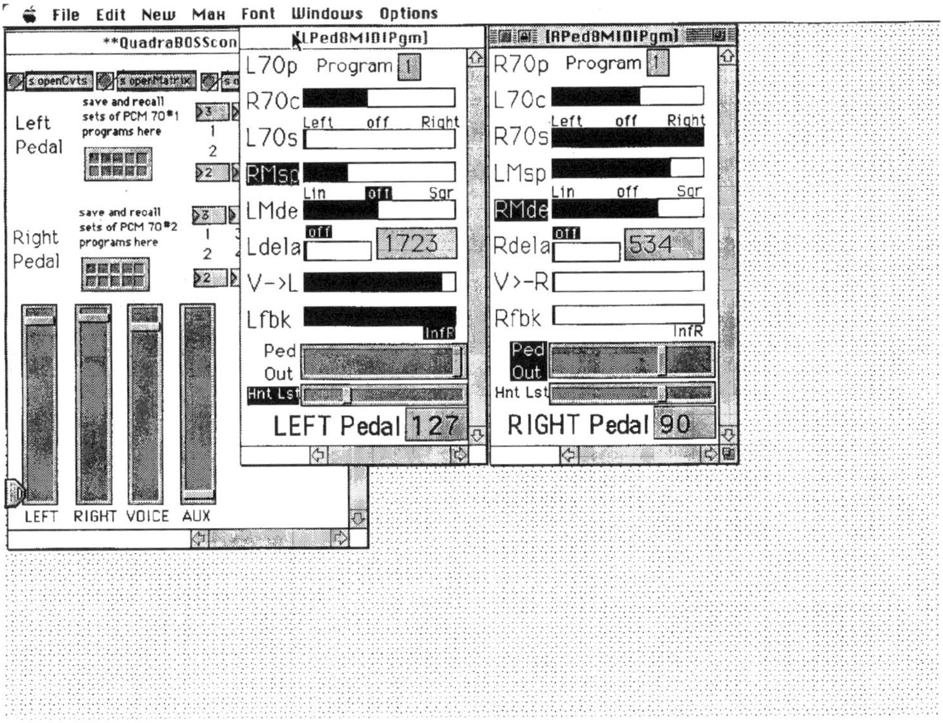

Figure 14.2

Computer screen read-out. Multifunction pedal programmed by David Gamper using MAX for the Expanded Instrument System. Each function is available by foot switch to a single pedal operated by a performer. The amount of each effect can be set and remembered by the program. A different function can then be accessed.

During a residency in 1990 at the Banff Centre in Alberta, I worked with computer programmer Cornelia Colyer to develop simulated super foot-pedal program. We used a custom digital interface for direct program control of the PCM 42s, constructed by David Ward of Pan Digital, Inc., and based on Shaw's circuit. I wanted to have a program that would simulate and then go beyond what I could physically do with my foot pedal. The resulting program was used in a piece for four performers called *The Lightning Box*. Sound output from each player was processed by the computer-controlled digital delays. The delay times were changed by the program, effecting transposition and pitch bending in a variety of forms and speeds that would have been impossible to accomplish with the slower foot pedal. The piece calls for the players, as they are listening to their sounds being repeated and modified, to respond to the modifications as well as to each other. Each player could also affect the sound to a limited extent with a foot pedal. Individual players' entrances and exits were cued by a computer-controlled lighting design as part of the score. The lighting design was programmed by Colyer with my advice.

THE DEEP LISTENING BAND

In 1988, Stuart Dempster invited me to record with him in an underground cistern at Fort Worden in Washington state. Dempster and I have shared an interest in unusual acoustic spaces for many years. Dempster's solo trombone recording in the *Great Abbey of Clement IV*[5] is an underground cult classic. The reverberation time in the Great Abbey of Clement is 14 seconds. The Fort Worden cistern has a reverberation time of 45 seconds. Dempster has clocked certain resonant frequencies responding to his trombone at 72 seconds. When I played in the cistern, my impression was that I had encountered the smoothest reverberant chamber ever. It was nearly impossible to distinguish direct from reflected sound. Panaiotis joined us with his voice in the recording session, which resulted in the compact disc *Deep Listening*.[6] We dubbed this trio the Deep Listening Band and took on the mission of performing in unusual spaces and trying to simulate these spaces with the EIS in our concerts. This also meant a continuing research commitment to evolving the EIS.

David Gamper joined the Deep Listening Band in 1990 and begin to help with research for the EIS. The experience of processing more than one instrumentalist presented a new challenge. Panaiotis and Gamper were the central controllers of the processed sounds. There was a growing need for all the performers to independently control their own processing, as I had done earlier. Panaiotis and Gamper experimented with widening performer control through a central computer, which was also being used to control specialized digital and analog signal processing. The customized interface constructed for the performance of *The Lightning Box* at Banff was used again to interface the veteran PCM 42s to the computer. This system was used in several Deep Listening Band concerts and for *Inside/Outside/Space* (1991), a piece commissioned by Nouvelle Ensemble Moderne of Montreal (Lorraine Vaillancourt, director) for fifteen players.

Inside/Outside/Space instructs the performers to listen for subtle acoustic phenomena—the effects caused by the sounds they are playing and the nature of the performance space. The performer causes and interacts with beats between notes, difference tones (tones resulting from the difference between two frequencies), phasing, and other variations. This continuous meditative interaction produces cloudlike textures that move throughout the hall, giving the audience a sense of expanding and contracting spaces. All of the sound is generated by the acoustic sounds of the performers in real time. Each player effects real-time transformations of his or her sounds with foot pedals supported by a computer program helping to access different functions. During a performance in September 1992 at the Glenn Gould Studio in Toronto, spatial distribution of the acoustic sounds was accomplished by Gamper and Panaiotis at the mixing consoles and by the conductor, who also used foot pedals.

The usefulness of this system was limited by the paradigm of centralized signal processing (although we provided somewhat more distributed control) and the lack of visual feedback for the performers.

FURTHER EXPANSION

In 1993, Gamper developed a configuration that distributed audio processing, computer control, and visual display of controlling parameters among all performers (see figure 14.2). In addition, he experimented with ways of allowing the performer to control many parameters with one controller. Although this configuration mostly eliminates the option of using individual processing units for more than one performer, it returns total control over the processing to the performer. In addition, it puts many of the interconnection options of all the processing units under real-time performer control. Without eliminating the possibility for each performer to send audio signals to each other, this configuration provides each performer with exclusive control over his or her own sound processing up until the moment the sound emerges from the speakers into the performance space. In an interesting regression, this configuration returns the PCM 42s to their original smooth and beautiful analog control and features them as front ends to the whole signal-processing chain. This decision arose from dissatisfaction with the "sound" of the PCM 42s under digital control. Development of a next-generation interface may finally allow the best of both worlds.

Areas of future EIS development include

- Improving hands-free performer controller options (we are currently stuck with foot pedals),
- Adding the capability to record and play back performer-control information,
- Exploring sharing performer control with computer control,
- Increasing interconnection options, allowing the use of more sophisticated audio and digital matrixing devices, and
- Continuing to discover beautiful ways to process live sounds in real time and to shape virtual space as a dynamic element of music.

NOTES

1. For example, see Schönberg, *Five Orchestral Pieces,* opus 16 (1909, rev. ed. 1949); and Webern, *Five Pieces for Orchestra,* opus 10 (1913).

2. See Pauline Oliveros, "Tape Delay Techniques for Electronic Music Composers (1969)," *Software for People: Collected Essays* 1963–1980 (Baltimore, MD: Smith Publications, 1984).

3. Pauline Oliveros and Panaiotis, "Expanded Instrument System (EIS)," *Proceedings of the International Computer Music Conference* (San Jose: ICMA, 1992).

4. See Pauline Oliveros with Panaiotis, *The Roots of the Moment* (composed 1987) (Therwil, Switzerland: HatArt, 1988; and Pauline Oliveros with Panaiotis, *Crone Music* (composed 1988) (New York: Lovely Music, 1989).

5. Stuart Dempster, *In the Great Abbey of Clement VI* (composed 1976) (San Francisco: New Albion, 1987).

6. Deep Listening Band, *Deep Listening* (composed 1988) (San Francisco: New Albion, 1989).

BIBLIOGRAPHY

Beranek, Leo L. 1962. *Music, Acoustics and Architecture.* New York: Wiley.

Helmholtz, Hermann von. 1954. *On the Sensation of Tone.* Rev. ed. New York: Dover. (Originally published in London in 1875)

Hutchins, Carleen Maley. 1978. *The Physics of Music: Readings from Scientific American.* San Francisco: Freeman.

Schönberg, Arnold. 1911. *Harmonielehre.* Leipzig: Universal.

DISCOGRAPHY

Deep Listening Band. 1989. *Deep Listening* (composed 1988). San Francisco: New Abion.

Deep Listening Band. 1995. *Deep Listening Sanctuary* (composed 1994) (Mode 46). New York: Mode Records.

Deep Listening Band. 1991. *Ready Made Boomerang* (composed 1990). San Francisco: New Albion.

Deep Listening Band. 1995. *Tosca Salad* (composed 1992–1995). (DL 3). New York: Deep Listening.

Deep Listening Band. 1990. *Troglodytes Delight* (composed 1990). Albuquerque: What Next?

Stuart Dempster. 1987. *In the Great Abbey of Clement VI* (composed 1976). San Francisco: New Albion.

Pauline Oliveros. 1968. "I of IV" (composed 1966), included in *New Sounds in Electronic Music* (32 16 0160). Odyssey Records (out of print).

Pauline Oliveros. 1979. "Bye, Bye Butterfly" (composed 1965), included in *New Music for Electronic and Recorded Media.* Berkeley: 1750 Arch Records (out of print).

Pauline Oliveros with Panaiotis. 1989. *Crone Music* (composed 1988). New York: Lovely Music.

Pauline Oliveros with Panaiotis. 1988. *The Roots of the Moment* (composed 1987). Therwil, Switzerland: HatArt.

15

"I Always Like to Go Where I Am Not Supposed to Be"

Rebecca Allen with Erkki Huhtamo

INTRODUCTION

The yet unwritten history of computer graphics and animation would be unthinkable without the contributions of Rebecca Allen. The Detroit-born artist and designer began to explore the possibilities of digital technology in the early 1970s, when most people hardly knew what a computer was. During her prolific career, Allen has tried her hand on a great variety of different approaches to computer imaging—still graphics and animation, logos for TV, music videos, video games, large-scale performance works, artificial-life systems, interactive installations. The driving force behind Allen's work is her passionate relationship with innovative technology. Where there is something new, most likely there is Allen as well, exploring yet unused creative possibilities.

The urge to be in the forefront of technological development has taken Allen to such highly regarded institutions as the Rhode Island School of Design, the Massachusetts Institute of Technology's Architecture Machine Group (the predecessor of today's MIT Media Lab), the Computer Graphics Lab at the New York Institute of Technology (CGL/NYIT), and most recently to the University of California, Los Angeles (UCLA), where she was founding chair and continues as professor at the Department of Design/Media Arts. In this interview, Allen talks about her career, her ideas about computer imaging, and her relationship with the various worlds she has been affiliated with—the research communities, popular culture, the video game industry, and the art world.

Erkki Huhtamo: How did you first get involved with computers, technology, and art?

Rebecca Allen: I became interested in technology as an art student at Rhode Island School of Design (RISD) in the early 1970s. I studied the art and technology movements of the early twentieth century, such as the Bauhaus, the Futurists, and the Constructivists. I was very interested in the idea of artists exploring the latest technology and trying to help society understand its relationship to technology. I was also intrigued by the idea of artists using new technology as a tool to create new art forms. As I began to think about the technologies of our time and their impact, I recognized that computers would become an important part of society and could be important tools for artists as well.

In trying to push the boundaries of traditional art, I became interested in the use of motion. Initially, I was studying graphic design—this was about 1972 or 1973—but I was not interested in static images. I wanted images to move, to come to life. My role models were not cartoon animators but Duchamp and other artists who had experimented with motion in the early 1900s.

A few things at RISD really inspired me. One was a lecture and film showing work by John Whitney Sr., a pioneer in computer graphics. I also had a professor, David Brisson, who talked about four-dimensional space. The idea of mathematics and multiple dimensions intrigued me. Then, in 1974, I saw a stereo computer-generated film that visualized four-dimensional rotations of hypercubes. It was made by Tom Banchoff and Charles Strauss, professors, respectively, in mathematics and computer science at Brown University, which was right next to RISD. This convinced me to explore computer animation.

EH: How did you come to use the computer yourself?

RA: I changed my field of study to film animation and video in order to explore the aspect of motion. I talked Charles Strauss into sponsoring an independent study; I made up a course called computer animation that I proposed to RISD through Brown University. My RISD teachers were very skeptical about the idea, saying artists didn't need to work with this kind of technology, but they agreed to let me go ahead, so I began to work with a computer called a Vector General. It didn't have a frame buffer yet, so everything was drawn as vector lines: that was the state of the art of computer animation then. Programmers had developed an interpolation program for 2D animation. I created animation using graphs and punch cards: a keyboard interface didn't exist then. This was how I made my first computer animation in 1974.

EH: Can you describe what it was like?

RA: Actually, I still have the drawings from it. I was interested in the idea of rotoscoping, a technique where you examine a film, a frame at a time, and transform it into animated drawings. I used the natural movement of the human figure, but I abstracted and stylized the imagery through animated drawings. I somehow acquired a number of erotic films from the 1950s, so I transformed a film of a woman lifting up her dress as she danced around. That was my first computer animation. I did it because I was struggling with the sense that the computer was antihuman.

I wanted computer art to be something sensual and slightly erotic. This seemed the antithesis of what a computer was supposed to be.

EH: So it seems to me that the computer appeared to be the perfect tool to realize the kind of things you were already interested in.

RA: Right, at that point in my life I really wanted to define where to go as an artist and a designer. My dream was to create my entire animated work independently without following the industrial, assembly-line structure that had been established for traditional animation. As an artist working with state-of-the-art technology, I could, hopefully, have some kind of an impact on its evolution. I probably also liked the idea of being a woman working in a field that was very male. It still is that way, but certainly back then it was much more so. I could be the bad girl in that field.

EH: How did people react to a female student entering the computer laboratory and wanting to explore digital imaging?

RA: I think they were amused. When I did my first piece, the one with the woman lifting up her dress, that upset some of the computer science students. I guess I was oblivious to their reactions because I was on a mission to explore the artistic side of the computer.

EH: You spoke about your first computer animation. What happened next?

RA: After I graduated from art school, I worked as a graphic designer for a furniture design company, but it became clear to me that I didn't want to be a graphic designer. I really wanted to explore technology. I spent a good deal of time looking around for places where anything was going on with computer animation. I called the NYIT computer graphics lab (they had just started), MIT's Lincoln Lab, and Bell Labs (where Ken Knowlton was working). I also knew about MIT's Architecture Machine Group (Arch Mach) and Ohio State University. There were so few places to go at that time.

I attended MIT as a special student enrolled in a course that was co-taught by Nicholas Negroponte. It was one of the very early computer graphics courses. A graduate program was being developed, but it wasn't yet open for admissions. After completing the class, I continued to spend a lot of time in the Arch Mach lab. In the meantime, I applied for the graduate program and was accepted. I am forever grateful to Negroponte for taking a risk and allowing me to enter his program.

EH: Was there any kind of a network connecting the people interested in computer graphics and animation at that time?
RA: There was a small research community of computer scientists who were somewhat familiar with each other's work. There was only a handful of people interested in computer art.
EH: Many people have emphasized the role of SIGGRAPH (the leading computer graphics conference) as a foundation for such a network. Did you attend any of the early SIGGRAPH conferences?
RA: The earliest was the one in Chicago in 1979. That's when I saw some of the computer animation that the Computer Graphics Lab at NYIT was generating. That was exciting.
EH: What was the importance of SIGGRAPH for you during those early years?
RA: SIGGRAPH exposed me to a larger community. The MIT Architecture Machine Group had a strong community as well, but Arch Mach was more concerned with issues of interface, interactivity, and human-machine communication and less concerned with computer rendering and animation. At SIGGRAPH, I was exposed to other directions. I became more interested in three-dimensional worlds. I recognized that interactive computer graphics was very limited. You had to work with very simple shapes and colors, like you see in the early video games. I was willing to give up interactivity for higher resolution and more complexity. I wanted to make something that would draw me into a 3D virtual world. This was the area I entered after leaving MIT.
EH: What was your role at the Architecture Machine Group, and what projects did you participate in?
RA: I was initially involved in a Defense Department project involving mapping and the credibility and believability of images. Most of a summer was spent in libraries studying topics such as perception, illusion, and image recognition. The project that followed was called the Aspen Movie Map. I was part of a small team that traveled to Aspen and recorded all the streets and buildings of the town. We also collected graphic data and recorded audio interviews and environmental sounds. At that time I was living with a musician/audio engineer and was familiar with audio recording techniques. In order to experiment with binaural audio, I built little microphone headsets that recorded binaural sounds that captured a strong spatial sense of the environment.

I worked on another project called Personalized Movies, also known as Movie Manuals. We were exploring ways to use interactive multimedia for education and training. This led to my thesis work, which involved the development of a system that transformed live-action film to stylized animation. The thesis was titled, "Computer Rotoscoping with the Aid of Color Recognition." As you can see, I followed up on my earlier interest in art in motion.

EH: What was the impact of MIT and the Architecture Machine Group on your work?

RA: We explored the design of human-computer interfaces; we used gesture control, voice recognition, eye tracking, speech synthesis, and graphics. This turned me on to a number of new directions. It was an incredible learning experience about all the things that were possible with technology. I worked with a fantastic group of people. It was like a floodgate that opened: I was exposed to so many innovative ideas in such a short period of time. It proved to me very early on that interactivity and interface design were very important areas and would become more so in the future.

EH: After having participated in prototypical interactive projects like the Aspen Movie Map, why did it take you so long, until the 1990s, to begin working with interactivity again?

RA: As a visual artist I wanted a certain quality of color, detail, and resolution that interactivity couldn't provide. I wasn't that excited about interactivity using laser discs, creating premade movie segments that you could branch to. And computers were too slow to render sophisticated graphics in real time. Even in the late 1980s, when virtual reality was all the rage, I found it too ugly. One could not control the motion, the movements were jerky, the frame-rate slow, objects were too simple and without texture. Creative ideas could not emerge within such limitations. It was not until I began to see what was going on in video games in the early 1990s that I thought it might be possible to work with real-time 3D without compromising quality. The minute I could get a certain level of complexity in real-time rendered 3D images and produce fluid movement, I was ready to get back to interactivity. It's amazing that it took almost fifteen years to get to that point.

EH: Why did you move to the New York Institute of Technology (NYIT)?

RA: At MIT, I had focused on ways to use technology but spent less time creating my own art with these tools, so I wanted to get back to doing something special with my own work. I wanted to experiment with three-dimensional form, space, and movement, so I accepted a position as researcher at the lab at NYIT, which at that time was the best facility for 3D animation research. My time there started in the summer of 1980 and lasted until the end of 1986. It was an interesting time. Just as I was starting at NYIT, George Lucas, who was interested in computer animation, lured a number of people from the NYIT lab to California to start a computer graphics division. (This group eventually became Pixar.) When this happened, Alexander Shure, the head of CGL/NYIT and our patron, decided to switch the focus of his lab from 2D cartooning to 3D animation. I came in as part of the second generation.

EH: What were the guiding ideas behind the activity of NYIT at that time?

RA: It was a very rare and unusual situation. Shure had built the top computer-animation research facility. We never had absolute agendas; there was never a requirement to create software products in a certain amount of time. He had hired some very talented people and encouraged them to push the edge of computer graphics and animation. At that time no commercial software existed, so we had to first develop software in order to do anything—drawing, painting, modeling, animation, rendering. It was a wonderful, free environment where everyone worked passionately and created some amazing work.

EH: What kinds of projects did you work on at NYIT?

RA: There was a large effort to create a totally computer-generated feature film called *The Works.* Lance Williams who was one of the researchers wrote the script and directed the project. A number of us were hired to work on this film, but it eventually became clear that it was just too difficult to complete this project. The rendering alone would have taken over ten years based on the speed of the machines at that time. One project that followed was a pilot for a half-hour TV series called *3DV.* It featured a computer-generated host and hostess and was shown at SIGGRAPH. Two directors, Annabel Jankel and Rocky Morton, saw it and were inspired to create Max Headroom, the "computer-generated" TV announcer, who, of course, wasn't computer generated.

While working on those projects, I was able to create my own work. I didn't want this work to be shown only at SIGGRAPH, and the art world was hardly interested in computer art, so I sought out projects that could be seen on mass-media channels that would allow me total artistic control in the design of the work. The first commercial work I created in 3D, with the technical expertise of Thad Beier and Paul Heckbert, was the opening sequence for the CBS science show called *Walter Cronkite's Universe.* This was a very early use of 3D computer animation in a TV opening, so it was really an experiment on CBS's side as well. I won an Emmy award in the category of outstanding individual achievement.

The next film I made was *Steps,* which was inspired by Oscar Schlemmer and the Bauhaus theater. Schlemmer's performers wore costumes made of geometrical forms; he considered them the "motors for his costumes." As an animator, I didn't need actors inside of costumes. I could take simple geometrical forms and make them move in a way that was fundamentally human.

After *Steps,* I worked with the choreographer Twyla Tharp. She had completed a piece called *The Catherine Wheel,* which I had seen on Broadway. Tharp saw my work and asked if I could create a computer-generated St. Catherine for the film version. I found this to be very appropriate; a saint has human qualities but is superhuman. In a way, the computer is seen as superhuman as well. My computer-generated figure of St. Catherine appeared before the live performers and showed them the way. The figure was rendered as a deconstructed wire-frame model, which gave it a sketchy, transparent look. David Byrne, a former colleague at RISD, created the music for this piece.

EH: What was your relationship with photorealism, which was for so many computer graphics pioneers their primary goal?

RA: In 1980, it could take an hour or more to render one frame of a fairly simple model. You had to think very carefully about what you were modeling and rendering. People would spend forever building realistic models, but that is not the same as developing an art form. I had no interest in re-creating reality with the computer. It is a good technical challenge, but I was more interested in making abstract films. It was motion, not realistic image rendering, that attracted me.

Throughout my time at NYIT, I never modeled anything. I would just take existing 3D models that were lying around. We had a model of a woman, built by Ed Catmull in 1975, that was modified so the head and limbs could articulate. It was very difficult to generate humanlike motion, but I was obsessed with the desire to get this female model to move. This related to what I had done initially with computer animation, working with the human female form. I wanted to put the human, the female, into the computer. That has remained as a theme in my work—to analyze human motion and understand the subtleties, frame by frame, of what is going on with the human body.

EH: Your relationship with popular culture, especially pop music, was very intense in the 1980s. What kind of projects did you create?

RA: Soon after *The Catherine Wheel* (1982), I was approached by Lynn Goldsmith, who was a photographer of famous people. She was recording a strange album titled *Will Powers* with a variety of well-known musicians. Island Records was producing the album, and Lynn wanted to do something unusual for the video. During this time, at the beginning of 1983, MTV was just beginning. They were happy to show weird and interesting videos that also had good music, so I agreed to direct a music video. I again asked for creative control, partly because I was an artist but also because our tools at NYIT were really research tools and it was very scary to do productions that had deadlines. I had to be able to change designs and ideas and artwork based on whatever was happening with the tools, so I really needed a lot of freedom.

My first music video was called *Adventures in Success* (1983) with music by Sting, Robert Palmer, and Lynn Goldsmith. It was a pop art piece about forms of self-improvement. I wanted to experiment with the integration of live action and computer animation and use 2D animation software developed at NYIT to generate 1950s-style animated imagery. An eerie image of a singing, three-dimensional rotating mask created an optical illusion and formed the thematic icon of the work.

The video received a strong showing on MTV and other music video programs. At that time, I thought this could be a new venue for visual artists who create short linear pieces. It didn't work out that way in the end, but in the early days it was great because my work received substantial international exposure. It also integrated the idea of pop culture and computer graphics, which was something I was very interested in doing.

EH: From your point of view, what went wrong with MTV?

RA: Music videos quickly became commercial marketing tools for the music industry. That brought in committees of people who would dictate the creative direction of the work—not a very appealing situation for a visual artist.

EH: Your experience and thoughts were shared by some video artists, like Dara Birnbaum and John Sanborn. In the beginning, they also thought MTV was a new medium for artists, a home for video art.

RA: On the other hand, I had always aligned myself with the art world, and yet they have been so slow to accept "computer art" and reluctant to understand how to critique it. People still say, "The technology is so new. It takes a while for it to mature as an art form." But video art appeared almost immediately after portable video technology was available, and the art world quickly embraced it. The same has not been true of art using digital media. I'm disappointed that the art world is afraid to embrace this new direction. I suspect that this new form of image making calls into question some fundamental principles of art, such as its relationship to popular culture and complex technology and the economics of the physical, original object.

EH: Your rotating masks from *Adventures in Success* have become media cultural icons that have been used in many contexts since then, metaphorically merging popular culture and computer culture.

RA: Yes, that has been satisfying. I made another music video called *Smile* (1983). I worked with break dancers and again drew from a Bauhaus theme. I was somewhat tired of modern dance, and on the streets of New York the break-dance movement was beginning. This was really fresh and exciting. This piece became very popular, especially in the club scene, because it was about dance, and computer graphics work well on large public displays.

EH: You also did other kinds of work for the club scene.

RA: There was a large club in New York called the Palladium that was opened in 1985 by Ian Schrager and Steve Rubel. It was a huge space that held about 7,000 people. Ian and Steve believed that the art world was where the action was. The art world was revitalized by Keith Haring, Kenny Scharf, Jean-Michelle Basquiat; there was a wide range of fresh artists who were breaking certain traditions of art. These artists and others were commissioned to create work for the club. I was commissioned to do a piece on video that would describe this new environment. The work was about the birth of a new environment, beginning with the formation of the basic elements of nature and ending with a spot-lit dancing tree. Fractals and particle systems were used, which were beautiful techniques that we were developing to simulate natural movement and form.

This project was very exciting because, although some of these artists were part of the fine-art world, all the work was designed for a nightclub, an important expression of pop culture. On that piece I worked with the musician Carter Burwell, who has gone on to score many feature films.

EH: How did you come to work with Kraftwerk?

RA: Kraftwerk contacted me in 1984. They were working on a new album and wanted a unique music video. I was thrilled to work with them because Kraftwerk was a group that was very interested in pushing computer technology. They not only made music with it but wrote songs about the role of technology in our culture. In addition to being artists and architects, they were very plugged into pop culture. As you know, they have influenced a large number of musical styles—pop, hip-hop, rap, house, techno, ambient. They had mannequins made of each band member, and during their live performance they would, at times, sit in the audience and let the mannequins perform for them, questioning whether real humans needed to be on the stage. This question would often occur to me when people expressed their fear that computer-animated characters would replace humans.

Kraftwerk led very private lives, but they allowed me to be part of their secret circle. I told them I wanted to bring them to life as virtual forms on video. The process of building and animating human faces was very laborious and complex, but, fortunately, Fred Parke, the man who developed the original process for 3D computer-generated human facial animation in 1974, was working at the lab. With Steve Dipaola, Robert McDermott, and Peter Oppenheimer, Parke's research was further developed. Kraftwerk sent us their mannequin heads and we began the process of bringing them to life. Since I was simulating their visual presence, Kraftwerk created a simulation of my voice to say such things as "musique nonstop." They ended up taking a very long time to finish their album, and I moved on to other projects. Finally, in 1986 they were ready. I had already decided to move out to Los Angeles, but I knew I had to finish this work.

During that time I also worked with Devo, utilizing some of the earliest morphing software, and with Peter Bauman, formerly of Tangerine Dream, who had a company called Private Music. I enjoyed connecting with musicians who were exploring new ideas and new technology.

EH: Certain parts of the *Musique Non Stop* video, such as the heads, have been separated from their context and used by other artists like Nam June Paik. They have been seen in magazines, in art installations, and on television, becoming icons of technoculture. Do you mind the fact that they have begun to live a life of their own?

RA: Once your work is on television, you can expect to see it all over the place. I knew it when I was doing that. I was trying to get to the essence of what computer animation really was, to develop a style that really read as computer animation, so I purposefully used the wire frame and the faceted model even though you could have made it much more realistic. I was influenced by cubism and Tamara de Lempicka. We used some of her work as a model for the color and lighting. The idea of a faceted model was so beautiful to me, much more beautiful than just a smooth model. I was pushing very hard to expose the computer as much as I could on that piece and hoped it would be popular.

It is in the collection of the Centre de Pompidou and other places, so in some parts of the art world it has received recognition. The disappointing thing is that the art world doesn't expose it that much, so I don't mind getting exposure. Of course, I mind when my name isn't credited. Nam June Paik asked if he could use some of my work for an installation at the Whitney Museum for little monitors on the side. I was surprised to see it was the main focus of the piece. We worked out an arrangement, but the museum still wanted it to be exclusively Nam June Paik. Paik tried to insist that my name would be up there with his, but that didn't fit their model. Paik was a known video artist; this was computer art, and they didn't know what to do with it.

EH: In the mid-1980s, you decided to leave New York and move to Los Angeles. Why did you do that?

RA: It was time for a radical change of scenery. I worked so hard on the Kraftwerk piece; I thought it would never be completed. I was burned out and needed a new perspective, and I sensed that the golden age at NYIT might be coming to an end. The lack of nature and mild weather in New York was getting to me, so I decided to move to L.A.

EH: Were you attracted by the film community in Los Angeles?

RA: I had never really wanted to work in a commercial or special-effects company, but I liked the fact that there was a large computer graphics community there. I knew people in the film and the music industries that could collaborate on projects. There was a lot of creative energy in L.A. As I was making plans to move, I was invited to teach at UCLA, and that looked like a good option.

EH: Was it easy to reestablish your production in Los Angeles?

RA: The technology that I was used to working with was always very expensive, so I had to be connected to a facility and have some kind of special arrangement. In the 1980s, commercial computer animation cost between $3,000 and $10,000 per second. In 1987, soon after I moved, I was commissioned by Rebo High Definition Studios to design a work on HDTV, and I created a piece called *Behave.* At that time, the field of artificial life was just emerging, and a good friend of mine, Craig Reynolds, had developed beautiful flocking algorithms for flying 3D models. With artificial-life software, one could set up procedures based on simple rules that would bring computer models to life, and they would behave in ways that appeared to be natural and familiar. I knew that this would become an important direction for my work, so I proposed a project that would integrate live-action and 3D animation using Craig's A-life software.

EH: Around the end of the 1980s and early 1990s, you ended up creating several works in Spain. How did this happen?

RA: In L.A., I had met the Spanish artist and curator Montxo Algora, who was convinced that computer art was the future of art. When he founded Art Futura Barcelona in 1990, he invited me to produce work for the festival. I had already created a piece called *Steady State* for a Spanish television series called *El Arte del Video.* Montxo suggested that I would work with La Fura dels Baus, a well-known performance group. I had expressed an interest in working with them after attending one of their performances: their work was very raw, physical, and aggressive. While the audience stood throughout a large open space, La Fura appeared from all directions, performing to intense live music on fast-moving mobile stages. Their strong physical presence expressed primal desires and was often combined with water, fire, food, animal entrails, blood, paint, and explosions. The audience was forced to move and interact; this was the ultimate interactive three-dimensional experience.

La Fura dels Baus is group of nine men, and our collaboration was the first time they had worked with a woman, so they decided this piece would be about women and include woman performers. We created a piece titled *Mugra* and experimented with ways to integrate electronic technology and ephemeral media with their very raw, heavy, industrial kind of performance. We settled on placing seventy video monitors throughout the space that hung just above the audience's heads and were at times activated with pneumatic motors. I created a series of images that complemented the physical performance.

In 1991, Art Futura awarded me with a second commission. For this, I worked with a unique LED display system developed by Bill Bell and created two light installations. During that year, I also received a commission from the city of Barcelona to create a piece about Barcelona. Spain hosted both the World Expo in Sevilla and the Olympics in Barcelona in 1992, so there was a lot of activity in Spain and money for the arts. Working with the Spanish computer-animation company Animatica, I created a piece called *Laberint* that included performances from members of La Fura dels Baus and music by John Paul Jones, a former member of Led Zeppelin. I was also commissioned to create three pieces for the opening of the Seville World Expo and segments of a large-scale live multimedia performance called *Memory Palace.*

For *Memory Palace,* I designed a piece using the highest-quality flight simulator. I was interested in virtual reality and real-time 3D graphics, but I wanted very high-quality images, so I contacted a company in England called Rediffusion Simulation, which was known to be one of the best flight simulator companies. Working with their flight-simulator pilot, I choreographed camera movements through simulations of different parts of the world. The edited film was presented across three giant screens. Being in a flight simulator is an incredible experience; you really experience the sense of movement. You can feel nauseous when things fly too fast or in the wrong direction. I wanted to give a large audience the sense of being in a simulator by projecting it as a large-scale panorama.

EH: During this period you were really sort of adopted by the Spanish.

RA: Yes, they gave me such strong support. I also liked the fact that Spain is on the edge of the European community—as they say, the third-world country of Europe. There is an energy and freedom that comes from that. But in 1992, after the World Expo and the Olympics, the money for art projects dried up; the fiesta was over.

EH: Then came one of the most surprising moves of your career. You became a videogame developer.

RA: After completing work with the flight simulator, I wanted to pursue the idea of real-time, high-resolution, 3D virtual worlds, but I didn't want to continue negotiating the use of multimillion dollar machines. So I decided to enter the video game world. I wasn't a game player, but some of the coolest 3D technology development was taking place in the video game industry. And there was my interest in pop culture: this was where young people were going; they lived in video games and quickly understood their interactive language. I wanted to learn more about this to better understand this side of pop culture while actually making games.

Many colleagues were surprised and disapproved my move to a game company. The first game I completed was *Demolition Man,* based on the movie. I did something unusual for me. I created a violent shoot-'em-up game. I was a misfit there, in one sense studying that culture but simultaneously participating in it. And after a couple of years, I began to get tired of working with very difficult technology only to build commercial products.

EH: Was your "adventure" in the game industry motivated by financial reasons?

RA: Certainly I liked making money for a change, but it was really an artistic pursuit. The game people know many things that the academic researchers don't know about. The academic VR researchers were not paying attention to games, and, like the art world, the academic research community considered popular forms of media to be trivial and inferior.

I always like to go where I am not supposed to be. I was very much a part of the computer graphics research community, but I did not see why I shouldn't go to this very commercial "low" art form and find out what was driving it.

I also thought that maybe I could change some of the video games and make them more artistic, but 3D games were so expensive to build that the industry did not want to experiment. They wanted to play it very safe, imitating *Doom*-like games that were already popular. When I first started at Virgin Interactive Entertainment, there was a lot of potential for new and different ideas, but it changed quickly; I lost my interest because I knew there was no way to really express my ideas.

EH: After you retreated from the game industry, what happened next?

RA: After two years at Virgin, I was offered a position at the University of California (UCLA) as a professor and chair of a new Department of Design that was converting to digital media. This was good for it allowed me to get back to my artistic work. I accepted the offer and immediately proposed and received a grant from Intel Corporation. With a team of computer science and design students, we have developed a kind of game engine, a system called Emergence. A major feature of this work involves artificial life. I felt this was really the right direction for computer animation, and it also connected with my initial interest in human and natural motion. Instead of only articulating the body or face to form expression, I could choreograph the movements and behaviors of individual characters and groups. We developed a behavior scripting language, so that one can define the relationships and behaviors, personalities and desires of characters and objects and set them into motion.

The first prototype interactive art installation was exhibited in 1997 at *Art Futura* in Madrid. This was followed by a more developed version titled *The Bush Soul.* My interest now is creating interactive installations using this new system. Currently we are developing a unique form of voice input and applying a force-feedback joystick so that you can feel the energy and other touch sensations of the artificial life forms. I want to get beyond the keyboard and mouse interface.

EH: Now that you look back, is it possible for a computer artist to make an impact on culture?

RA: I'm a bit disappointed because when I started back in the early 1970s, I was sure that within ten or fifteen years computer art would be in the forefront. I also thought the art world would change and that the idea of the single, original art object exhibited in a museum would go through radical changes. It is still not happening. I also thought that art created with new technology would be distributed in a way that is similar to music distribution—where you could purchase art through a variety of channels. Experimental music may not sell a large amount, but it is available in the record store. That hasn't happened to computer art and may not happen soon. Yet I believe that in the future it will become obvious that the exploitation of digital media in art is very important and that certain important works have been created over the years and will finally be recognized.

The art world is stuck in a rut; it has to change. I went to art school in the early 1970s when conceptual art was strong. That was also a very confusing time for the art world because if there were no art objects, then the whole structure—museums, galleries, curators, buyers, and patrons—would not make sense. The art world has been struggling with this and has tried to hold onto the traditional forms. Maybe computer art came at a time when the whole way we view art had begun changing, and we still haven't come to terms with that yet.

EH: The attention of both the business world and the young hacker world is focused on networking. How do you see the possible impact of the Internet on digital imaging and your own work?

RA: Certainly the broad use of the Internet is extremely important to our evolution. I pay a lot of attention to it, though very little work has inspired me so far. I am considering how I want to approach the Web. The technology isn't quite there, in that the bandwidth isn't quite capable of displaying the fluidity of movement or complexity of image that I desire, but I am counting on that to change soon. What I don't like about the Internet is that it encourages you to sit in front of a computer, usually alone and in your own personal space. I want to give people reasons to go outside of their homes. I am still interested in getting people to physically come in contact with one another and to experience an immersive projected environment. I want to see the interface of the computer develop radically, and I am still uncomfortable with a form of communication that further excludes our physical presence.

Currently I am really excited about artificial life and rich real-time worlds that are alive and interactive. I continue to focus on behaviors and relationships, including our relationship with the computer. These are ideas that have concerned me all along—relationships between humans, the meaning of human presence in a virtual world, and the fact that we are going to relate to artificial-life forms more and more. I have always been intrigued with the sense that the computer is my partner, and artificial life further personifies this.

16

Algorithmic Art, Scientific Visualization, and Tele-immersion: An Evolving Dialog with the Universe

Donna J. Cox

Art attempts to find in the universe . . . what is fundamental, enduring, essential.

—SAUL BELLOW

ORGANIZATION OF RESEARCH FROM 1981 TO 1999

I borrow from psychological theorists and scientific philosophers[1] to organize my technological explorations in art and science into three developmental stages of work. Each stage is examined for its (1) methods and technical innovations and (2) content. These explorations suggest patterns in conscious processes. The underlying link between the evolution of consciousness in artist, individual, and society is an assumption of my writing.

Magic-Mythic Stage

In this early stage of development, the individual is more egocentric and body conscious.[2] Symbolism and individualism play central roles.

Methods: Individualism—Algorithmic Art

In the early 1980s, I worked alone and developed software to create art.[3] My collection of software tools included programs to interactively manipulate color and digital values in images that were scanned using photometers.[4] These interactive tools provided capabilities to create composite images, perform mathematical functions, and map the color characteristics of today's computer paint programs. One algorithm produced statistical matrices that could be interactively manipulated, resulting in complex, interleaving patterns.

To amplify the final size of the computer image, I developed a collection of "compulages."[5] The final artworks were installational collages of photographic sections that increased the size of the displayable computer image.

I employed elements of chance and mathematics as part of the creative process. Many digitized photographs were manipulated with randomness and controlled interplay. Algorithmic, electronic art—where the computer becomes a kind of creative partner—characterized my artistic process during this stage.[6]

Content: Symbols—Goddess and Metamorphosis

The creative work seemed like a magical experiment with the computer. While interacting and exploring algorithms, symbolic and mythical images emerged from the computer screen. *Butterfly Mask* (series, 1983–1985) was created during such an experiment where I discovered the image within the algorithm. One perceives *Butterfly Mask* among alternative interpretations of face, mask, or butterfly. This image is a Rorschach experience and conveys Jungian cross-cultural symbolism relating to the flight of the soul.[7]

I also created a series of digitally manipulated self-portraits that composited facial expressions with houses from childhood memories.

A reemergence of the goddess, a symbol for feminine power, occurred in the women's movement.[8] Feminism brought with it the resacralization of the female body in reaction to research that disclosed that most women lack self-acceptance of their bodies.[9] The reassertion of a female image that is controlled by women is one of the characteristics of feminist body art.[10]

I was keenly aware of feminist ideas concerning the body and was in the process of redefining my self-image.[11] *Agony & Ecstasy* (figure 16.1, see inset image in lower right corner) is a full-body self-portrait being transformed in a mythic Faustian struggle between good and evil. This colorful, crucificial image questions the emotional state of the objectified female.

The myth of metamorphosis is an artistic genre and informs much of the work during this period.[12] *Metamorphosis*[13] (1983–1985) is a ghostly self-portrait of face and raised hand, symbolizing peace, interleaved in a swirl of complex mathematical patterns. Myths and symbols are embedded in the collective consciousness of individuals and culture. Freud, Jung, and Piaget reached essentially the same conclusion: all the world's great mythologies exist today in each of us, in me and in you.[14]

Mythic-Rational Stage

A world of calculated reason incorporates myths as playfulness. Self-image becomes less important and abstracted. Collaboration and understanding the roles of others are essential to the work.[15]

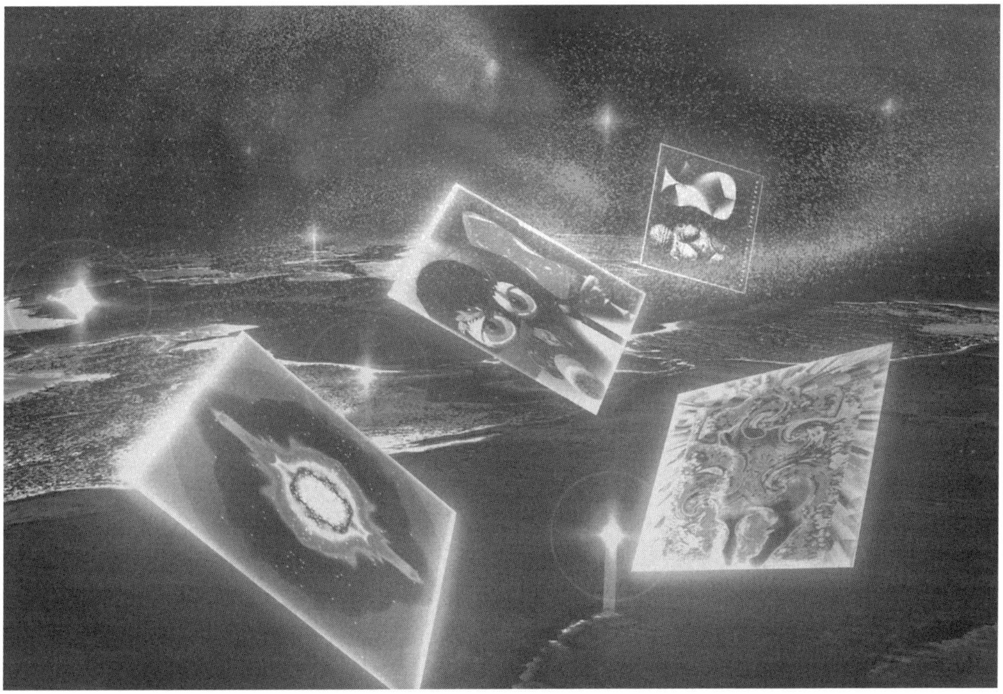

Figure 16.1
Donna Cox, *Old Familiar Is-ness,* 1998, cibachrome, 24 by 20 inches, *SIGGRAPH 98 Art Show.* A collage of the artist's earlier works that was inspired by a poem she wrote in 1998. *Lower left: Colliding Neutron Star. Center: Venus and Milo. Upper right: Venus in Time. Lower right: Agony & Ecstasy.* Starry background composited from colliding galaxies; earth-plane from *Garbage.*

Methods: Collaboration, Scientific Visualization, Information Design

In 1985, I went to the University of Illinois, Urbana–Champaign, as assistant professor and began a long-term affiliation with the newly established National Center for Supercomputing Applications (NCSA). I developed some of NCSA's first computer graphics software and began working with scientists to visualize supercomputer data.

The process of collaboration contrasted with my former method of working alone to create computer art. In 1987, I formulated the concept of interdisciplinary Renaissance teams involving artists, scientists, and technologists to solve problems in scientific visualization.[16] This research included collaborations with many disciplines including astrophysics, entomology, and mathematics to create methods, visualize data, design software interfaces, and produce graphical imagery and animations.[17] These projects and processes have been described in detail elsewhere.[18]

Collaboration with Kodak engineer Richard Ellson resulted in the "glyph" technique to visualize multidimensional data.[19] Glyphs are abstract graphical symbols that visually represent numerical data. This research became some of the pioneering work in computer graphics information design for supercomputer and data visualization. Many of these techniques have been imbedded in off-the-shelf software from NCSA and published in textbooks.[20]

One of the most popular data visualizations from NCSA was developed in collaboration with another NCSA research artist, Robert Patterson. The *Visualization Study of the NSFnet* shows the enormous growth of the national network and backbone. I sketched an initial backbone and client network design on paper; then Patterson and I collected data from the network and modeled, designed, and rendered the images and animations using Wavefront Software.

The art of scientific visualization is a highly conceptual process that requires more rational design methodology compared to the algorithmic process described in the former magic-mythic stage. I became more involved with the software design, data organization, and image rendering and less involved with hands-on software development. The magic of using the computer disappeared into the "black box" process, and "as you program in ever higher-level languages, you know less and less precisely what you've told the computer to do!"[21] While the magic was gone, the mythic forms remained part of the collaborative content.

Content: Venus, Web of Life, Gaia

Working collaboratively with scientists revealed new visions, an enormous web of life unfolding through the integration of art and science, where bridges between disciplines are as important as the nodes. Interrelationships with others in the creative process were paramount to the research here.

Scientific visualization is a very conceptual and rational process where symbols represent real data. The visualization of the supercomputer simulations seemed more real than my individualist artwork. Yet, "real simulations" is an oxymoron and reflects key elements in postmodern thought: simulation is the map that has become the territory,[22] and such visuals are icons of truth to the public.

We need a visible past, a visible continuum, a visible myth of origin to reassure us about our ends, since ultimately we have never believed in them.[23]

A myth offers the best of what it has when it ceases to dictate over reason and is released into the realm of possibilities.[24] Myths offered playful possibilities in a transformed context. Many of these cosmological simulations appeared to me and my colleagues as Rorschach impressions of male and female forms. I often exhibited astrophysical images using alternate titles for the work. For example, I entitled an astrophysical jet *Cosmic Sperm*,[25] and *Colliding Neutron Stars* (figure 16.1, inset image in lower left) appeared as Georgia O'Keefe female flowers and Judy Chicago feminist art.[26] People seem to have a propensity to identify male and female icons in these astrophysical forms.

There are potentially not only an infinite number of *pathways* in a brain but also an infinite number of *symbols*. As was pointed out, new concepts can always be formed from old ones, and one could argue that the symbols that represent such new concepts are merely dormant symbols in each individual, waiting to be awakened.[27]

Collaborative work with George Francis, mathematician, and Ray Idazsak, computer programmer, resulted in the raw material needed to build archetypal forms from mathematics. *Venus on Glass*[28] presents geometric surfaces as figurines displayed on marble pedestals under glass. These activate into a protagonist character in the narrative, humorous animation, *Venus & Milo* (see figure 16.1, center inset). The Venus eats and burps chocolates in a synthetic postmodern art museum and mathematically transforms while interacting with Milo, the museum janitor.[29]

Scientific forms, animations, and mathematics provided new ways to realize transformations. *Venus in Time* (see figure 16.1, upper right inset) juxtaposes an ancient archetype to a postmodern supercomputer Venus. This juxtaposition implies many levels of transformation and alludes to change from representational form into abstraction. An individual hand created the paleolithic Venus, while a collaborative team generated the supercomputer postmodern Venus.

At (Art)^n, artistic collaboration is the primary mode of working and provides a collective synergism for invention. Ellen Sandor acts as creative producer and director and organizes, coordinates, and participates with her creative team. *Human Papilloma Virus*[30] was created in collaboration with Stephen Meyers and (Art)^n, representing female sexuality and issues surrounding cancer-causing HPV. Likewise, *A Tribute to Sadie Elmo*[31] relates to female breast cancer. (Art)^n collaborations are some of the most important in the contemporary art world. Such interdisciplinary collaboration potentiates important educational directions.[32]

Many of the visual icons produced for science became extremely popular with the public and are well-documented in media.[33] *Visualization Study of the NSFnet* creates visual connection between the Internet and the dendritic web of life. Likewise, *Garbage*[34] delivers an animated environmental public-service announcement; however, its visual message also relates to Gaia[35] and Mother Earth.[36] The mythic-rational work incorporates rigorous techniques in scientific visualization as well as symbolic forms that extend beyond rational thought.

Rational-Universal Stage

Global collaboration and complex interrelationships are paramount to the creation of this networked and universal work. Other perspectives and horizons open before the imagination.[37]

Methods: Global Collaborations, Virtual Teams, Tele-immersion

From 1994 to 1996, I took sabbatical and leave of absence to be associate producer for scientific visualization and art director for the NCSA/PIXAR segment of the IMAX movie *Cosmic Voyage*, which premiered August 1996 at the Smithsonian Air and Space Museum in Washington, D.C.[38] This movie is currently playing in IMAX and OMNIMAX theaters around the world and was nominated for an Academy Award in the documentary short-subject category in 1997.

This movie shows an unprecedented fifteen minutes of IMAX computer graphics images. A world-renowned Smithsonian Science Advisory Committee approved each scene. While I was involved in most scientific visualization discussions, my primary responsibilities were to produce approximately five minutes of computer animation using scientific simulation data. To accomplish this task, I organized and coordinated a global Renaissance team[39] to create visualization sequences showing the formation of the early universe. This team included cosmologists, graphics researchers, computer artists, and technical directors from PIXAR Animation Studio, NCSA, Santa Barbara Studios, San Diego Supercomputer Center, and several major universities.

Frank Summers, of Princeton University, simulated galaxies forming in the early universe using NCSA's Power Challenge Array, requiring all processors for six weeks. Chris Mihos and Lars Hernquist, of the University of California–Santa Cruz, simulated two spiral colliding galaxies (see figure 16.1, upper background), using over 900 hours of C-90 Cray supercomputing. We network transferred over 120 gigabytes of raw simulation data to NCSA, where it was stored and visualized.

Loren Carpenter, PIXAR senior research scientist, created a very fast particle renderer.[40] Erik Wesselak, NCSA, created software to interface the raw data to the renderer, enabling Robert Patterson and me to control aesthetic parameters for the images (including color, softness, and depth). We created over 100 gigabytes of high-resolution (4028 by 3002 pixels) images for IMAX film recording.

We designed the scenes while developing a virtual-reality choreographer. Robert Patterson, Marcus Thiébaux, and I cocreated Virtual Director™ and demonstrated the prototype at the SIGGRAPH '94 VRoom exhibition.[41] Patterson and I conceived of and designed the Virtual Director, and Thiébaux designed and implemented the software.

To choreograph scenes for *Cosmic Voyage*, Virtual Director was developed into a valuable voice- and gesture-driven virtual-reality system to interactively navigate, record, edit, and play back three-dimensional camera paths through data in the CAVE™, a multiperson immersive 3D stereo display room.[42] The galaxies appear suspended in the 3D stereo virtual space. Patterson choreographed and I helped to design choreographies for *Cosmic Voyage*. The virtual camera output coupled to the PIXAR renderer to create digital images for the movie.

Virtual Director provides remote virtual collaboration and enables users to be represented as avatars[43] and meet in virtual space via network connection (figure 16.2). The user's head and hand motions are magnetically tracked and conveyed as the motion of the avatar's head and hand in virtual space.

Stuart Levy, NCSA, works with Patterson and me to expand Virtual Director capabilities with other powerful software in the Grid.[44] We have collaborated with Glen Wheless and Cathy Lascara at Old Dominion University to integrate Cave5D[45] with Virtual Director, resulting in Collaborative Virtual Director™ (CVD), which we use to remotely connect, visualize, navigate, and record interactive virtual sessions.[46] Multiple users can interact over large distances using a variety of devices and visualize a variety of data, including environmental hydrology and atmospheric science. Using the vBNS network,[47] we teleimmerse[48] in a virtual space while each user sees identical data from his or her personal point of view (see figure 16.2) and hears through digital voice transfer.

Content: Grand Visions of the Cosmos

Global teleimmersion is the technological extension of basic human sensory experiences that can also expand mental imaginations. We project representations of ourselves as incarnated, disembodied, archetypal, gendered symbols (avatars); we fly; we play; we exchange information; we view environmental data as glyphic symbols in a virtual space; we cooperate.

Such virtual interaction and symbolic communication has profound implications. Remote virtual collaborations across international networks will transform information exchange among social systems. Evolution

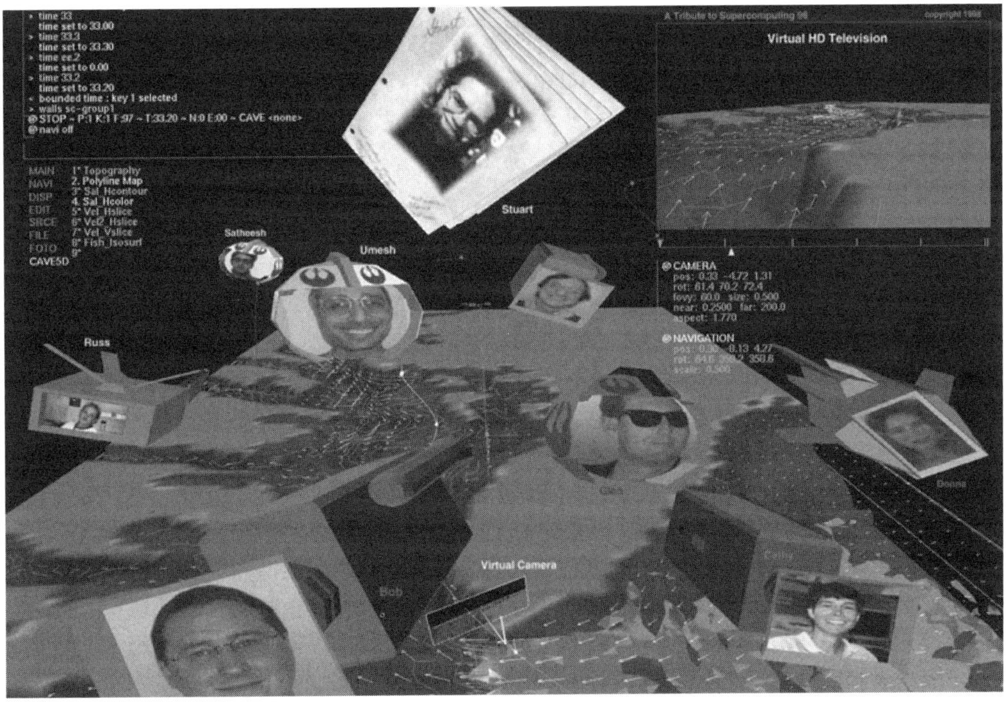

Figure 16.2

Donna Cox, *Virtual Team Group Portrait,* 1998. Image generated with collaborative software: Virtual Director by Cox, Robert Patterson, Marcus Thiebaux, Stuart Levy; and Cave5D by Glen Wheless, Cathy Lascara. Many users, represented as avatars, remotely collaborate across the United States and meet in virtual space. Each user can independently navigate and view identical data while sharing the virtual camera, television, and other information. Available at ⟨virdir.ncsa.uiuc.edu/virdir.html⟩.

does not always take the form of unrelenting competition; cooperation on the global scale is how life organizes and advances.

The view of evolution as a chronic bloody competition among individuals and species, a popular distortion of Darwin's notion of "survival of the fittest," dissolves before a new view of continual cooperation, strong interaction, and mutual dependence among life forms. Life did not take over the globe by combat but by networking.[49]

Cosmic Voyage is a story of the relative scale of evolving structures from the microcosm to the macrocosm; it provides visual imagery to demonstrate how humans fit into the grand scale of the universe. We collapse billions years into a few minutes, warping time and scale to provide a vision of creation. The births of galaxies provide powerful visual stories about the beginning of the cosmos and can change how people think about scale, time, and the universe. To "hold infinity in the palm of your hand, and eternity in an hour"[50] is possible when using computer graphics, and yet such visualizations have the capacity to transform human thought and consciousness.[51]

CONCLUSION: CONSCIOUSNESS, FINAL FRONTIERS

Advancing stages of consciousness imply an arrow of time, a movement forward. Yet quantum physicists do not view linear time as an unchangeable law of the universe.[52] Time is constructed from human perception. In a nutshell, the quantum paradox derives from the fact that the fundamental equation in quantum mechanics is symmetrical in time; hence it is our measurements, our interventions that give direction to time.[53]

Extrapolating from this paradox, consciousness might also be viewed as an active participant that affects probabilities in the slices of space-time from the material universe.[54]

Great speculations about quantum theory and the evolution of consciousness have little to do with what we can actually demonstrate in our daily lives—or possibly understand. Limitative theorems of mathematics and computation suggest that a high-level view of a system may, in fact, not be defined or explained from a low-level view. Our brains may preclude the complete understanding of ourselves and developments of consciousness on the grand scale of the universe.[55]

Consciousness evolves, children grow, the adult matures, society changes, and technology marches onward. Birthing and dying are equal partners in these evolutionary processes.[56] A recent poem relates personal experience to issues of space, time, consciousness, and the cyclic nature of existence. Figure 16.1 is informed by this poem.

Excerpt from *Than This*

   Oh, sly, rhythmic awareness
beating all your drums.
   Undulating fingers nod,
surfing eternal roles,
   Then disappearing helical threads,
Retracting those heartfelt souls.

Old friend, daughter, mother,
   old acquaintance, brother, lover,
. . . All siblings in-kind.
Old familiar is-ness,
You've been everything to me:
   Anchors of laughing candles,
Intermingling crests of regret,
   Revealing memories unborn,
Confessing tearful forgiveness.

We've so-journed, so-long together,
   arriving again and again,
Beyond all appearances of
   evolving atomic finesse,
Discovering no slice is better to claim,
No-thing . . . No-time . . . Than This.[57]

NOTES

1. Ken Wilber, *Sex, Ecology, Spirituality* (Boston: Shambala, 1995) (Wilber extends Piaget's cognitive theory and contends that the adult advances to a new conscious awareness that is more global, less egocentric); Pierre Teilhard de Chardin, *The Phenomenon of Man* (New York: Harper Torchbooks, 1956).

2. Wilber, *Sex, Ecology, Spirituality,* 165, 185.

3. I wrote software in C programming language, working in a scientific environment.

4. Donna Cox, "Interactive Computer-Assisted RGB Editor (ICARE)," *Proceedings of the Seventh Annual Small Computers in the Arts Symposium, 9–11 October* (Philadelphia: SCAN, 1987), 40–45; Cox, "Using a Supercomputer to Visualize Higher Dimensions: An Artist's Contribution to Science," *Leonardo,* 21, no. 3 (1988): 235–236.

5. I coined the term *compulage* in 1983 to describe the process of reconstructing a matrix of enlarged digital sections from the computer image and reassembling the sections to create a large-scale installation. I explored printing on various archival papers and cloth.

6. Brian Reffin Smith, *Soft Computing* (Wokingham, England: Addison-Wesley, 1984), 72; Frank Popper, *Art of the Electronic Age* (New York: Abrams, 1993), 86, 102, 121.

7. Barbara G. Walker, *The Woman's Dictionary of Symbols and Sacred Objects* (San Francisco: Harper, 1988), 245, 371.

8. Riane Eisler, *The Chalice and the Blade* (San Francisco: Harper & Row, 1987); Eleanor W. Gadon, *The Once and Future Goddess* (San Francisco: Harper, 1989), 146, 149–150, 228, 257, 285, 341.

9. Naomi Wolf, *The Beauty Myth* (New York: Anchor Books, 1991), 184–185 (women's apparent obsession with beauty is a reflection of the pressures of culture); Hilde Bruch, *Eating Disorders* (New York: Basic Books, 1973) (degraded body image correlates to eating disorders in well-documented medical studies); Susie Orbach, *Fat Is a Feminist Issue* (New York: Berkeley, 1978).

10. Gadon, *The Once and Future Goddess,* 294; Mary Kelly, "Re-viewing Modernist Criticism," *Art after Modernism,* ed. Brian Wallis (New York: New Museum of Contemporary Art, with Godine, Inc., 1984), 98.

11. Vivian Gornick and Barbara Moran, eds., *Woman in Sexist Society* (New York: New American Library, 1971).

12. Donna Cox, "Caricature, Readymades, and Metamorphosis: Visual Mathematics in the Context of Art," in *The Visual Mind,* ed. Michele Emmer (Cambridge: MIT Press,

1993), ch. 15; Cox, "The Tao of Postmodernism: Computer Art, Scientific Visualization, and Other Paradoxes," *Leonardo* (Suppl. issue, ACM SIGGRAPH '89) (July 1989): 7–12.

13. *Metamorphosis* is one of a series and a compulage. Androgynous and cross-cultural symbols are transformed into self-image.

14. Wilber, *Sex, Ecology, Spirituality,* 220.

15. Ibid., 224.

16. A Renaissance team was first defined in Cox, "Using a Supercomputer," as collaboration among small groups of artists, technologists, and scientists.

17. Ivars Peterson, "Twists of Space," *Science News* (24 October 1987); cover, 259, 264–266; Margaret Neal, "More Than Science, More Than Art," *IEEE Computer Graphics and Applications* (November 1988): 3–5; Donna Cox, "Collaborations in Art/Science: Renaissance Teams," *Journal of Biocommunications* 18, no. 2 (1991).

18. Cox, "Using a Supercomputer"; D. W. Onstad, J. V. Maddox, D. J. Cox, E. A. Kornkven, "Spatial and Temporal Dynamics of Animals and the Host-Density Threshold in Epizootiology," *Journal of Invertebrate Pathology* (January 1990).

19. Richard Ellson and Donna Cox, "Visualization of Injection Molding," *Journal of the Society for Computer Simulation* (November 1988): 184–188.

20. Peter Keller and Mary Keller, *Visual Cues: Practical Data Visualization* (Los Alamitos, CA: IEEE Society Press, Manning Publications, 1993), colored illustrations and text by Cox, 51, fig. 1-9; 58, fig. 2-7; 61, fig. 2-10; 147, fig. 11-1.

21. Douglas R. Hofstadter, *Gödel, Escher, Bach* (New York: Vintage Books, 1979), 306.

22. Cox, "The Tao of Postmodernism."

23. Jean Baudrillard, "The Precession of Simulacra," *Art after Modernism,* ed. Brian Wallis (New York: New Museum of Contemporary Art, with Godine, Inc., 1984), 253–281.

24. Wilber, *Sex, Ecology, Spirituality,* 239.

25. *Astrophysical Jet* series (1986–1987) was created in collaboration with Michael Norman. See Cox, "Using a Supercomputer." This animation was shown at the *SIGGRAPH '87 Film and Video Show*. Presentations as *Cosmic Sperm* were made at the *Simulations/Dissimulations: A Symposium,* Art Institute of Chicago, Chicago, IL, 5–7 November 1987. Presentations included Jean Baudrillard, Michel Segard, Joan Truckenbrod, Ed Emshwiller. *CRASH (Computer-Assisted Hardcopy),* was presented 23 October to 23 November 1988 at the Wright Museum of Art in Beloit, Wisconsin.

26. Gadon, *The Once and Future Goddess,* 320–321. The *Neutron Star* series (1986–1987) was often interpreted as flowers by viewers and reiterates the forms of vessels, vulva, and female morphology.

27. Hofstadter, *Gödel, Escher, Bach,* 384.

28. Fred Ward, "Images for the Computer Age," *National Geographic* 175, 6 (June 1989): 719–751 (Cox quote, 750; Venus, 721, 739).

29. Cox, "Interdisciplinary Collaboration Case Study in Computer Graphics Education: *Venus & Milo,*" *ACM SIGGRAPH Computer Graphics* 25, no. 3 (July 1991): 185–190. *Venus and Milo* won first place in the art and entertainment category at Nicograph, 1990.

30. (Art)^n is available at ⟨http://www.artn.com⟩. *Human Papilloma Virus* was originally created by Cox and Stephen Meyers and later made into a PHSCologram. *The Politics of Pleasure/The Nineties: A Tribute to Robert Mapplethorpe* (1990) incorporated the third edition of the *Human Papilloma Virus* PHSCologram into a computer sculpture.

31. Donna Cox, *Tribute to Sadie Elmo,* 1988, original computer image, PHSCologram made in collaboration with Ellen Sandor, Dan Sandin, Ray Idaszak, and George Francis. This image tributes my grandmother and all who have died with breast cancer.

32. Cox, "Scientific Visualization Collaborating to Predict the Future," *Computing and Communications in Colleges and Universities EDUCOM Review* 25, no. 4 (Winter 1990): 38–42. Based on Cox's keynote to EDUCOM '90, October 14–17, Atlanta, Georgia (keynotes also by Jimmy Carter and Steven Jobs).

33. Judith Turner, "An Artist Whose Work Reflects and Extends the Most Dominant Forces in Our Culture," *Chronicle of Higher Education,* 11 October 1990, B5–B7; "Backbone Bore Brunt of Internet Traffic," R&D (August 1996): 25.

34. Cox et al., *Garbage* (1991), computer graphics PSA that shows proliferating garbage, houses, and cities and warns people to recycle; appeared at *SIGGRAPH '91* and on broadcast television.

35. Lynn Margulis and Dorion Sagan, *Microcosmos: Four Billion Years of Evolution from Our Microbial Ancestors* (New York: Touchstone, 1991), 265 (Gaia is the superorganism system of all life on earth); Rick Fields, with Peggy Taylor, Rex Weyler, and Rick Ingrasci, *Chop Wood Carry Water* (New York: Putnam's Sons, 1984), 230.

36. Gadon, *The Once and Future Goddess,* 228–232.

37. Wilber, *Sex, Ecology, Spirituality,* 174.

38. "*Cosmic Voyage,* a science educational film, was funded by Smithsonian Institute and Motorola Foundation and produced by Cosmic Voyage Inc. I was involved from the fund-raising stages to final production. Mel Slater, vice president, Motorola, was a champion for the film. IMAX film contains ten times the emulsion area of standard 35 millimeter film. See also Sara Latta, ⟨http://141.142.3.131/Pubs/access/96.2/Cosmic.html⟩;

39. A Global Renaissance team is a much larger and more remotely located group than a Renaissance team. See Cox, "Using a Supercomputer."

40. PIXAR eventually packaged Star Renderer into StarMan™.

41. In 1992, Robert Patterson and I conceived of the Virtual Director, and a very early version was created using a Fakespace Boom. Marcus Thiébaux, an EVL student, joined our team in 1993.

42. Tom DeFanti and Dan Sandin, UIC, developed the Cave Automated Virtual Environment (CAVE),™ a multiperson, room-sized, high-resolution 3D video and audio environment. Real-time computer graphics are rear-screen projected in stereo onto three walls and floor. An operator wears a head-tracking device, and participants wear stereo glasses to view images in 3D. The operator contols applications with a hand-held tracking device called the *wand.*

43. The term *avatar* evolved from Indian mythology and describes the symbol of an agent in virtual-reality space. Figure 16.2 shows customized avatars in Virtual Director.

44. The NCSA organized a National Computational Science Alliance to prototype a national technology grid.

45. Cave5D is a configurable virtual-reality application framework that was developed by Alliance principal investigators Wheless, Lascara, and Bill Hibbard, University of Wisconsin–Madison. See also ⟨http://www.ccpo.odu.edu⟩.

46. Available at ⟨http://www.ncsa.uiuc.edu/VR/grants/virdir⟩. Figure 16.2 collaboration was demonstrated at Internet2 in Washington, D.C., April 1998, and at SC 98, November 1998.

47. Very broadband network system (vBNS), 620 mbit. See ⟨http://www.vbns.net⟩.

48. Available at ⟨http://www.evl.uic.edu/spiff/cavernsoft⟩; Mike Goslin and Jacquelyn Ford Morie, "Virtopia: Emotional Experiences in Virtual Environments," *Leonardo* 29, no. 2 (1996): 95–100; Roy Ascott, "On Networking," *Leonardo* 21, no. 3 (1988): 231; William R. Sherman and Alan B. Craig, *Working with Virtual Reality* (San Francisco: Kaufmann, 1999).

49. Margulis and Sagan, *Microcosmos,* 17.

50. William Blake, "Auguries of Innocence," *Blake: Complete Writings* (Oxford: Oxford University Press, 1972), 431.

51. Michael Murphy, *The Future of the Body* (New York: Putnam's, 1991).

52. Darryl Reanney, *After Death: A New Future for Human Consciousness* (New York: Avon Books, 1991), 19–48; Danah Zohar, *The Quantum Self Human Nature and Consciousness Defined by the New Physics* (New York: Morrow, 1990), 30, 34, 37–39.

53. Ilya Prigogine, "Science, Reason and Passion," *Leonardo* 29, no. 1 (1996): 42.

54. Reanney, *After Death,* 26.

55. Hofstadter, *Gödel, Escher, Bach,* 696–698. Hofstadter suggests that death is perhaps the greatest contradiction of all, according to Godel's theorem, which implies that it is totally impossible for humans to comprehend what it is like to "be no more." Since we hold all that we know in our mental descriptions of who we are, it is impossible for our self-model to understand nonexistence.

56. Freeman Dyson, *Infinite in All Directions* (New York: Harper & Row, 1989), 294.

57. Excerpt from Cox, “Than This” (1998). This poem was written in Oklahoma after my mother was diagnosed with cancer. An excerpt was printed in the *SIGGRAPH Electronic Art and Animation Catalog* (Orlando: SIGGRAPH, 1998), which documented the SIGGRAPH ’98 art show, July 19–24, 1998, where *Venus in Time* and *Familiar Is-ness* were also exhibited.

17

My Autobiographical Media History: Metaphors of Interaction, Communication, and Body Using Electronic Media

Agnes Hegedüs

As young people (no matter in which creative field one is active), we start with an almost empty page about ourselves. As time goes on and different activities define themselves through our interests, it becomes clear that the initially seemingly endless potentialities of creative expression are a utopian illusion and that we become focused on a few personal ideas and obsessions that we repeatedly address in various forms. This is what becomes identified as our "style" and also constitutes a particular expertise.

After fourteen years of creative work, ten of which were with electronic media, I recognize this emerging pattern of recurrent thoughts and their variations in my own work. While I often resisted the idea of the artist working with the same medium on reiterating themes and motives, my work has shown me how the concentric orbits of certain ideas come to define a field of interest and activity. In this process, the personally perceived imperfections of each consecutive creation propels the maker to try it again and again to find a way to better express a strangely coded singular message that preoccupies her mind.

A personal obsession with completeness and wanting to encompass everything led me on a journey through many different media. I first received my diploma as a ceramist. In the art school where I was enrolled (the Hungarian Applied Art Academy), I then had the opportunity to work with photography and graphic design. But my activities in these fields were forms of experimental work that were somehow problematic in those proscribed academic circumstances, and I was ready for a change. In 1987, one of life's chance events gave me the opportunity to begin to work with video at that same institution. I was among the privileged few who were accepted as students in the first media department to be established at an art education institution in Hungary. Following this, I was able to study abroad—first in Holland (AKI Art Academy, Enschede) and then in Germany (Städelschule Institute für Neue Medien, Frankfurt)—at more advanced and better-equipped media art institutions, thus opening an opportunity for me to begin to work with computer-assisted media.

As I try to understand the hidden logic of my development in those years, it seems to be a creative ambition to embody, with various media, the notion of a synesthetic *gesamtkunstwerk.* To begin to understand and creatively use the synaesthetic qualities of audiovisual media was a personal breakthrough for me. The fact that I could thereby begin to use the widest range of my creative abilities was also felt to be a way of breaking out of gender limits set by the art establishment. The frustrating sense of being caught in ghettos of artistic sensibility that I previously experienced as a ceramist, photographer, designer, and video artist was liberated in the practice of media art, through its particular (especially cooperative) modalities of production and its (especially interactive) modalities of reception by the general public.

METAPHORS OF COMMUNICATION

The original dice were the spinal bones of sacrificial animals that were used for prophesying.[1] Communication via computers embodies a basic strategy of encoding complex information into simple rules and operations. This process of abstraction was not unfamiliar to human thinking in the past and is evident, for instance, in the modality of various ancient games, which also encoded layers of information within the protocol of simple rules.

The essential random element in games allowed the machinations of a divine manifestation to enter into play with the players. That game was considered a microcosmic reflection of the macrocosmic game—that is, God's play with man. The element of competition in games allows the manifestation of divine providence. Two main categories of games can be distinguished from the point of view of necessary skills. One category offers a playground of numeric randomness that develops either linearly (such as board games played with dice) or in cycles (such as roulette) without making much demand on the players' intellectual abilities. The other category is games that have complex spatial or structural information (such as chess). Such games cultivate the player's ability to grasp and exploit their underlying patterns.[2]

My interactive installation *The Fruit Machine*[3] constitutes a significantly different category of gaming (and aesthetic experience). Its aims to configure a social dimension for the new digitally visual spaces of communication. A game of luck becomes a game of skill at remotely cooperatively manipulating virtual objects in represented space. While most collective games are competitive, *The Fruit Machine* differentiates itself by sublimating both the competitiveness and the winning. *The Fruit Machine* also carries an element of those games based on structured arrangements, but instead of an abstracted two-dimensional intellectual structure, it embodies its visual components in a fully articulated three-dimensional Cartesian space.

The basic intention of *The Fruit Machine* was to create an aesthetic work whose modality of interactivity resembles the video game but whose visual characteristics have more to do with Arcimboldo than Nintendo. The iconography of *The Fruit Machine* is a geometrical form that is divided into three parts that can be separately manipulated in the pictorial space by three separate spectators using three axis joysticks to control these objects' movements. The half-octagonal surface carries texture-mapped rows of fruit, evoking the well-known gambling machines. As in any game, there is a defined task, a goal to strive for, and of course a reward for success. The task here is for the three users to coordinate and control the movement of the three image elements. The goal is to bring them together so that the rows of fruit are joined correctly, and success is rewarded with an exuberant shower of virtual coins. Contrary to its ancestry, this new fruit machine is not a game of chance but a game of skillful coordination and cooperation between the three players.

My later installation *The Televirtual Fruit Machine*[4] (figure 17.1) is based on the functionality and visual iconography of *The Fruit Machine* but is modified with respect to a new situation where there are two geographically separated installations (such as Karlsruhe and Tokyo at its premier) that are telematically connected. The two dislocated or relocated players now embody the coordinated and cooperative aspects of the original work in a cyberspace of telepresent interactions.

Replacing the three semi-octagonal wheels, *The Televirtual Fruit Machine* has a spherical form that is divided into two parts that can each be manipulated in the televirtual space by the two remote operators using the three-axis joysticks. The outer surface of these shapes again carry images of fruit, but additionally a map of the world covers the inside surfaces. Joining the two halves joins both the fruit and the world, signaling a successful aesthetic and physical connection between the distant players, which is further confirmed when the rewarding shower of virtual coins appears in the local currency of each location. An ISDN link between the two installation sites enables a real-time videophone connection so that the remotely located players can both see and talk to each other while maneuvering the virtual objects in cyberspace.

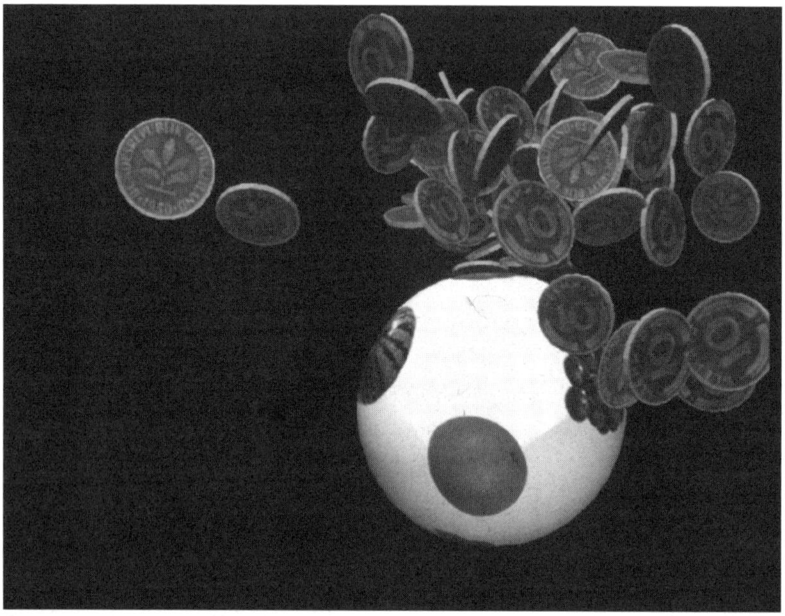

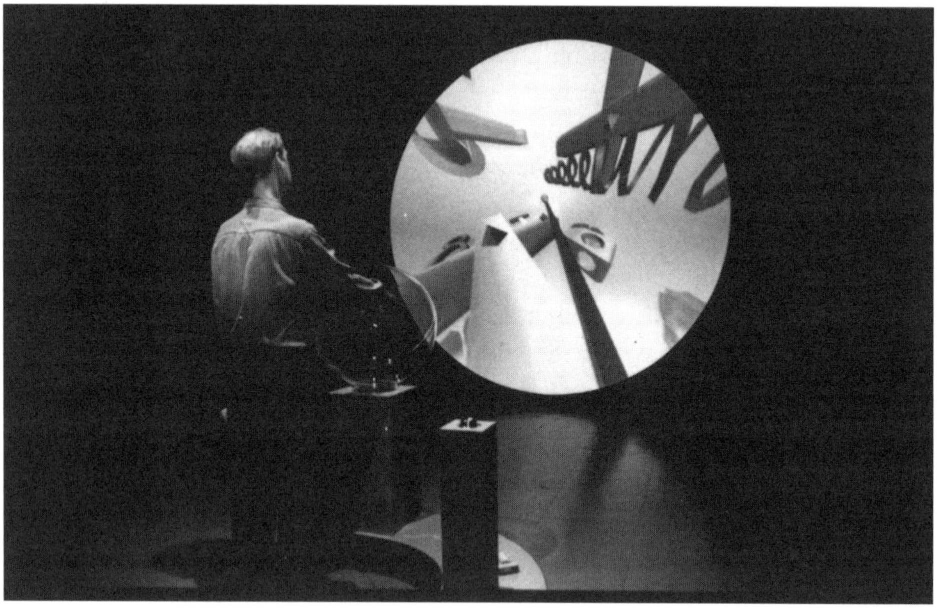

Figure 17.1
Agnes Hegedüs, *The Televirtual Fruit Machine,* 1993, interactive telecommunication installation connecting the ZKM Karlsruhe and the ICC, Tokyo. Software by Gideon May.

Figure 17.2
Agnes Hegedüs, *Handsight,* 1991–1993, interactive computer graphic installation. Software by Gideon May.

These components—the interlocking spheres, the fruity imagery, the physicality of the interactivity, and the communicational infrastructures—together evoke a particularly sensual dimension in this work. Margaret Morse describes the "cooperation with a skilled player as an almost erotic pleasure . . . playful choreography takes on a seductive pattern of approach and avoidance by anticipating the partner's moves."[5]

METAPHORS OF THE BODY

The hand has always been an elementary mediator between thought and action. And this is increasingly so in our technological era, where merely fingers on a keyboard can release enormous forces. The interactive interface design in my work reflects my fascination with the hand, for its gestural efficacy in a symbolic context, for its power to mediate between the body and machines, and for the sense of control and assurance it gives us in our handling of our surroundings. And hands become especially significant in our handling of the immaterial situations released by the new technologies of simulation. Here they are both the pragmatic and symbolic point of contact between our bodies and what is incorporeal. In the tactility of their actions, no matter how slight, our body maintains a physical context of experience and understanding.

Handsight[6] (figure 17.2) and *Between the Words*[7] are interactive installations where I have attempted to articulate this delicate relationship between the body and the virtual by means of a carefully considered aesthetic and conceptual strategy directed at the functionality and signification of the viewer's hands.

Handsight is constituted by a circular projection screen, a hand-held interface that has the form of an eyeball, and a transparent sphere with a hole into which the viewer can insert this interface and move it around inside. Holding the interface outside, the transparent sphere is represented as a virtual eye on the projection screen. When entering the sphere, the viewer enters this virtual eye through its iris and sees an image tableau, which is located within its virtual interior. Moving the interface within the transparent sphere, the viewer is able to explore all the features of this virtual image tableau from every point of view.

The intention of this work is to emphasize those salient aspects of virtuality such as telepresence and the reembodiment of the senses. It creates a transverse relation between the real and the virtual by correlating a physical object with its virtual representation. In *Handsight,* the interface becomes a symbolic object of perception. It takes two primary senses—sight and touch—and conjoins these two in a synaesthetic manner. The eye-shaped interface enables the viewer to locate himself as a virtual presence inside a virtual world. At the same time, this hand-held eye maintains the physical presence of the viewer within the transparent sphere that soberly circumscribes the boundary of that virtual space. This consolidation of the body and virtuality is further emphasized in the concentric layers of identification between the computer-generated eye, the hand-held model eye, and of course the actual eyes of the viewer. Unlike the journeys of Alice in Wonderland, who had to change her size to accommodate to the worlds that she entered, here the travel in a virtual world becomes an effortless journey by means of a simultaneous displacement of the viewer's body in both the physical and virtual dimensions.

Whereas *Handsight* makes a synaesthetic correlation between the hand and the eye, *Between the Words* makes the hands themselves the subject of such a correlation between the tactile and the visual. *Between the Words* constitutes a specific modality of communication between two persons where facial expression and hand gestures are its essential agencies. Two people face each other through an opening in a wall, where an optical system of semitransparent mirrors allows computer-generated virtual hands to appear in the space between them. On each side of the wall, there are holes where the viewers put their hands to manipulate multiaxis joysticks, which control these two pairs of virtual hands. Thus the viewer's real hands disappear within a technological system that then re-presents them in a virtual space. The merely functional directional movements of the viewers' handling of the joysticks translate into a peripatetic selection from a large vocabulary of prescribed hand gestures, which are contained in the work. As these virtual hands are interactively moved, there is also an algorithmic metamorphosis of the lines that constitute each hand, creating an interplay of abstract transitions between each gesture.

In the disembodied domain of televirtual networking, the fabrication of surrogate body languages is a cardinal endeavor. In the virtual, we are the instigators of communication but also the spectators of other elaborate communicational processes that express the singularities of the hardware and software morphologies that we have created. *Between the Words* substantiates these two levels of conversation, the direct and the mediated, and allows two viewers to engage in an improvised visual communication with each other via a metalanguage of gestural permutations. Distant from each other and yet face to face, these people interactively manipulate their surrogate virtual hands that are at the same time vitalized by their algorithmic aesthetics.

METAPHORS OF KNOWLEDGE, KNOWLEDGE SPACES

Digital technology produces technical apparatuses that deal with information in an atomic and fragmented state. These physical properties deeply influenced our conceptual and philosophical approaches to the world of information. We are given a new kind of freedom to use meanings disengaged from their original context and place them into other constructions. I am fascinated by the tension between cultural heritage and these discontinuities created by the new technologies. Each of my artworks has a symbolic significance in that they locate an emotional and intellectual moment in a personal history, which is then objectified.

Handsight is a form of "memory theater" of these symbolic objects from my past works. These are interwoven in the virtual space with another range of symbolic objects—those votive signs found in the traditional Hungarian "patience jars." Thus emerges an experience that belongs neither to myself nor to the past but becomes an autonomous (interactive) event in the present.

This somewhat eclectic convergence of manifold iconography is especially evident in the installation I have made for the ZKM Media Museum in Karlsruhe, the *Memory Theater VR.*[8] Conceived as a "memory theater" of pioneering virtual-reality technologies, conjoined to current artistic VR practice, and contextualized by a cultural history of relevant precedents, a didactic intention is amalgamated into a careful aesthetic construction. As if it were as grand *objet trouvé,* this work is almost wholly constituted by all of its quotations, and from the outset one can recognize the maternity of *Handsight* and the paternity of Jeffrey Shaw's *Virtual Museum*[9] in its basic configuration. A description of this can best begin by identifying its three separate yet closely interconnected spatial morphologies:

THE REAL ENVIRONMENT

The installation is built symmetrically around (and incorporates) two columns in the ZKM Media Museum. The external aspect is a 9-meter diameter cylindrical wall built up from rectangular panels that are as similar in appearance to theater flats—exposed wooden frames and hinged support struts on the outside and flat wooden surfaces on the inside. The entrance is an opening in this cylinder, above which (like an old cinema) there are neon and illuminated signs announcing the identity of the space *Memory Theater* and the title of the work *VR.*

The interior surface of this cylinder is painted in a rough theater style showing black and white images derived from well-known works of expressionist cinema (such as *The Cabinet of Dr. Caligari*), with an iconographical emphasis on shadowy staircases.

Inside, opposite the entrance, there is a large white smoothly curved projection screen attached to the wall. The cylinder has no ceiling, so the viewer is confronted with the actual museum ceiling from which two structural columns on either side come down to the floor. In the far center of the room, between the screen and the columns, there is a white pedestal. In the near center of the room, between this pedestal and the entrance, the user interface assembly is positioned.

THE SURROGATE ENVIRONMENT

The user interface assembly is constituted by a fully detailed transparent Plexiglas scale model of the real installation space—its framed cylindrical wall, projection screen, and white pedestal.

The interface is a hand-held device contoured around a photographic image of that often published NASA picture of a woman wearing a head-mounted display, her hair blowing in the "virtual" wind, and her outstretched "data-gloved" hand pointing into the virtual world. This device contains a 3D position tracking system,[10] allowing the viewer to use it to explore the interior (and exterior) space of the Plexiglas model, which, in turn (like a virtual camera), also controls the synchronous exploration of the computer-generated virtual environments that are projected onto the large screen in the real installation space.

THE VIRTUAL ENVIRONMENT

This is the space that is computer-graphics generated and projected onto the screen within the installation. The viewer is able to interactively navigate this space using the interface device.

The virtual space reproduces the architectural form and appearance of actual installation (and consequently of the interface assembly as well). It is a circular environment showing the two columns, the pedestal and the projection screen, all optically presented at one-to-one scale in relation to the real space. Looking upward in this virtual space, one also sees a full-scale representation of the real Media Museum ceiling that covers the installation. Only the interface assembly itself is left out of this overall simulation.

Using the interface device in relation to the Plexiglas model, the viewer has the following methods of navigating the virtual spaces. First, by moving the device within the Plexiglas cylinder, the viewer can explore the interior of a virtual environment. Second, by lifting the interface above the model, the viewer is offered a "menu" that allows the choice of four different virtual environments. Third, moving the interface around the exterior of the model, the viewer is shown a dematerialized wire-frame representation of the circular wall.

The specific design of this work allows the viewer to explore four different virtual worlds within the boundary of the one physical model of the space. When the viewer holds the interface above the model, its virtually represented ceiling is seen to be divided into the four segments of a compass—north, south, east, and west. Moving the interface figure down into the model through one or other of these four segments ("directions") will constitute a choice of one of the four different virtual worlds. At any time, the viewer can shift to another virtual world by moving the interface up out of the model and then choosing another direction to enter down through.

The installation offers many levels of correlation between the virtual spaces and the actual space that contains them. Specific and significant components of the real situation—the cylindrical enclosure, the columns, the ceiling, the pedestal, and the projection screen—are iterated in the virtual situation and given specific functionality there. The empty pedestal in the real room has different virtual objects put on it in each of the four virtual environments. The projection screen in the real room is used to visualize the virtual scenes—but within those scenes that screen becomes a film projection screen. The installation's cylindrical painted interior enclosure becomes the surface for various other images. And the columns and ceiling always remain intact to reaffirm this correlation despite the differences.

The four virtual worlds together constitute an idiosyncratic history and "cosmography" of virtual reality. They identify aspects of the broad range of VR strategies such as representation, reconstruction, replay, and re-embodiment and put the (now almost mythical or for others fatuous) existential ambitions of VR into a broader context. The following descriptions of the south and east environments give an introduction to the content design of these four worlds:

- *South* Here the virtual environment is a virtual museum of early VR technologies. On the virtual projection screen, there is a movie showing Jim Clark (founder of Silicon Graphics) using Ivan Sutherland's head-mounted display (HMD). On the pedestal, there is a 3D model of the logo of the first Internet browser by Mark Andreesen. The circular wall is textured with the typical scenery used in a military flight simulator. The interior space is occupied by a reproduction of the Scott Fisher team's 3D wire-frame model of their lab at the National Aeronautics and Space Administration, which was made to be seen via their head-mounted display. The original wire-frame representation of their "data glove" (sequentially gesticulating) also follows the viewer's point of view. On the walls of this NASA lab, there are three wire-framed "pictures"—Mark Bolas's 3D deconstruction of a Mondrian painting, a photograph of Morton Heilig's *Sensorama,* and a morphed animation of Scott Fisher using his HMD and data gloves. Three large 3D signs—*A, 2, Z,* taken from Jeffrey Shaw's *Virtual Museum*—also bounce randomly about inside this wire-frame room.
- *East* A place of imagination and fantasy. The ultrawide-angle and paradoxical optical arrangements used in the visualization of this environment evokes the child's-eye view of Tamas Waltzky's *The Garden.* On the virtual projection screen, there is the mirror passage sequence from Jean Cocteau's film *Orpheus.* On the pedestal, a golden piggy bank spews coins from the slit in its back when you approach it—a reference to my *Fruit Machine* and Jeffrey Shaw's *Golden Calf.* On

one half of the circular wall, there are drawings by kindergarten children, on the other half drawings by Antonin Artaud—interspersed with pinned illustrations from *Alice in Wonderland.* The floor is textured with a photo of various children's toys. Some components are modeled in 3D, such as a tiny frog (prince), which grows very large when the viewer gets close to it, as well as a rotating gyroscope (with its shadow), taken from my video work *Ise d'oil.*[11] There is a child's bed in which an image from Paul Sermon's *Telematic Dreaming* is mapped onto the sheets. The Tin Man from *The Wizard of Oz* is standing over the bed. Behind him a large animated block puzzle fits together images from fantasy comics and science fiction films (such as *The Invisible Man* and *The Golem*).

Knowledge spaces are embodied spaces. *conFIGURING the CAVE,*[12] made for ICC Tokyo, is an aesthetic and conceptual discourse on the theme of the lost and regained conjunction of body and space. The work takes as its starting point the historical perception of the harmony between macrocosm and microcosm and reconstitutes this equivalence in a technologically conditioned form, which reflects the dialectics of our contemporary perception of our location in the world. Algorithmic and representative imagery, supported by an evocative electronic musical score, create an open narrative structure that can be interactively configured and individually interpreted by each viewer.

At the center of the physical *CAVE*[13] is a wooden puppet, which is used by the viewers as an interface to control various transformations in the audiovisual space. Visitors are invited to play with this puppet, freely moving its body and limbs and head and in so doing modulating and exploring each of seven singular virtual worlds. The visitor can move from one world to the next by first covering and then uncovering the puppet's "eyes" with the puppet's hands. In each world, the interactive functionality of the puppet is differentiated, and the visitor gradually discovers the various ways that the images and the music react to the handling of its body. The following description of five of the seven worlds introduces to the specific forms and functionality of this work:

- *World 1 MATERIAL* Constellations of geometric forms move about in a complex yet organic symmetry. Starting in the distance, the action of the puppet causes them to come closer and enter the *CAVE* space. Some of these shapes carry fisheye photographs on their surfaces—pictures of a tearoom, the foyer of the Tokyo Opera City building, and the homes of the homeless in Sinjuku station. At the center of this organic constellation, a long wooden rod moves dramatically in and out of the *CAVE* space, sometimes saturating it with its interior colors.

- *World 2 LANGUAGE* Layered walls of changing texts, alphabetic characters, and hieroglyphics constitute the impression of a universal matrix of languages and information. On the *CAVE* floor, there is a representation of an ancient Chinese rubbing stone carved with text. This all is set against the background of an ancient Hebrew astrological map. All movements of the puppet's body and limbs dynamically shifts these text walls about in the *CAVE* space. Tilting the puppet upside down suddenly creates a vortex of alphabetic signs around its head.
- *World 3 MACROCOSM* Representations of the five Platonic geometric solids continuously move from the distance into the *CAVE* space. Movement of the puppet's limbs interactively deforms their symmetry. These geometric shapes are set in a visual space of three rotating concentric spheres. The first is the eminent surrealist map of the world, the second is a satellite picture of earth where the continents have been transformed into clouds, and the third, in the far distance, is a traditional astrological map of the heavenly bodies.
- *World 5 UNION* Two undulating image planes constitute this space: one is a satellite relief photograph of the Hiroshima geographic area, the other is a daguerreotype of two naked women. Tilting the puppet from vertical to horizontal causes the image of the two women to move downward and upward. Evoking some ancient creation myth, they merge in one extreme with the earth's topography, and in the other they become the firmament. When above ground level, this image of the women can be sensually distorted by movements of each of the puppet's limbs.
- *World 6 PERSON* In a fiery space like an alchemical inferno, a mirror image of the wooden puppet stands in the *CAVE,* casting shadows on the walls and floor. The reflected and shadow puppets echo movements of the real puppet. Seven objects dance in the fire without ever being consumed—a candelabra, a compass, a camera, a gyroscope, a

child's wagon, a blue sponge, and a cube on whose sides there are six symbolic representations of the human figure. When the puppet's limbs are folded, all these objects (except the sponge) converge with the mirrored puppet's body.

PRIVATE SPACES

Things Spoken[14] is my first work for CD-ROM. It reflects a desire to turn within myself and give form to an idiosyncratic interior world of freely signified objects. Instead of internalizing external phenomena and trying to locate myself in those tenuous spaces, in this work I set out to externalize the feelings and narratives I have extracted from my environment. In this way, I erect a virtual projection of myself, insofar as such a fragmentary assembly of things that I have deeply drawn unto myself can construe me.

Most people collect objects during their lives: these can be gifts, souvenirs, mementos, personal artifacts, and found things. Their significance for their collectors are usually contextual and personal: other people can only suspect what meanings are attached to them by the owners. *Things Spoken* presents a selection of objects that I have collected and that have been put together so that the viewer can make an interactive exploration of both their singularities and their possible (inter)relationships.

Fifty objects are displayed. Each has been digitized on a flatbed scanner, whose consequent transformation of the original objects is a form of aesthetic reconstitution characteristic for multimedia. Embedded in a machinal darkness, the objects reveal themselves insubstantially, idiosyncratically in the reflected red, green, and blue light of the scanning process. Appearing almost like ghosts, they seem like transcendental images of imagination.

The objects present themselves in an endless row that can be scrolled left or right on the screen. The viewer can sort these objects in various ways—by objective criteria (such as size, weight, color, or function) or by more subjective criteria (such as in the case of gifts, the gender of the persons who gave them to me). In this way, that "feverish" method by which digital archives can be reorganized according to any criteria is here applied in a manner that is as gratuitously personal as the objects themselves.

Each object is accompanied by my spoken explication of its significance to me, where I disclose the contextual and personal properties that led me to keep these often trivial things. I have also asked a number of friends to talk spontaneously about these objects, and what they have said constitutes a second parallel set of narratives to my own that the user of the CD-ROM can freely shift back and forth between. My friends have no personal relation to these objects, but because these persons are close to me, as are the objects, a triangulation of relationships is established that amplifies the ad hoc meaningfulness of these objects.

The third layer of interactivity offered to the viewer comes from within these spoken narratives. Specific words in each of the narratives are hypertextually linked to any reoccurrence of those words elsewhere. In this way, the viewer can instantly make links between objects and their associated stories. This leads to chance encounters between the objects and chance conjunctions in the narratives—an amplification of potential relationships that lets the viewer discover further layers of congruency and signification within this very personal collection of objects. It creates a structure that emulates human thinking, not in the sense of artificial intelligence but in the way people talk to each other, ideas follow each other in a conversation, and recognized similarities of feelings and thoughts connect people on an emotional level.

To conclude, I consider that this peculiar undertaking stems from a deep desire to shift away from the somewhat abstracted constructions of my previous works and confront the idiosyncratic configurations that everyday proposes—as if living itself was one grand found object. Derrida[15] makes clear that the actual activity of archiving tends to kill the things that are archived. Assembling my things in *Things Spoken,* I effectively disassemble myself to configure my own mortality. Identity, child and mother, man and woman, relationships of people close and distant, privacy and openness: all is confounded. I feel like I'm forcing memories to remember their emotional tatters, enforcing the absurd conjunction of things to expose their (ir)real transcendental (in)coherence.

NOTES

1. Nigel Pennick, *Games of the Gods: The Origin of Board Games in Magic and Divination* (Boston: Weiser, 1989).

2. Pennick, *Games of the Gods.*

3. Agnes Hegedüs, *The Fruit Machine,* 1991, interactive computer graphic installation, software by Gideon May, collection of ZKM Mediamuseum Karlsruhe.

4. Agnes Hegedüs, *The Televirtual Fruit Machine,* 1993, interactive telecommunications installation connecting the ZKM Karlsruhe and the ICC Tokyo, software by Gideon May.

5. Margaret Morse, *The Televirtual Fruit Machine,* in *Hardware Software Artware, Confluence of Art and Technology* (Karlesruhe: Cantz Verlag, 1998).

6. Agnes Hegedüs, *Handsight,* 1991–1993, interactive computer graphic installation, software by Gideon May.

7. Agnes Hegedüs, *Between the Words,* 1995, interactive computer graphic installation, software by Sebastian Egner & Adolf Matthias, construction by Bossinade Lightworks.

8. Agnes Hegedüs, *Memory Theater VR,* 1997, interactive computer graphics environment, software by Gideon May, construction by Bossinade Lightworks, collection of the ZKM Mediamuseum Karlsruhe.

9. Jeffrey Shaw, *The Virtual Museum,* 1991, interactive computer graphics installation, collection of the ZKM Mediamuseum Karlsruhe.

10. Polhemus, Fastrak.

11. Agnes Hegedüs, *Ise d'oil,* 1988, video.

12. Agnes Hegedüs, Bernd Linterman, Jeffrey Shaw, and Leslie Stuck, *conFIGURING the CAVE,* 1997, interactive virtual-reality installation, collection of the ICC (NTT InterCommunication Center) Tokyo.

13. *CAVE* visualization system developed at the University of Illinois.

14. Agnes Hegedüs, "Things Spoken," *ArtIntAct 5,* CD-ROM magazine (Cantz/ZKM 1998).

15. Jacques Derrida, *Achive Fever: A Freudian Impression* (Chicago: University of Chicago Press, 1996).

18

Reflections on Some Installation Projects

Judith Barry

When I first began making installations, I wondered what it would be like to render the image world inhabitable. How would the body make sense of these spaces? What would it be like for the body if it or its agents were suddenly thrust into a three-dimensional image-derived space? In an early transitional work done for a Coca-Cola Corporation convention (1980), I tried to use the structure of the company's advertising campaign—"Bob Hope sharing a Coke in the developing world with the indigenous population"—as the linchpin to animate the Disney-ish eat-a-thon that was the raison d'être for this event. So instead of bland food-town displays, I proposed showing how the developing world really lived at the time that Coca-Cola gained worldwide hegemony. Not surprisingly, my designs for these dioramas were censored by Coca-Cola, but this piece led me to begin to scrutinize advertising and corporate display as an articulated ideology about space.

Over the years, I have expanded on these initial investigations from a variety of perspectives. In what follows, I briefly describe several installations that illustrate some of the approaches I have taken to spatialize different types of image experiences through installations. This is not a linear description of my work but more an attempt to delineate some of the questions that have interested me. These include notions about the city and its architecture, public and private space, rural and agrarian landscapes, first- and third-world relations of capital, the flow of information and its capital accumulation. This list is not programmatic but reflects the way I have thought about how contemporary experiences and their representations might be interrogated.

One of the major issues that has illuminated my thinking has been the notion that it has become increasingly difficult to "image" the present because so many of the necessary relations that govern our passage and behavior in urban space occur along the primarily invisible circuits of information flows. As these flows determine more and more the nature of our interactions, it has become more difficult to actually "image" the present. On some level, this is an effect of our increased access to technologies. A quick repast of the "pomo" debates about the "loss of the real" (best described by Baudrillard, among others) and of the fears that virtual reality and the simulacra will replace "real lived experience" should remind us not that there is no real "out there" any more but that the relationships between our senses of what is real and can be made manifest and what is real but has no physical form has been occluded. For example, how do we represent this space of "flows"? One way this might be thought about is to confront our relation to images. In fact, given the problems in determining what is real, we might ask "what are images," and what possible relations can and do we have with them? How do we perceive the changing paradigm shifts that digital computing has rendered in the image world? For example, where is the original, or how many originals can we have? The commonsense answer says that images inhabit the place between the object and the viewer. We know this from the analytical work by semioticians, but it is increasingly in the realm of "imagining"/"imaging" that the transformed nature of our impression of this condition can be represented. As Paul Bishop discusses it, this problem is one of both access *to* and representation *of* an interdiscursive space in which our relationships to images is negotiated. To some extent, this problem in access is attributable to a kind of loss of "imaginable depth," a collapse of imaginary space, because of the confusion produced by both the ability to simulate images and the multitude of visual images that we encounter. This perceived loss can result in an inability to delve deeply into the psychic structure of imaginary space, so that we remain on the surface of an image and do not penetrate into a deeper realm that would connect us to the underlying metaphorical nature of experience and hence forge new understandings of meaning. Bishop argues that this situation leaves us ungrounded, unable to distinguish between images, the effect of their ever-shifting grounds, and our inability to locate ourselves within them.

What might be a remedy for this sense of a loss of ontological ground? One possibility would be to embrace this destabilized terrain and to acknowledge this condition as a given using a Deleuzean model that calls for ever-shifting notions for determining a ground within an undefinable reality. Yet another might be to attempt to locate competing representations within these ever-shifting grounds with the aim of interrogating the conditions that produce these effects. Examining these competing representations is not easy and is destabilizing, but these are the conditions of the present, and it is these relations—described from multiple viewpoints, examined from numerous points of view—that need to be described, particularly if we want to understand the conditions that produced them.

To give a more concrete example of this in the following discussion, I try to describe the ways I have used notions about technology as a frame of reference to interrogate various issues.

Sometimes I think the term *technology* has become like the term *psychology*—a word that eschews meaning unless rendered visible through the very specificity of its use. At a conference on the millennium (another impossibly difficult concept to characterize), I was asked why it seemed that women artists of my generation (Kruger, Holzer, Simmons, Charlesworth, Sherman) used technology in their work, while the men (Schnabel, Salle, Longo) used painting, a less technologized form. At the time, I answered that I thought there was some confusion between the media's need to create categories and make gender characterizations and the actual histories of this moment. I cited several prominent male artists who also use technology in their work, including two of the previously mentioned men. Yet I was bothered by the exchange and realized that the men were significantly less associated with gender politics than the women and that perhaps gender politics was really what was at stake in describing these seemingly different relations to technology. The art world is still a place in which gender is visibly at issue. Perhaps more so here than in other cultural arenas because the authorial presence must be gendered and by this association guarantee the authenticity of the work. And while I began to show publicly as an artist by representing overtly feminist issues—first in performance and later in single-channel videos—my perspective evolved to situate issues as an active participant and as an often unacknowledged term within many other questions, including questions of technology.

Technology is such a broad term that it is difficult to define it for the purposes of a conversation about women *artists working with technology-based art forms* or even to categorize and distinguish among the various ways it could be defined. Certainly, I have used technology, in the most general sense, as a kind of instrumentality (as in transforming the natural world through some human intervention) in all of my projects. However, I have often sought to interrogate multiple and competing technologies both from the standpoint of their ideological implications and through an examination of the forms that they might take within the specific subject matter my work.

Nearly all of my installation projects use technology to render their manifest content in a visible way, quite often eschewing any built evidence. For example, when you turn on the lights and turn off the slide, film, or video projectors in many of my pieces, from *In the shadow of the city . . . vamp r y* (1982–1985), to *Rouen: intermittent futures/touring machines* (1993) and *Voice-off* (1998–1999), there is often nothing to see, nothing to experience, as the pieces come into being only as the technology works. Whether this might be seen as a transparent use of technology I cannot say, as I assume the viewer is aware in a commonsensical manner that technologies in various forms are producing the experience of the work. Yet that is only one way that I deal with technology.

In many pieces, I have used an understanding of how certain technological effects of representation illuminate aspects of "the technological" that are deeply inscribed in daily life while making specific relations possible. For instance, *Maelstrom: Max Laughs* (1988) is an exploration of information technologies and the changes in daily life that we have experienced as our private and public sense of space, our experience of place, and our ways of negotiating these spaces are transformed by information–communication, storage, and retrieval. For *Maelstrom: Max Laughs,* I defined these technologies very broadly to trace language made visible (as a visual trope) from the beginning of writing to its apotheosis—when the computer interface replaces our need for any overt representations of narrativized subject positions. For part of the installation, the viewer is taken on a 3D computer-animated, motion-control ride through new spatial paradigms, where the *I* and *eye* conjoin and where the physical body is kinesthetically invoked. In reproducing the viewer in this space, it was my desire to show how fundamentally simulated and yet simultaneously how convincingly real our sense of these spaces is, even as we are fully aware that our bodies cannot physically go to these places where our eyes have taken us. Further, what produces this ride and what is woven through the installation itself is a minihistory of graphics and graphic design as a spatial paradigm that both undermines and catalogs how design as a system of competing technologies works on the viewer. For example, in one section, which presents the paradoxical effects of abundant information—not as knowledge but as a storage and access problem—the viewer falls through a space into what seems to be a bottomless abyss, only to reemerge as a data stream reproducible in a computerized form.

If I were to describe all of the different graphic techniques that were used in the production of just this section of the project—live-action shots, 2D and 3D animation, blue-screen and other computer techniques—and attempted to show how they are used to produce various subject effects, it would be a rather long essay that would not allow the reader to really experience the effects I am describing. However, in foregrounding all these technologies overtly and visually, the viewer comes to some realization and understanding of the complexity of technologies, their ideologies, and the various subject positions as she or he is viscerally effected by them.

I am often asked how or if my work resists or alters the instrumental, commercial, or positivistic attitudes built into technology—a kind of "Are you for or against technology?" question. For me, the issue of the ideological impact of technology is a difficult one, as in many instances the technology has a different effect depending on how it is used and who is using it. The emancipatory properties of a word processor used by middle-class intellectuals versus the ability of the same machine to enforce a system more invasive than Taylorism, because it has the inherent ability to monitor key strokes, is just one example of how the same technology can be applied in different situations for a variety of ends. I think that more than resisting or altering any of the attitudes around "the technological," I try to make technology in its many guises visibly part of a larger ideological context so that how it works and what it means in a particular instance is inscribed directly in the work. In *Imagination, Dead Imagine* (1991) (figure 18.1), five video projectors hidden in a 10-foot minimalist cube describe a three-dimensional head that I have added to Robert Morris's famous mirror cube in an attempt to return the body to minimal art. That these images are electronically produced, defaced, and finally delivered "clean" only to be sullied again leaves no uncertainty for the viewer that this head, while perhaps at first seemingly human, is in fact technologically generated. As such, it is both alterable and restorable through the use of a slow "wipe" that by definition both seemingly cleans the image away and restores it again and again to its original, pristine surface. It is the agency of this "wipe" that renders this particular situation visible.

Another question that circulates endlessly around artists' use of technology is the question of transformation: How much can we or should we be able to transform technology through our use of it? And how does this technology use us? In a sense, these questions are also another way to

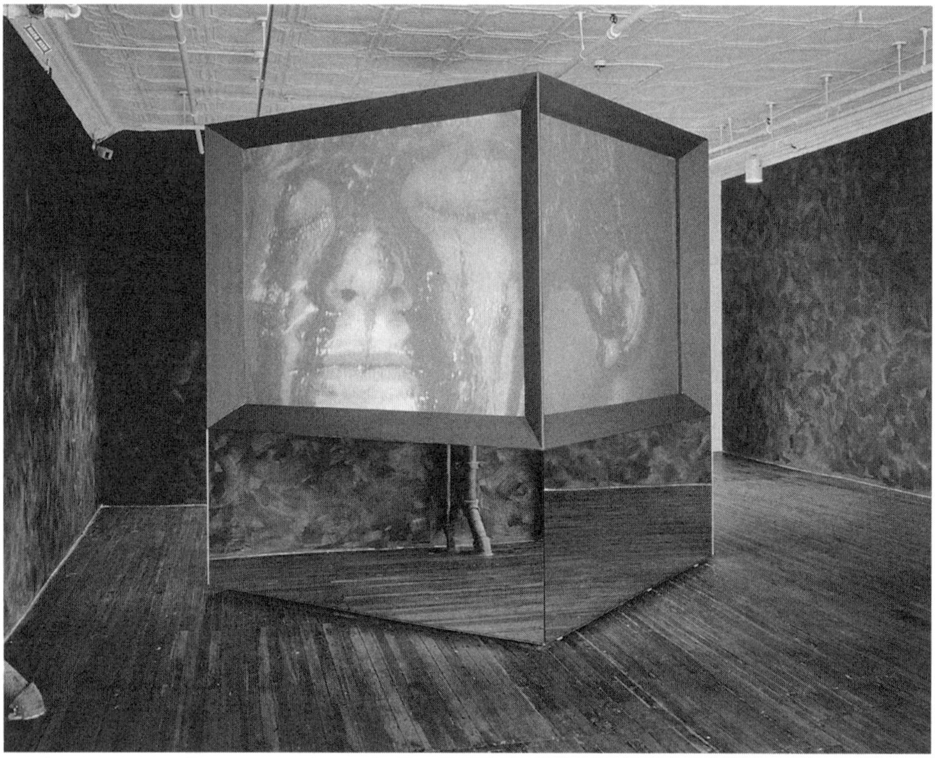

Figure 18.1
Judith Barry, *Imagination, Dead Imagine,* 1991, five-channel video projection/sculpture. Collection: La Caixa, Barcelona, Spain. Photo courtesy: Nicole Klagsbrun.

describe interactivity. I think anyone who employs technologically derived programs (applications) has the possibility of potentially effecting how we use technology in terms of figuring out something new to do with it. I have cynical friends, not of the art world, who consider Photoshop and Director the two greatest artworks of recent years, and while I don't agree with this, I do see these programs as producing a number of possibilities, especially in terms of the accessibility and interactivity. I think these programs are truly interactive, more so than what is often claimed to be interactive—that is, branching systems where you choose among several preordained paths. In terms of my own work, I have done what I consider interactive works where the viewer has input into the choices that can be made, but I have done them often in an old-fashioned way, as in *Ars Memoriae Carnegiensis* made for Carnegie International in 1991. Here the viewer was invited to create an imaginary museum using some of the ancient mnemonic techniques of the memory theater. By practicing the memory exercises included in the set of nine cards that were given free to each visitor to the museum, the user could build a museum in her or his mind as a memory theater, saving for future recollection whatever she or he deemed worthy of exhibiting. Once stored properly using mnemonic techniques, this "museum" could be revisited whenever the viewer wished to recall it. As one critic noted, this installation in an exhibition where each installation was bigger than the next was perhaps the largest of all and yet it took up no physical space.

Recently, I realized a four-channel immersive interactive video and sound narrative in collaboration with the artist-writer Brad Miskell. *Speedflesh* (1988) takes place in a 360-degree point-of-view (pov) theater. The povs of these four characters can be accessed by dialing a "radiolike tuner" that allows you to simulate *tuning into* the last five minutes of their lives by moving from one character's point of view to another across the dial. You are in the center of a world that is moving away from you at a rapid speed. A woman floats toward you and then away, elongated through the anamorphosis that is produced by the projection. Her movements seem to conceal and sometimes reveal the four characters whose lives are caught and displayed in the band of light that closes in around you. As a viewer, you can engage with the narrative by *tuning in* using the wheel that is at the center of the space. You experience the narrative through the eyes of each character as each character's pov is projected

360 degrees around the cylinder using three projectors and video-wall technology to split the image.

The structure of the work, the 360-degree projection, has antecedents in the history of the anamorphic image, a history that figures throughout the narratives of representation and also marks the Neo-Platonic period and recalls the power of hidden iconographies in describing new social orders. On one level, in using the anamorphosis as the form and content of the imagery of *Speedflesh,* we were attempting to show that even an apocalyptic moment poses other histories, which can be seen if you can decode them but will not necessarily be exhausted by this process.

Although we tend to associate interactivity with the computer, video games, and new technologies, the idea of interactivity between an artwork and the viewer has a long history, from the previously mentioned mnemonic devices to postmodern discourses: through engaging with the forms and symbols in an artwork the viewer discerns meanings. In my work, I often use narrative codes and architectural displacement to effect an interactivity on the part of the participant, which produces some of the same effects as more computer-based interactivity but is perhaps less obvious. For example, *in the shadow of the city . . . vamp r y* uses a double-screen projection surface that bisects a room, forcing the viewer to walk around it to see it. This displacement is underscored by the narrative fragments that comprise the story, producing a desire for normal cinematic narrative closure and the simultaneous realization that our desire for images is not easily sated as the viewer paces around the screen.

In *Whole Potatoes from Mashed* (1993), a project that contains computerized interactivity, part of the hermeneutics of that work involves figuring out how its various prerecorded sound tracks are mixed with on-site recordings of what you as a viewer say in that space. Here you find yourself seated, campfire style, as you read the pages from what seems like a dictionary strewn around the rug of a living room. As your conversation is recorded by the talking fiber-optic sculpture that is the main visual element in the work and then suddenly played back, you may be reminded of the publicness of communal spaces and the ways that technology creates communities even as it produces alienation.

Rouen: intermittent futures/touring machines (1993) requires another kind of interactivity on the part of the host institution whenever this four-channel video and sound, fiber-optic sculpture is installed. Initially commissioned by the city of Rouen for the inauguration of its National Urban Institute, this installation tours to other cities in Europe. To make sense of locating an urban institute in Rouen (definitely not an urban center), I cowrote a guide book for a new imaginary city of Rouen that has suddenly discovered that it is the center of the world. When this sculpture tours, the sponsoring institution reads this book and maps the Rouen sites onto the host city by making a video document that identifies these new host sites in that city. The piece becomes site specific once again as it interrupts the previously recorded three-channel video of Rouen with images of this new host city.

The question of interactivity in installation and sculpture also circles back to questions of how the body might be thought. My approach has been affected by my interest in sports, dance, architecture, and film studies. I went to school in the late 1970s, when questions of representation, gender, and subject positions were being articulated across a multitude of disciplines. Although initially I thought that performance art was a way for me to combine all of this, by the early 1980s I had begun making installations. In many of these works, I have tried to articulate a kind of synaesthesia for the participant, merging the body within the spaces constructed by the environment and allowing the viewer a variety of "subject positions." As the viewer moves through the space, the work comes to life.

What is at stake when we invoke the idea of technology and the body and question its disembodiment or reembodiment? Ideologies of technology speak of the imaginary relations that human subjects make with various modes of production on a continuum that is less and less tangible. I think we are undergoing a metamorphosis in terms of what constitutes the physical body and what we might include under its various regimes, particularly in terms of what we (however you think of a unified body and its parts) have to take responsibility for/negotiate with/live by, and so on. We are all cyborgs, as Donna Haraway has noted, and there are many questions about our cyborg bodies that remain unasked, let alone answered. It used to be that the writer created a fiction out of an external world that was reality, but now reality is seen as a fiction, and the writer must make sense of the fantasy that is the so-called real world.

By way of an illustration and as somewhat of an open-ended conclusion to some of the issues I raised in the preceding paragraphs, I want to conclude this text with a brief description of several other installation projects that address these questions.

For *Adam's Wish* (1988) (figure 18.2), a one-minute continuous video loop initially designed to project into the 60-foot dome of the World Financial Center, I asked the question, "Why do corporate spaces eschew any overt corporate iconography?" By allowing the dome projection to function as a tromp l'oeil investigation of various historical spaces in relation to the notion of a worker and questions of public and private space, I was able to humorously raise this question directly within a corporate environment. During the course of the escalator ride that provided the timing for the piece, the viewer was positioned in numerous *tromp l'oeil* spaces where suddenly parts of the overall image cohered, underlining how in corporate America, those who control space are never seen, while those who are quite literally "out of control" are made to fly.

In *The Work of the Forest* (1991), I used the notion of the panorama as a whirling room to structure the sense of interiority that characterizes art nouveau space. By providing multiple viewpoints through and above the three screens that comprised this video panorama, the viewer attempts to negotiate the troubled histories of art nouveau architecture, African objects, and the Belgian Congo, realizing that these histories are different and definitely competing.

Model for stage and screen (1987) uses a simple light and fog device to update a sixteenth-century anatomy theater by turning the viewer into a projector. The piece produces retinal effects in the viewer in a chamber where two disks are suspended over an illuminated floor. As you enter the chamber, your eyes experience retinal excitation caused by the light and fog in this space, and as you leave, you become a projector as you see the complementary color on the white walls of the antichamber. This piece demonstrates that vision is out of our conscious control and that the body is often the effect of external stimulation that acts on it.

(Home)icide (1993), a collaborative project with the architect Ken Saylor, took up issues of architecture and space directly within the redesign of one of the apartments that make up Le Corbusier's *L'Unite* in Firminy, France. It asked whether, in a "House of the Present," our living environments adequately reflect the ways we live: in particular, do they represent the discourses that make up our daily lives? In this project, the body was treated as a production of the discourses that mark and interrupt

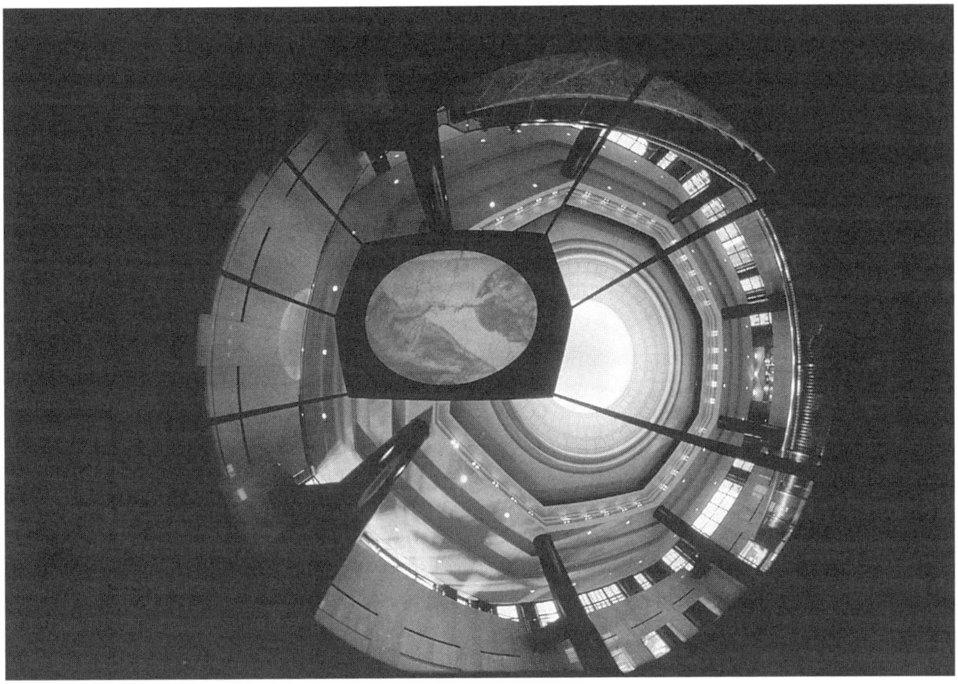

Figure 18.2
Judith Barry, *Adam's Wish,* 1988, single-channel video projection. Private collection, New York. Photo courtesy of Rosamund Felsen Gallery, Los Angeles.

the configurations of the traditional home, ever-changing as it is shaped by competing information. We sought to redesign the cell that Le Corbusier saw as reproducing the worker in a way that was reflective of all the contemporary functions of the home. Functionalism was turned upside down in a computer program that morphed lifestyle questions into multipurpose spaces, rendering the space itself a whim. Not surprisingly, this was exhibited on a computer, which constituted the main part of our project.

Hardcell (1994–1995), a collaborative project with the artist and writer Brad Miskell, treated the body as a site of recombinant computer debris, which over time seemed to cohere into a living organism somewhere on the continuum between human and machine. Using some of the minimalist sculpture tropes of accumulation where the viewer must discover the relationships between the various elements in the installation, *Hardcell* at first looks like nothing but dead tech at a dump site. Broken computers, circuit boards, bits of motors, and other pieces are piled knee-deep in a room. As you spend time with the installation, you begin to notice that parts of it resemble a crate and that this crate is the site of an organism that gradually becomes more animated the more time you spend with it, ultimately cohering into something that seems to have sentience.

This interest in animating technology continues in two recent projects. *au bout des levres* (1993–1996) is a group of five sculptures that use projection and animation to examine the relationship between the limits of language (as in what might be said or written and what cannot be said or written) and the insistence of the visual (as in what cannot always be described within language but insists on coming into being) and its possible representations in the visual realm. Each sculpture presents the viewer with a set of references whose rebuslike meanings become clear as the viewer interacts with the structure of the work and reflects on the shifting meanings portrayed by the piece as it responds in different ways—through video projections and through various mechanical and sculptural devices. Each piece is referential, referring to a common form such as an urn, and each is also not what it seems. For instance, the piece that looks like an urn is made of hundreds of feet of what appears to be hair but is actually spun plastic. As it turns, it presents you with a figure, but just as you draw near enough to see it, it disappears, only to reappear again as you move away.

The terror and possibility in the things not seen is a reflection on some of the relationships that might be posed by rethinking contemporary notions of landscape by using sophisticated computer techniques to form a five-sided video sign. In this project, the concept of landscape functions as a series of questions that frames the dialogs between what used to be called *nature* and *culture.* These dialogs, while referring to the long traditions of perspectival space, art history, natural history, and geography, also now reflect the largely invisible relations of the flows of capital and technologies of representations. When I began to work on this project, first shown at Insite '97 in San Diego, what was most striking was how different the cities of Tijuana and San Diego appear, even though they share essentially the same geography. The project evolved to become an interlocking series of five billboard-size video projections underscoring the fluid and interdependent relationships between the two cities. In rethinking *landscape* in contemporary terms, I used photography and animation techniques to pose antimonical relations as an ever-changing collage of possibilities. While each of the five-sided T-shaped signs is complete unto itself, the animated motion of the video propels the viewer around the sign, provoking a set of responses that the viewer can question. Rather than fixed and immutable, this landscape shows the possibilities inherent within the contemporary environment, where both what is seen and what is not seen have powerful consequences.

19

Do While Studio

Jennifer Hall and Blyth Hazen

HORRERE VACCUA

> "It was 1974, and the math department had just set up their first computer. On a Friday before a long weekend, I began to read a manual that came with the system. Figuring out enough to get a line of text to appear on the screen, I then sent it to the printer using another instruction. I waited a while, and when nothing happened, I found a command to repeat the instruction. What I did not realize was that the system needed to compile my command by sending it to another location over telephone wire. Still nothing, so I left. Tuesday morning, Professor Hochkins opened the door to find streams of perforated paper filling the room—each page covered with the words I had typed days before: HORRERE VACCUA. I had unknowingly created my first telecommunication artwork.
> —JENNIFER HALL, FOUNDER, DO WHILE STUDIO

BACKGROUND ON JENNIFER HALL

My media education began very early. As a child, I bounced back and forth between my grandfather, who descended from a clan of Hudson Valley School artisans and made his living by creating flawless copies of famous paintings, and my mother, who was a television producer in New York City. I remember when one of my grandfather's paintings was sold by some unsavory dealer as an original. Fate had it that then, at eight years of age, my job was to paint his name in lead on the canvas before it was gessoed. Being the honorable forger that he was, my grandfather could then guarantee that his original copy of an original master was secured forever in the annals of art duplication history.

My mother was the first woman producer of television commercials. In the seventies, I worked as her production assistant. From the Jerry Lewis Telethon to the research and development of Crazy Foam, I was immersed in the professional art of mass illusion. I became the agency's young mistress of edible props—painting white highlights on ears of corn and searing BBQ stripes on steaks with a soldering iron. Interpreting the dance of real and fake or truth and lies became a focus for me. I will always relish the experience of my first shoot. On the production set for a dishwashing detergent commercial, I turned to my mother, "Mom, what does virtually spotless mean?" "Shhhhsh!" she whispered. "It means it has spots!"

It is not surprising that my early artwork responded to appropriated content. At an age when I had barely outgrown crayons, I was collecting unused video footage and audio jingles from the postproduction studios—discarded artifacts from industry, repurposed as art. I also built tiny theater sets filled with scaled models of useless and absurd gadgets. I carefully photographed them to appear large and actual. By the time I went to art school, I had a portfolio of video footage that I didn't shoot and documentation of odd devices that didn't really exist. I brought the first video camera to my campus but was unsure what it meant to me as an artist. While in graduate school at the Massachusetts Institute of Technology, I used tracking devices from cruise missile warheads to generate data from dancers. It was there, among the labs in which new media emerged, that I first recognized the authority of technological reappropriation.

The greatest resource at MIT was the collaboration possible between people. Biologists, software engineers, artists, and theorists were joined in common research by the funding hip. In 1985, I started Do While Studio to forward what I learned about cross-disciplinary artwork at the Center for Advanced Visual Studies at MIT. Independent from the trappings of a larger institution, I prepared an artist studio space for technology collaboration. Soon after, other members of the MIT community joined—including Joan Shafran, Rob Haimes, and Laura Knott—to implement the first independent, nonprofit organization on the East Coast dedicated to the fusion of art, technology, and culture.

DO WHILE STUDIO

Located on South Street in downtown Boston, Do While Studio[1] is flanked by old neighborhoods constantly reinventing themselves. Today the street is inhabited by upstart Web design and multimedia companies, all looking to catch the newest wave of transnational commerce. We are strategically placed among these organizations, offering a human-scale perspective on the research and development of new media.

"Do while" is a conditional command from early programming languages. It instructs the computer to continue to execute a routine until another specified event occurs. At Do While Studio, it also suggests a cultural construct—requiring automated ways of doing things to be interrupted by human reflection. Do While Studio is a small and focused community offering an alternative to the way technology is assimilated in day-today art practice. It offers a place to think about art outside of academic, military, or commercial research models. We ground our collaborative works in the development and critical appraisal of digital technology—always in concert with traditional forms of artistic expression such as painting, sculpture, poetry, choreography, storytelling, music, and design.

Projects at Do While Studio follow in the rich tradition of the artist as tool maker. From the grinding of pigments to innovations in ancient casting and stone-cutting techniques, artists have been developing and altering the technology of their day to fit the nature of individual expression. As new tools emerge, new working methods require moving within multiple software packages, hardware platforms, and networks. To succeed in this complex environment, professional artists need a community of informed peers with different perspectives on the development, implementation, and social ramifications of emerging technologies. Artists at Do While Studio are currently working with engineers, management consultants, physicists, neurologists, and anthropologists. These kinds of creative partnerships are at the heart of art making at the studio.

> Alliances with industry are a natural for not-for-profits such as Do While. We can provide ideas and solutions that cannot happen within the confines of a company. The goal is to create a unique symbiosis between the art environment and the industrial environment that is beneficial to them both.
>
> —JOAN SHAFRAN, PRINCIPAL MEMBER, DO WHILE STUDIO

Notions of community, diversity, and accessibility are being retooled in the wake of emerging technology. Technical innovations—in particular, telecommunications—appear to defy boundaries both physical (neighborhoods, cities, towns, states, nations, and continents) and cultural (race, ethnicity, gender, art, and science). Artist and audience are tethered by economic and political access to the global grid. Hardware and software interfaces define aspects of these interactions, so we must be aware of who builds them and why. End users and creative problem solvers contextualize evolving technology, offering valuable insight to the developers of imaging, information management, and networking systems.

Do While Studio has historical ties to twenty-five years of New England art technology innovation. Dozens of artists, educators, and students have come to the studio to engage in dialog and collaborate on a variety of projects. The studio provides residences for individuals looking to develop their own art making in this environment. During this time, artists work in a variety of contexts ranging from research and the planning of large-scale projects to the development of artworks. The projects outlined below are examples of work done at Do While and often take several years to move from research to funding and finally to implementation.

Out of Body Theatre[2]

Principals: Jennifer Hall, Mat McMaken, Meg Young

One of the first large-scale projects completed at Do While Studio was the *Out of Body Theatre* (1986–1992). The focus of the theater was a computer-controlled robotic marionette that would interpret, rather than replicate, real human movement by using motion-tracking technology. Sensors, motors, balers, and other hardware were mounted on a full-scale reproduction of a human skeleton suspended from the ceiling. Radio frequency signal processors interfaced with MIDI controllers to input and export marionette motion, projections, lights, and sounds. The marionette would track, assess, and reassemble according to the chaotic patterns of random input within her immediate environment. At one point during the construction of the installation, a gull flew over the gallery skylight, casting a shadow into the input range of the marionette. She spent the rest of the afternoon trying to replicate the gesture. Comparable to a convoluted Icarus, the attempts to create humanness had been eluded by her desire to reach beyond and fly.

From the Storm[3]

Principals: Jennifer Hall, Dr. Steven Schacter

From the Storm (1992–1995) was a collaboration between artists who were living with temporal lobe epilepsy (TLE) and neurologists who were studying the disorder. Artists at Do While Studio and Dr. Steven Schacter of Boston's Beth Israel Hospital were researching the notion of human perception from different areas of expertise. Together they created a collaboration that showcased the artwork and life stories of twenty-seven artists throughout the world who were living with TLE. Along with a high-resolution multimedia database of visual art and video documentation, this project was presented at the Contemporary Museum in Sydney, Australia, during the 1994 International Neurology Association Conference. Since then, the multimedia archive has been distributed to museums, conferences, and heath-care facilities throughout the world.

World Wide Simultaneous Dance[4]

Principal: Laura Knott

> On the Internet, *World Wide Simultaneous Dance* will become its own site-specific performance. The project holds out the possibility of bringing together established and evolving technologies to create a completely new form of art. But perhaps most important, *World Wide Simultaneous Dance* offers the opportunity to build a community of people who think it would be beautiful and important to know that at some point in our lives, there were people dancing all over the world.
>
> —LAURA KNOTT, CHOREOGRAPHER

World Wide Simultaneous Dance (1998) was envisioned as a uncomplicated and poetic gesture—people all over the world dancing together during the same thirty minutes. The development team needed a way to connect the participants using current and internationally accessible telecommunication technology—a much more difficult task.

During the two years of project planning, development of technical options for live connections exponentially increased, but access for most artists still lagged. In response, the project sponsored technology for participants who could not gain access otherwise. In many countries, it was more complicated to pass a digital camera through customs than it was to access the Internet. In the end, many participants connected via concurrent video contact, others connected by asynchronous chat, and some later posted descriptions and sent images of their dance using e-mail and postcards. In June 1998, on one evening for part of one hour, individuals and groups from all over the world danced together.

Selections from the Algebra Drawings[5]

Principal: Blyth Hazen

> For several years my oil paintings have taken a turn toward chaos and a return toward order. I would lay down a series of marks on a canvas until the surface was a mass of disordered marks

> and color. I would then proceed to create form, structure, and movement by the selective painting out and addition of new marks. I always felt that I was growing the content of my work rather than portraying it. Readings on complexity theory and the common patterns underlying mathematics, biology, physics, and the social sciences furthered this inclination. A desire to work in an environment where this type of interdisciplinary conversation was encouraged brought me to Do While Studio in 1996.
>
> —BLYTH HAZEN, PAINTER

Standing in front of *Selections from the Algebra Drawings* (1998), the viewer sees a drawing emerge on the screen. Inspired by numerous conversations at Do While, the project began by creating a database of marks similar to the ones recurrent in Blyth Hazen's drawings and paintings. Then she constructed an algebraic algorithm to mimic the order and placement of marks onto canvas. A certain element of randomness was added to the process that would create a unique yet similar "drawing" each time the program was run. Several different versions of the algorithm—each with a slightly different family of marks, opacity, and variables of sequence—was developed. She used a touch-sensitive interface to allow the viewer to begin or end a drawing or choose a new family of marks.

Selections from the Algebra Drawings is an intuitive perusal into emergent patterns. The ability to use this drawing engine to create variations on a specific category of mark makes possible a new way of looking at one's own drawing technique. Through participation, one may pay homage to the natural order of such things as birds in flight, the firing of neurons, and perhaps traffic on the interstate.

Acupuncture for Temporal Fruit[6]

Principals: Jennifer Hall, Marc Locasio, Blyth Hazen

Directed by Jennifer Hall, many artists and engineers participated in the interfaces, robotics, and sculptural elements of *Acupuncture for Temporal Fruit* (1999) (figures 19.1 and 19.2). The installation consisted of twelve glass pods that responded to the viewer's presence and subsequently administered acupuncture to the tomatoes within. The more a viewer watched, the faster the tomatoes would turn to an acidic fluid caught in bags hung from each pod. The tomatoes were sealed within the pods and

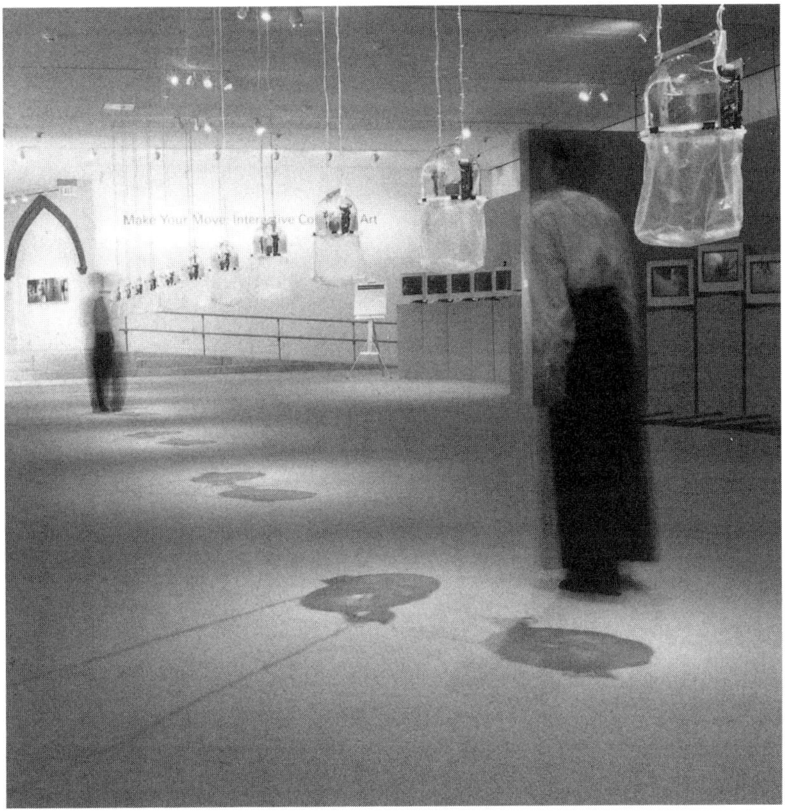

Figure 19.1
Jennifer Hall, Documentation of *Acupuncture for Temporal Fruit,* 1999, Decordova Museum. Installation with twelve pods suspended from the ceiling in a straight line. Photo by Sam Ogden.

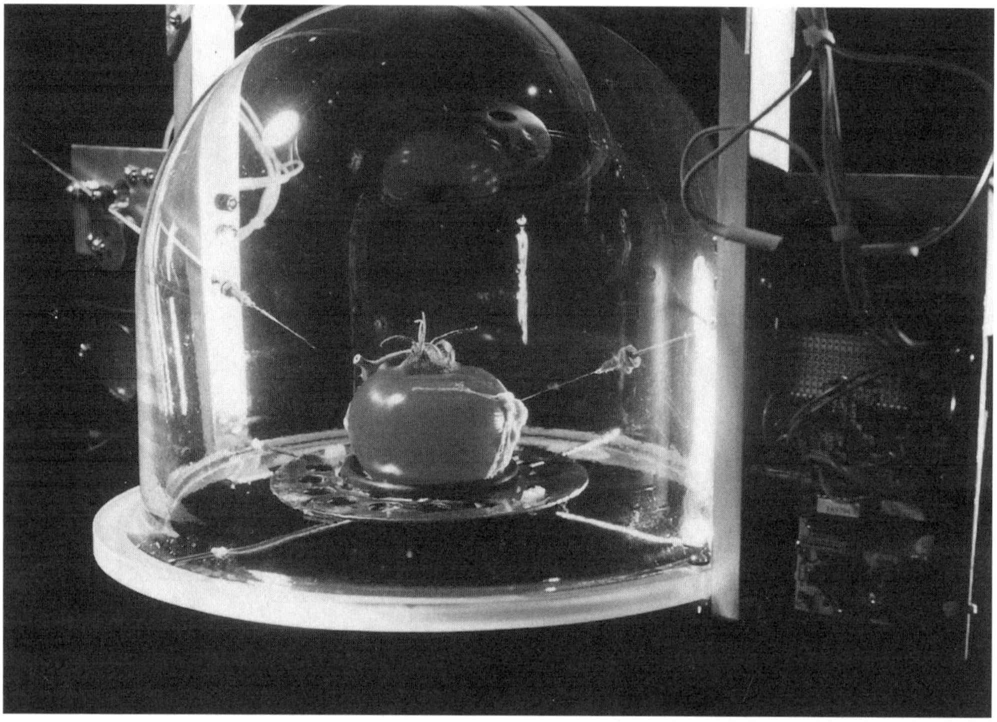

Figure 19.2
Jennifer Hall, Documentation of *Acupuncture for Temporal Fruit,* 1999, Decordova Museum. Details of one pod with a set of sensors, circuits, motors, and needles. White mold is growing from the needles' hole, and black mold is growing from the stamen of the fruit. Photo by Sam Ogden.

became unique ecosystems, creating their own environments of different molds and decay. In an ironic twist, the technology that drove this installation simply dropped away from importance, leaving the viewer focused on the long needles jabbing into the yielding flesh of decomposing fruit.

This artwork was constructed one circuit at a time, the soldering version of a country quilting bee that produced almost 100 handmade boards and 24 robotic devices. Teams of artists, engineers, musicians, and health-care professionals worked together for three months to build the components. Through the windows of Do While Studio, the projects progress was regularly monitored by people on the street. It was not unusual to see the same faces from the Financial Center at lunch time or visits from families headed to Chinatown on the weekend. Those who watched had both a fascination about physical construction and a visceral response when they were operated. The installation evoked many conversations about issues such as medicine, genetic engineering of food, and experimental interfaces for the body.

> Here, the high-tech world of robotics and surveillance intrudes on the natural world with disturbing consequences. . . . As we pass through the gallery, our every movement echoed by the twitching needles, we are acutely aware that standing astride the cutting edge is not necessarily a comfortable place to be.[7]
>
> —MILES UNGER, *NEW YORK TIMES*

DO WHILE VISION

Research addressing the shifts in artists' lives, the impact of global-scale artwork, and the cultural implications of technology are tremendous motivations for new artistic work and are highly valued at Do While Studio. In response to the advent of global communication, artists are looking with fresh perspectives at their work on a regional level. They are describing their own cultural identity in more relevant and personal terms and are working toward understanding the ramifications of broadcasting that identity beyond the domains of village, state, and continent.

Participation in the new global dialog demands the ability to navigate through massive quantities of information while reassessing notions of space and time. We find ourselves wading through a tremendous sea of disconnected ideas. Artists have responded to the information complex by developing a new working paradigm—connecting chunks of data. We have become concerned not with a specific piece of information or a particular image but with the changing relationships of elements within an overall structure, with customized views, and with the coherent transmission of ideas from one place and time to another.[8] The data that is collected and the information systems that explore and reveal the data shape our social perception. This perception is the essence—the raw material that artists call on for timely and seminal contributions to our community.

We are living in a culture where content is compulsively pushed and pulled through media as though the act itself is sacred. This digital productivity is the perfect bedfellow for a commerce-driven culture that demands that available media spaces be filled as soon as possible. Content providers can't be found quickly enough, and the new educational industry of preparing young artists for this task can easily become glorified pop culture, segregating the art from critical discourse. It is now technically easy to appropriate content utilizing new media, and the making of such imagery is no longer a thought-provoking act. If visual material is indeed essential to the new communication tender, then we must rethink our educational strategies.

CONCLUSION

The "do while" model is an interruption of the cultural obsession to simply absorb, produce, and distribute. When artistic focus is maintained through all aspects of technology research and production, ventures into personal expression, industry consultation, or community outreach will undoubtedly shift the track of emerging technologies.

We believe that artists are the historical purveyors of cultural reflection. To maintain this role, face-to-face research and development in the early stages of technology is essential to our work. Do While Studio continues to attract artists from around the world seeking the opportunity to consider the social ramifications of technological expansion and to imagine roles as the creators rather than the consumers of new media.

Horror vacui translates from the Latin as "horror of the void." Technology has given us a virtual universe of empty spaces to fill and new tools to organize and share information. Our need to fill this space is fired by an unsettled feeling placed on us by the expanding digital world. Do While Studio exists as a place to pause and reflect in the collective panic, to copy, collate, and staple the data that transfigure and become us.

NOTES

1. For up-to-date information about Do While Studio, visit our Web site at ⟨http://www.dowhile.org⟩. We also maintain a Silicon Graphics multimedia archive of the studio's first decade, which is available to educators and historians by contacting the director at ⟨jenhall@massart.edu⟩.

2. For more information about *Out of Body Theatre,* visit the Web site at ⟨http://www.dowhile.org/physical/projects/theatre⟩.

3. Participating artists include Jeannette C. Abulafia, Noreen Adams, Maureen Burhenn, Bill Burke, Eric Clarke, James Colella, Richard Davis, Isabelle Delmotte, Ben Doress, Fritz Fairhurst, Todd Gieg, Jennifer Hall, Bonnie Jean Hollywood, Lee Hope, Barbara Kennedy, Mica K. Knapp, Florence T. Kundsen, Edward P. McMorrow, Greg Scaff, Walter G. Sine Jr., Graig Smith, Jacqui Streeton, Carolyn Stuart, Diana Tsacalis, Jim Weiner, Paula Wooley, and Carol J. Whyte. Principal collaborators: Jennifer Hall and Steven Schacter. For more information about *From the Storm,* visit the Web site at ⟨http://www.dowhile.org/physical/projects/storm⟩.

4. For information about *World Wide Simultaneous Dance,* visit the archives at ⟨http://www.wwsd.org⟩.

5. *Selections from the Algebra Drawings* continues to evolve as an interactive kiosk located at Do While Studio. Visit samples of this work at ⟨http://www.dowhile.org/physical/people/hazenb.html⟩.

6. In-kind donations for *Acupuncture for Temporal Fruit* were provided by Poloroid Corporation; design and constuction support provided by Ellen Abdow, Sue Bennett, Ethan Berry, Linda Bourke, Cindy Brown, Kathy Desmond, Keith Donaldson, Laurie Lindop, Anne Loyer, John McCormack, Sue Oppie, Lynda Shift, Bethany Shorb, and Doug Sweeter; additional support provided by Nick Capasso, Brad Goyer, and George Peet. For additional information on *Acupuncture for Temporal Fruit,* visit the Web site at ⟨http://www.dowhile.org/physical/projects/acupuncture⟩.

7. Miles Unger, "Boston Cyberarts," *New York Times,* Sunday, April 25, 1998, 35.

8. Credit to Rob Haimes, Do While board member, for helping to focus our mission over the years. Other board members includes Matthew Belge, Time Carlson, Terra Freidricks, Jeff Garmel, Jennifer Hall, Hubert Hohn, Beth Kanter, Anne Koufman-Frederick, Susan M. Lewis, Joan Shafran, and Meg Young.

20

Tech Work by Heart

Brenda Laurel

This chapter is about doing humanistic work in the context of computers and HCI. I'm hoping to cover some of the questions and answers that have come up for me in that quest over the last twenty-one years in the interactive entertainment business. Personal stories are not necessarily the best examples, but they are the examples I know best, especially tales from Purple Moon and our adventures with girls' software in the United States.

At Atari in 1980, I was managing the software planning and marketing group for the new Home Computer Division. Atari had become enormously successful with arcade games—from Pong to Asteroids—and it had also established a lucrative business with a little game console called the Atari VCS that could play versions of these games at home. The home computer—the Atari 400 series—was a brand-new product, and the company needed to position it in the minds of the public.

It's interesting to note that no one had yet succeeded in creating a coherent idea of what a personal computer was in terms of popular culture. Apple, Commodore, and before them Intellivision and Cybervision and other obscure little systems had dabbled with allowing people to write BASIC programs or play Hangman, but there was not yet a suite of applications or capabilities that really defined a home or personal computer and differentiated it from a video game console. My group was experimenting with everything from staples like Visicalc and word processing (40 columns, all caps) to concepts in self-improvement and learning. My imagination had been infected by Alan Kay, who had just joined Atari as chief scientist. I learned about his work with Smalltalk and his commitment to and respect for children. I became enamored of the idea of computer literacy—and the belief that empowering children to use computers would vastly increase the evolutionary potential of humanity.

Armed with this bright optimism, I wanted to see what was really going on with kids—Atari's audience—and what they were thinking about computers. I was fond of visiting arcade parlors—places where boys and

---

Portions of this chapter were previously published in the CHI '98 Conference Proceedings, which is copyrighted by ACM. Portions of this chapter were also published in Brenda Laurel, *Utopian Entrepreneur* (Cambridge, MA: MIT Press, 2001).

young men were playing arcade games, Atari and others—to watch them and strike up conversations if I was lucky. Arcade parlors in those days were pretty much the same as they are today—dark, stuffy, and appallingly smelly, with the constant deafening din of simulated aircraft, shuttlecraft, submarines, weapons fire, explosions, and screaming—the noise testosterone makes. I got used to it, and the boys got used to me. I'm sure they thought I was sort of strange: I never played, only watched and tried to get them to talk to me (even at a self-confident twenty-nine, I did *not* want to make a fool of myself in front of all those young men).

One day I was hanging around the arcade watching a little guy about ten years old play one of the newer games. Like all the boys, his whole body was involved in the action, and his concentration was razor-sharp. When a game ended, I strolled over to him and introduced myself, and then I asked him if he liked the game.

"Oh, *yes.*" An enthusiastic response.

"Do you think you'll play it some more?"

"Yes," again.

"So tell me," I said. "Does it make you want to learn how to program computers someday?"

He looked at me blankly. "What do you mean?"

"Well, it's a computer. Somebody programmed it. Wouldn't you like to know how to do that—to make some games of your own?"

He laid his little hand on the machine. "Lady, this is *not* a computer," he explained, as if to someone mentally impaired. "Computers are *smart.* This is just a stupid video game."

Well, it was indeed a stupid video game, and even though Atari was trying to differentiate their home computer from their VHS unit, it seemed that upper management wanted us to do a whole lot of stupid video games instead of most of the cool empowering things we were working on, which of course in my humanistic fervor I resisted. Don't get me wrong. I enjoyed stupid video games as much as the next girl, but computers meant something different to me. "Don't you see the possibilities?" I wanted to shout to the folks at Atari. "Kids could learn things! Everyone could communicate much better! People could balance their checkbooks and organize their recipes! Artists could make great new works! Personal computers could change the world!"

"Yes, yes, dear."

When I finally found myself fired from the Atari Home Computer Division, I called my by-then friend Alan Kay and asked him for a job at Atari Research. He and Bob Stein hauled me into the lifeboat. It was nearly two years until the corporate types figured out that I was still working at Atari, and boy were they mad. But by that time, I was ready to embark on a new humanistic quest, thanks to Alan.

I've used the term *humanistic* a couple of times now, and I guess it's time I define what I mean by it. The most satisfying expression of humanism I know is D'Alembert's *Preliminary Discourse to the Encyclopedia of Diderot.* D'Alembert's translator, Richard Schwab, said that the *Discourse* "breathed a confidence that man, through his own intelligent efforts, could transform the conditions of human life." D'Alembert succeeded in defining a philosophical space between the rationalism of Descartes and the empiricism of Locke and Newton. This is a tough line to walk—to believe in the primacy of sensation and observation as how we know the world and at the same time to employ a methodology that submits these observations to reason and, sometimes not so obviously, to ethical consideration.

The truth is that humanistic work is values-driven work. It is work that you are doing because you think it's a good thing to do. The Enlightenment humanists fell at different points on the continuum between universal truths (including ethical values) and empirical investigation. So do humanists today. Many will deny that they have values and claim that their work is entirely objective. But this is to ignore the glaring single value that is the very heart of humanism: a belief that humanity's power to shape its own destiny through the application of knowledge and reason is a good thing. In other words, whether we admit it or not, we humanists make the implicit assumption that we can do good and therefore that we can know what is good to do.

I want to describe some of the work that we did at Purple Moon to illustrate some of the points I'm trying to make. But before I skip entirely over the intervening decade of personal history, I want to observe that doing what I consider to be humanistic work is something I think I had to work up to, first through rhetoric, then through research, and then through a really arduous education in business.

By 1986, I had become totally frustrated by my inability to produce any products I could consider humanistic in the world of computer games.

I had finished a Ph.D. in interactive fantasy and had created a lot of product designs that publishers generally couldn't see a market for. I turned my attention to writing critical theory, making rousing speeches, and creating virtual-reality art that would be seen by only a few hundred people. I think of those times as a period of closet humanism. I felt that if I couldn't make something that could touch a lot of people, I might be able to make things that would influence other makers who could step up and change the world. To a certain extent, I think I succeeded. But it is really easy to hide from the challenge of doing real products when you define yourself as marginal—a theoretician, a sideliner, a commentator. I was sending the postcards, but I wasn't taking the trip. At the same time, my enduring love affair with popular culture increasingly alienated me from the artistic elite. I felt more and more like a homeless intellectual.

In 1992, I met David Liddle, who was just forming Interval Research Corporation with his partner, Paul Allen. David had read my books, and I had heard about David's work at Xerox PARC. In the process of exploring whether there might be a fit for me at Interval, we discovered that we had a strong common interest. Both of us were extremely curious as to why there didn't seem to be any computer games for little girls. We both knew that exposure to computer games gave boys a level of comfort and familiarity with the machine that girls generally did not share. Neither of us knew of any reason that girls would be intrinsically less interested than boys in computers or computer games, and both of us were deeply puzzled about why no one had been able to make something that worked for them. David's summary of the missed business opportunity was, "There's a six billion dollar business with an empty lot next door." Most important, we agreed that if this were an easy problem, someone would have already solved it. In sum, the problem had all the characteristics of a good research problem—puzzling, consequential, and complex. So I signed on at Interval and began a four-year research project that led to the formation of Purple Moon.

Our first goal was to define the question we were trying to answer. It seemed too narrow and trivial to ask simply, "Why hasn't anyone made successful computer games for little girls?" This question has some ready-made answers. Computer games as we know them were invented by young men around the time of the invention of graphical displays. They were enjoyed by young men, and young men soon made a very profitable business of them, dovetailing to a certain extent with the existing pinball business. Arcade computer games were sold into male-gendered spaces, and when home computer game consoles were invented, they were sold through male-oriented consumer electronics channels to more young men. The whole industry consolidated very quickly around a young male demographic—all the way from the gameplay design to the arcade environment to the retail world—and it made no sense for a company to swim against the tide in all three of these areas at once. Even though the occasional computer game like Pac Man was a hit with girls and women, scoring sometimes as many as 25 percent female players, conventional wisdom was set and remained fundamentally unchallenged. Whenever a "girl" title was attempted, it was launched all alone onto the shelf without adequate marketing or retail support, and the inevitable failure easily became a proof that girls would not play computer games. Even as late as 1994, major game companies steered clear of the potential girls' market because they feared that being seen as doing things for girls would alienate their male audiences. By the way, our research showed that—initially, at least—their fears were indeed well founded.

When we began our research at Interval, we simply wanted to find out what it would take to motivate a little girl to put her hands on the computer and become comfortable with it. That was our core value because we knew that comfort with technology would eventually tend to broaden a girl's range of choices in both education and work. We finally agreed on this question: "For children ages seven to twelve, how is play influenced by both age and gender?" This question enabled us to investigate differences and similarities, culture and biology, skills and preferences, framed in the context of children's pleasure and enjoyment.

This was a humanistic approach. Once we identified the foundational values and questions, we employed a methodology based on empirical observation and a dialog with nature—in this case, the nature of boys and girls in the United States (a limitation dictated by our resources). As a research corporation, Interval did not have existing products or services that it was trying to bolster or justify, which freed us to look fully and clearly at our subject. As a private enterprise, Interval was largely immune to the political risk that inquiries into gender difference entailed in academic and other institutions (especially in the 1970s and 1980s). To the end of doing good research, David and I and the rest of the team agreed to check our preconceptions and political agendas at the door. Some people found this requirement really irritating.

We could have approached our work at Interval and later at Purple Moon in ways that would have satisfied some of those folks. Some feminists would have been satisfied if we had avoided the crassness of market research and proceeded aggressively on an agenda to educate and transform girls according to their particular values. Hard-nosed business realists would have been satisfied if we licensed a strong branded character or fiction property that had already proven itself to be powerful with our target audience. Indeed, when we started our work in 1992, the business realists thought they "knew" that girls would not play computer games, so there was nothing we could do to make them happy except to continue to refrain from trying to spend their money on a product that was bound to fail.

The truth is that people do a lot of talking about girls and not enough listening. No one is neutral about the roles played by women and girls in American society or probably in any other. People have strong ideas of what girls are like, what they need, and who they should be. These voices drown out the voices of the girls themselves.

Being a preteen girl is rather like being in that corridor of radio silence that a spacecraft passes through when it reenters the earth's atmosphere. American culture nourishes girls very poorly during this time. An executive of one of the world's leading toy companies told me that there are approximately ten times as many toys marketed to boys as to girls in the age range. Girls are generally too old for dolls and too young for serious fashion and cosmetics. Until quite recently, television and film did not target them because the studios "knew" that girls will watch material designed for boys but that the converse was not true. The primary cultural artifacts that girls engage with in this period of life are books, movies, and music.

Between the ages of eight and twelve, girls are absorbed in the process of self-construction—both external social identity and deeper internal self-awareness. The materials with which they construct themselves are the materials at hand—primarily provided by popular culture. Yet it seems to me that the few materials we do provide them are only superficially related to the actualities of their lives.

A single, shining exception is American girls' participation in sports, up 800 percent over the last twenty-five years since the U.S. Congress mandated equal funding for boys' and girls' athletic programs in schools.

For the first time, girls have a way to experience and value their bodies that is different from the idea of attractiveness.

How girls fare at the task of self-construction predicts to a large extent how well they will brave the storm of adolescence—the time when their self-esteem is at greatest risk. Girls who have trouble at this passage may experience depression and setbacks well into their adult lives. Yet it seems that at the very moment we should be listening to them, seeing them, and offering them back an affirmation of who and where they are in life, we instead tend to place them in a life stage—"preteenhood"—that is no place to be at all.

During the course of my research at Interval and after that at Purple Moon, my Interval team and I were graced with the partnership of Cheskin Research in our goal to hear and see girls. We began with a thorough literature survey of research in areas that we thought might have some light to shed on our question (something that computer game companies have rarely done)—areas like cognitive and developmental psychology, play theory, neurophysiology, and gender studies. In our second stage, we interviewed a hundred adults—experts from academia and industry as well as adults whose lives are spent on the ground with children at play—on topics that seemed most promising. In our third and largest stage of work, we conducted interviews with about eleven hundred children from all over the United States, initially both boys and girls and then focusing exclusively on girls.

In addition to this work, we employed survey data from another ten thousand children and conducted separate interviews with over five hundred parents. We had strong quantitative findings. For example, the leading reason girls gave for disliking traditional video games was not that they are violent or competitive but that they are boring. Girls tend to find the characters entirely unsatisfying—so weak that you can't even make up good stories about them. Girls are typically unmotivated by mastery for its own sake but demand engaging and relevant experiences from computer games. Both boys and girls see video game machines as for boys, and they see computers as gender-neutral.

But for me, the real understanding came with our qualitative work. We knew that if we were going to create things that were truly relevant to girls, we needed to know every single thing we could find out about them. We asked them to talk to us with their best friends at their sides for at least an hour on topics ranging from play preferences to gender signals sent by toys to technology to the issues in their lives that concerned them most. We asked them to take pictures of things that are important in their lives. Among the most common life issues were problems with friends, problems with siblings, favoritism, divorce, and social pressure to do things that aren't quite "right." Girls took pictures of their friends, collections of all sorts of things, sporting equipment, arts and crafts, pets, bedrooms, trophies, and stuffed animals. Even if I hadn't had three preteen girls at home, this research would have made me feel that I could walk around inside of the life of a preteen girl in the 1990s—which is a whole lot different, in many ways, from my preteen life in the 1950s and early 1960s.

After the interviews were finished, we spent several months consolidating our findings and then transforming them into design principles to use in developing products. After a stage of advance development inside of Interval Research, we formed our company, Purple Moon, and launched three businesses—interactive CD-ROMs, the Purple Moon Web site, and an array of Purple Moon merchandise. All of these products were linked by a diverse cast of characters drawn from the lives and experiences of girls and worlds of imaginative play that we explored through our conversations with them.

Our first two Friendship Adventures, *Rockett's New School* and *Secret Paths in the Forest,* were both in PC Data's top fifty entertainment titles during the 1997 holiday season with sales at approximately ten times our original forecast. From launch until February 1999, the Web site served over 300 million pages, with about 240,000 registered users who visited us at least once a day and viewed an average of thirty-five pages per visit. These girls collected about five million virtual treasures and sent each other nearly ten million postcards. Over the life of the company, we launched seven more CD-ROM titles, including more Friendship Adventures, creativity products, and the first-ever line of sports games for girls.

I consider our work to have been a cultural success in the sense that it touched the lives of millions of girls and gave them fresh views of girlhood and a new portal into the technology. But our business failed. It failed for reasons that had to do with investors' expectations, market conditions, and some weakness in strategic planning. The short story is that the investors pulled their support, and we were forced to sell the business to Mattel, where it lives in some form today.

Purple Moon's heroine, Rockett Movado, was unique in her ability to see the possibilities open to her and to make conscious choices about what to do next. Our products allowed girls to try this skill and to see how things might turn out differently depending on what moods, attitudes, or choices they bring into the next moment. I am taking the same approach to my career now. A few choices made differently would have made Purple Moon a financial success. I have learned a great deal, and I'm not dead yet.

Although the end was traumatic, thanks to Purple Moon I have a new kind of satisfaction in my life—one that comes from having tried to make an intervention at the level of popular culture. Probably one humanist in thousands actually succeeds in making a great big dent in things, but there are some excellent role models. I think of Gene Roddenberry, whose *Star Trek* television series continues to appall the artistic and cultural elite but has made strongly progressive depictions of literally every significant social issue in the United States since 1967 embraceable by popular culture. I wonder how much genius Stephen Spielberg has wasted trying to convince his critics that he can do "serious" work, where that is somehow defined as "not for mass consumption." Because his films appeal to millions, his values are somehow invisible. When he finally emits a film like *Schindler's List* that appeases his critics, obscure academics step in to grab their fifteen minutes of fame by questioning his politics and historical accuracy.

And so it goes. A humanist who attempts to do popular work will always be attacked by the elite who claim his political turf and who insist that people who do values-driven work must by definition be marginalized. These folks believe that popular culture is intrinsically poisonous. If it is popular, it is bad for you; if it is bad for you, it is probably popular. There is a great deal of inevitability about it. If you do market research, you are probably fundamentally crass and exploitive. This position is possible because the values at the heart of humanism have still not been understood.

The advice that has guided me best is simple. Know who you are working for: be clear about whom you wish to please. The goal is to introduce new genetic material into the culture without activating its immune system. Good research and strong values go hand in hand. Today, humanistic work in our field is possible, and it is necessary.

21

Imagine a Space Filled with Data . . .

Monika Fleischmann and Wolfgang Strauss

The concept of human interaction with space changed in the twentieth century from conquering space to dissolving space, and this reconceptualization has been brought about primarily through the new means of transport that have become available. The term *navigation* signifies the definition of and adherence to a course and is derived from the Latin *navigare,* which can be translated as "steering" or "traveling." The same symbols are used on the Internet as in real space—though virtual navigation involves the *reconfiguring* (that is, production) of a time process. The old metaphors of explorer or navigator have remained the same: the language of the conqueror is apparent to everyone on the Internet. As artists, we explore aesthetic strategies with communication processes to influence and transform the development of the community of the future.

This chapter discusses some developments of the eRENA project (electronic Arenas for Culture, Performance, Art, and Entertainment), which is one of thirteen European long-term research projects. The eRENA consortium brings together internationally known digital artists, experts in multiuser virtual reality, artists, social scientists, and networking experts from significant European institutions. We are both research artists, and we direct the project at GMD (the German national research center for information technology located in Sankt Augustin, Germany). Monika also is head of MARS (Media Arts and Research Studies), an interdisciplinary group of computer experts—artists, architects, and social and media scientists at GMD.

GOALS AND CHARACTERISTICS OF THE eRENA PROJECT

eRENA focuses on developing information spaces in which all participants can be mobile and socially active. By addressing questions of communication and representation research, the MARS lab focuses on the situation of a mixed-reality "stage," which is an abstraction of everyday life.

Our approach to mixed reality is to interconnect the real and the virtual to produce a new framework for communication and interaction possibilities. Real physical space is filled with the virtual space, and the user's exploration of virtual space is connected both to real space as well as to other users' experiences. This spatial arrangement of real and virtual is the means for creating a situation that connects the participants with each other.

Linking Real and Virtual Space

The basic concept of our mixed-reality stage installation is the linking of real and virtual space—a room filled with data. The "room" stands for physical interaction space, but the "furniture" of data is virtual and stands for an information space. It is a spatially organized information space in which data are revealed through users' movement in the combined real-virtual space and through interaction with other users.

Our concept is realized as an interactive field of sounds and words—the *Murmuring Fields,* a virtual soundscape consisting of four voices that are triggered by users' movement and emitted into the physical space. As movement in physical space causes sounds and as sounds are heard in the physical space, the resulting impression for users is that of an invisible field of sounds existing in the physical space and revealed through their actions.

Visitors to *Murmuring Fields* are engulfed in a sound picture that is combined from controversial statements made by media thinkers Vilem Flusser, Marvin Minsky, Paul Virilio, and Josef Weizenbaum and from sound samples. The protagonists control the sound stage by their positions on stage. The layers of language and sound of each participant intersect and weave around, thereby generating an environment of digital storytelling. That interactive conversation is augmented by a visual map, projected on different semitransparent displays like net skins.

Implementing the Mixed-Reality Stage

The MARS group developed an electronic multiuser stage environment (eMUSE) as a platform for multiuser interaction and communication, interface, rendering, and display organization, in shared physical and virtual space. It provides an operating environment for mixed-reality spaces based on VRML 2.0 like *Murmuring Fields.* The mixed-reality sound stage is connected through a vision system—a camera-based tracking system capable of locating several humans moving freely in a room.

The mixed-reality concept is based on the following situation: one or several participants in the physical space are simultaneously present in an information space that is made visible only through participants' actions in the real space. The overlapping and integration of these two spatial situations—physical space filled with virtual information space—create a new reality of perception. The virtual material is to be arranged and rearranged through the user's movement and interactions with other users and is not a predefined 3D scene to be observed and navigated.

The Mixed-Reality Situation as Interface

To realize the mixed-reality stage concept, the whole system—the medial staging of action—needs to be understood as the interface. The interface is not only the connecting of man and machine but also the action-oriented situation in which the visitors become involved. This concept of "situation as interface" acknowledges the distinction between two different levels of an interface environment: (1) The connection of participants with each other through the situation created by the environment and (2) the technical devices needed to realize an unobtrusive underlying computer system.

To achieve these goals, different perception levels need to be layered and related in a coherent structure. The interface environment functions through the interplay of following elements:

- *The vision system* A fundamental point of the mixed-reality stage concept is connecting a participant's bodily sense of being in physical space with a sense of being in virtual space at the same time. To achieve this sensation, we use an optical tracking system to connect participants' movement in real space with navigation in virtual space. The data provided by the tracking system control the position and movement of the user in the virtual space, parts of which are "displayed" accordingly in appropriate locations of physical space. This supports the perception of a virtual space that reacts to users' actions as an integrated part of the physical space in which they are situated.
- *Visual representation of content* The visual elements of the virtual environment serve as place holders for sounds and an orientation aid for the participants. Without this visual reference, establishing the relationship between one's movement in physical space and the sounds triggered in the virtual space becomes much more difficult.

- *Audio representation of content* A participant's movement in physical space controls the creation of the corresponding trace avatar in the virtual space. Trace avatars of all participants trigger sounds and words in the virtual space that are emitted in the real space. The words are statements from contradictory media thinkers about the future of our digital culture taken from interviews with the authors in 1992. The words inspire further movement and lead to a mutual play of the participants in producing sound patterns and dialogues together.
- *User representation: trace avatars* The trace of one's own movement—the trace avatar—is the orientation field. Participants communicate through their virtual traces and determine thereby the action space as well. The trace avatar becomes a graphically visible interface and a medium of communication. It shows the current position of the participants and triggers the sounds touching the sound objects. This "virtual touch" and the triggered sounds initiate an audiovisual composition between participants.
- *Physical interaction space* In some systems, navigation in virtual space is achieved by directly manipulating some physical device, but the mixed-reality stage cannot be navigated without the existence of an appropriate physical space. The physical space becomes an essential part of the interface. As a participant becomes immersed in the play of movement and sound, his or her awareness of being in space and of interacting with others becomes increasingly a bodily one.

None of these five parts alone can be called the interface. Instead, the interface is the whole situation of the environment and presence of the participants that connects the participants with each other as well as with the underlying computer system. The eMUSE system itself, along with the virtual environment and the vision interface, demonstrates the "situation as interface" paradigm. Figure 21.1 illustrates how this concept of the situation as interface is realized.

With eMUSE, we have built an instrument for the mixed-reality stage, an instrument for theatrical play and learning to communicate in networked space. This is not a stage to be observed but a stage to be entered. In our search for an adequate design, we see the design principles of "form follows function" confronted with a design framed by rules—"form follows structure."

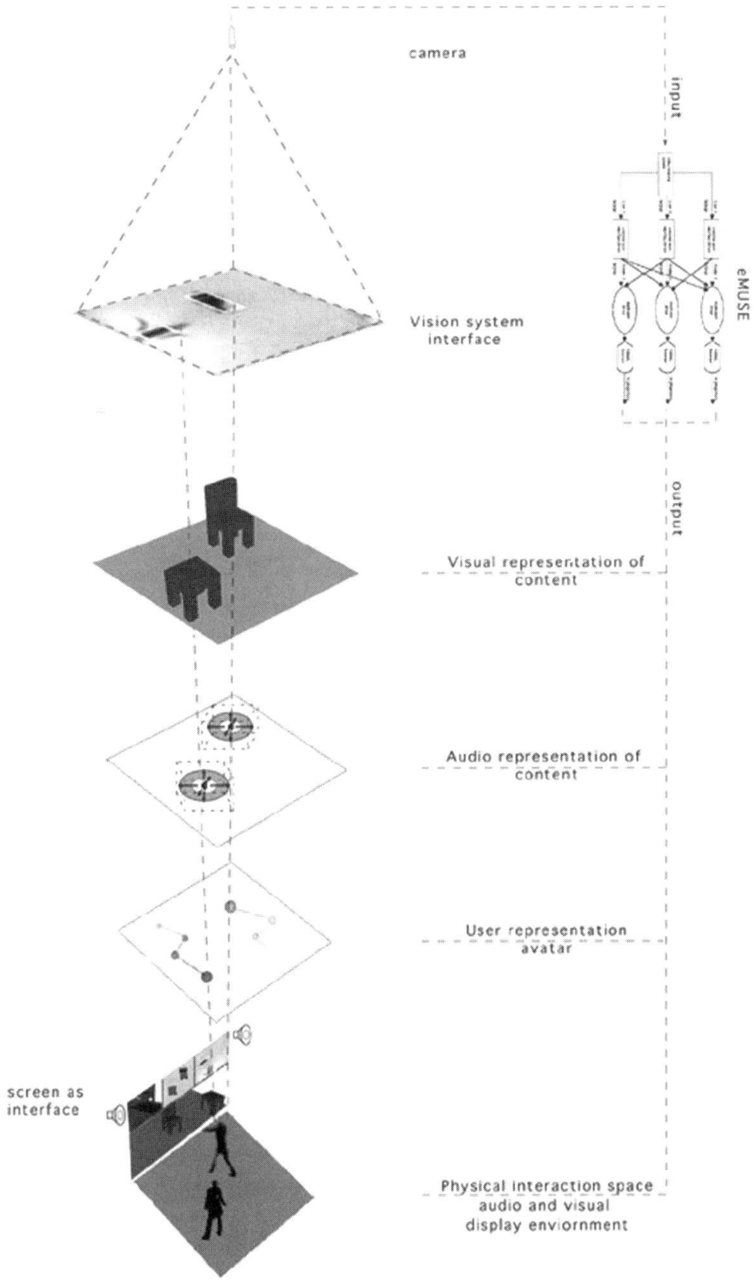

Figure 21.1
Monika Fleischmann and Wolfgang Strauss, The Mars Group, layers of the mixed-reality stage in *eMuse,* 1999. An electronic multiuser stage environment as a platform for multiuser interaction and communication, interface, rendering, and display organization in shared physical and virtual space. It provides an operating environment for mixed-reality spaces.

Murmuring Fields Staged as a Mixed-Reality Performance

We were invited to present and discuss *Murmuring Fields* at Fidena '99—an international theater, puppetry, and new-media festival held annually in Germany. The presentation took place in the theater of the Museum Bochum on 25 April. The two performers Maya Brosch and Martina Leeker—together with the authors as directors—examined the interactive stage and its possibilities.

Traditional forms of movement, such as sign language or gestures, have no meaning for this stage, since the vision system is not based on gesture recognition. Instead, the vision system registers the simple stretching movements of the body and its movement through the space. Rehearsals therefore serve to theme the limitations that arise from viewing the performance and to contribute to the development of specific methods of practising physical expressions for the media performance.

The performers are dressed in white and look like mobile projection surfaces against the black stage background. The performers pick out individual image components—sentences or words—that are heard as a sound at the same moment. The activated virtual sound object becomes audible and at the same time visible on the body or in the space.

Comments from the Performers

In performer Maya Brosch's opinion, various levels overlap during the performative interaction of participants with the installation: the desire to deliver a performance interacts with the desire to have a synaesthetic experience. The mixed-reality concept presents difficulties when it comes to implementing choreography: "Because it's not about me presenting something great—it's about me perceiving and experiencing something. It requires people to listen to the sounds triggered by their movement. If this listening exercise is successful, it creates a very fragile type of perception that is independent of the outside world. It's as if I'm moving through sounds and these sounds are tangible to my body" (figure 21.2).

For performer Martina Leeker, a key experience in *Murmuring Fields* is the influence of tactile elements on orientation in the sound space. She uses the hooped skirt as a "tool that makes possible certain movements that belong to the sounds." She sees the lace hem of the skirt "as something very tactile, like lots of small fingers or hairs . . . like an animal that makes very small movements and touches sounds as it does so."

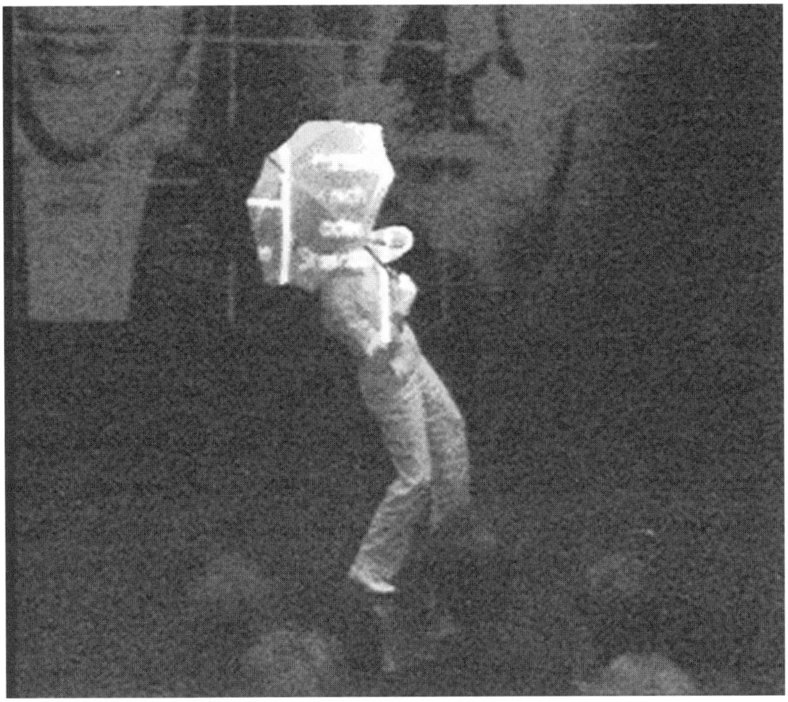

Figure 21.2
Monika Fleischmann and Wolfgang Strauss, 1999. *Murmuring Fields,* a mixed-reality performance space.

The real actions of the performers cause avatars to appear on the different displays. The performers interact with their own virtual traces as though the traces were imaginary partners. The result is an overlapping of the real and the virtual, which not only opens up a new level of communication but also redefines the notion of a stage and a performance in terms of the theater. The participants are linked together in real time on the Internet and on the stage.

CONCLUSION

Interactive media are time-based. This means that the action occurs in real time and that the performers bring their own rhythms and their own subjective concepts of time into the action. They can stop or alter the course of events at any time and therefore structure perception themselves. In interactive systems, the body reforms itself.

The discoveries made through the mixed-reality stage in terms of patterns of behavior go far beyond the bounds of the theater and could perhaps be used in a more highly developed form in the field of psychiatry. What type of contact can be made there? How is the virtual space created? What makes it a communication space? The moment is the unit of time that must be perceived within its situation. Networking means communicating in the present.

REFERENCES

Billinghurst, M., and H. Kato. "Mixed Reality: Merging Real and Virtual Worlds." In *Proceedings of the First International Symposium on Mixed Reality (ISMR '99)* (261–284). Berlin: Springer Verlag, 1999.

Fleischmann, M., W. Strauss, M. Thomsen, and J. Novak. "*Murmuring Fields:* The Making of e-MUSE." Design. *I3 Magazine* (October 1998), EU ESPRIT projects.

Fleischmann, M., W. Strauss, and C.A. Bohn. "Liquid Views—Rigid Waves." In *Visual Proceedings SIGGRAPH '98.* Orlando, FL: SIGGRAPH.

Fleischmann, M., W. Strauss, W. Heiden, K. Sikora, T. Sikora, and J. Speier. "The Virtual Balance: An Input Device for VR Environments." In *La Lettre de INA, Interfaces 97,* Proceedings number 123, Montpellier, May 1997.

Fleischmann, M., and W. Strauss. "Images of the Body in the House of Illusion." In *Art@Science.* New York: Springer, 1997.

Krüger, W., et al. "The Responsive Workbench: A Virtual Work Environment." *Computer* (July 1995).

Marvin, S. "Telecommunications and the Urban Environmen: Electronic and Physical Links." In Peter Droege, ed., *Intelligent Environments: Spatial Aspects of the Information Revolution* (179–197). Amsterdam: Elsevier Science B.V., 1997.

Strauss, W., et al. Staging the Space of Mixed Reality: Reconsidering the Concept of a Multiuser Environment." In *Proceedings VRML99: Fourth Symposium on the Virtual Reality Modeling Language,* 23–26 February, Paderborn, Germany (93–98). New York: ACM, 1999.

Strauss, W., M. Fleischmann, J. Novak, and M. Thomsen. "VRML-Based Extended Galleries: Reconsidering the Concept of a Multiuser Environment." i3-eRENA 1st year deliverable, Bruxelles E U–Research Project Report, May 1998, 1–30.

Winkler, T. "Creating Interactive Dance with the Very Nervous System." In *Proceedings of the Connecticut College Symposium on Art and Technology,* 1997.

Yelistratov, V., W. Strauss, and M. Fleischmann. "Two Approaches for Intuitive Navigation in Virtual Environments." *Proceedings of GraphiCon–99, the Ninth International Conference on Computer Graphics and Vision, Moscow, Russia, 26 August—1 September 1999.* Moscow: Russian Computer Graphics Society, 1999.

22

Landscape, Earth, Body, Being, Space, and Time in the Immersive Virtual Environments *Osmose* and *Ephémère*

Char Davies

Osmose (1995) (figure 22.1) and *Ephémère* (1998) (figure 22.2) are immersive and interactive virtual-reality environments.[1] These works are known for their embodying interface, painterly aesthetic, and themes of nature. They are the most recent fruits of an artistic project that I have been engaged in for almost twenty years and that has encompassed painting, film, and making three-dimensional computer graphics and animation.

The impulse behind this project has been to communicate an intensified experience of being embodied in the space-time of the living world. *Osmose* and *Ephémère* are my attempts to distill and amplify the sensations and emotions of being conscious, embodied, and mortal—that is, how it feels to be alive here now among all this, immersed in the vast, multichanneled flow of life through space and time. In these works, I seek to remind people of their biological, spiritual, and psychological connections to the natural (rather than human-made) environment and of the regenerative source and mythological ground of those connections.

For nearly a decade, I attempted to communicate this sensibility through the medium of painting. The practice of painting and its mode of apprehending the world has significantly informed my entire approach to VR. Evidence of the themes of *Osmose* and *Ephémère* can be found in my work as early as 1975, and their visual aesthetic appears as early as 1981, when I turned my attention as a painter to investigating my own extremely myopic eyesight. In doing so, I was initiated into an alternative experience of space whereby "objects" had apparently disappeared and where all semblance of solidity, surface, edges, and distinctions between things, including figure and ground—all the usual perceptual cues by which we objectify the world—had dissolved. In their stead was a sense of space without sharply defined, separate objects in empty space and with an ambiguous intermingling of varying voluminous luminosities and hues. Within this spatiality, there is no split between the observer and the observed. The withdrawal of the sense of sight—of the visual acuity that dominates the human relationship with the world and is tied to the Cartesian paradigm—allows another way of sensing to come forward, one in which the body feels space very much like that of a body immersed in the sea.[2] This alternative mode of perceptual spatiality has profoundly influenced my work.

In the mid-1980s, I exhibited a series of paintings called *Espaces Interlacés* (Interlaced Space). This body of work attempted to communicate a

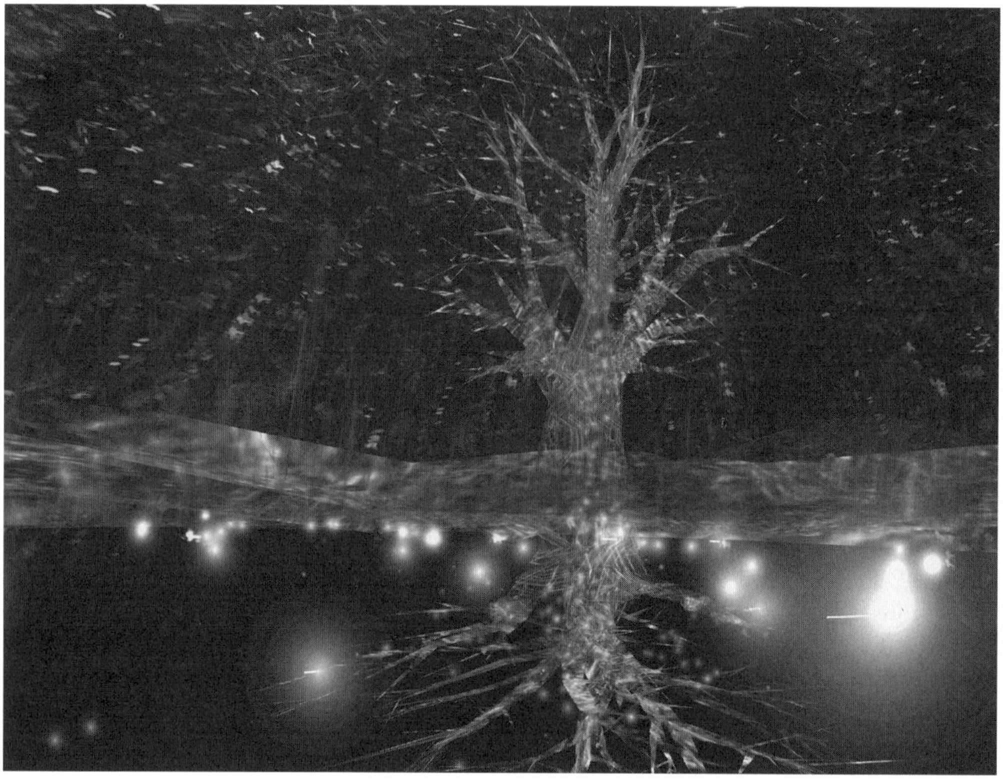

Figure 22.1
Char Davies, TreePond real-time frame capture from *Osmose,* 1995. Char Davies/Immersence Inc. & Softimage Inc.

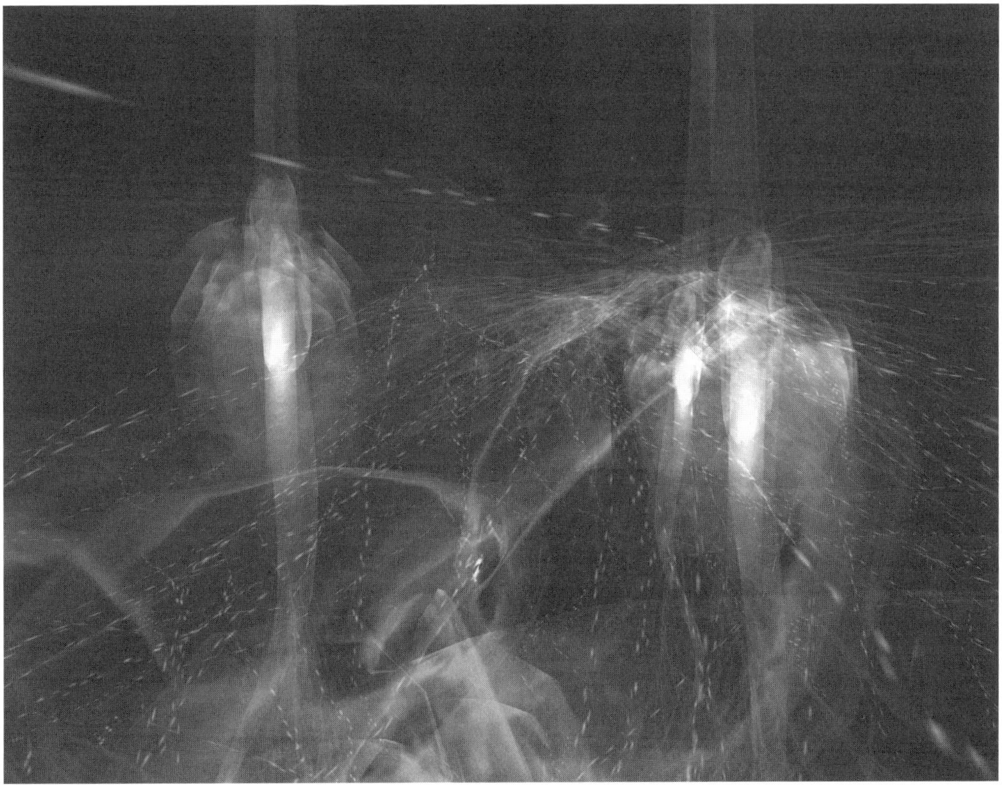

Figure 22.2
Char Davies, Seeds real-time frame capture from *Ephémère,* 1998. Char Davies/Immersence Inc. & Softimage Inc.

subjective experience of the intermingling of interior self and external world, of body and nature. To this end, I developed painting techniques based on the application of layers of oil pigment alternating with the dissolving and sanding away of certain areas, using the resulting semitransparent and semiabstract/semirepresentational effect to create spatial and conceptual ambiguity. In this work, I also sought to reconstruct the sensation of being encircled by horizon, of being sensuously, spatially enveloped. The two-dimensionality of the painterly picture plane, however, ultimately posed an insurmountable limitation to the achievement of this goal. Consequently, I abandoned the medium of painting for that of 3D computer graphics—a medium that offered the possibility of creating in virtual three-dimensional space on the other side of the picture plane.

In late 1987, I became a founding director of a start-up computer graphic software company, Softimage, becoming its first director of visual research. Learning to use the young software by cowriting its first manual, I eventually resumed my own work in the form of three-dimensional computer graphic images composed and rendered as stills. Between 1990 and 1993, I produced a series of 3D stills titled *The Interior Body,* dealing with metaphorical correspondences between body and earth, the same themes that were present in the earlier paintings and that reoccur in *Osmose* and *Ephémère.* As an artist who had already developed a particular visual aesthetic style, I instinctively and immediately bypassed the usual 3D computer graphic techniques. These images were created by making 3D cg models and placing them in three-dimensional space, as if constructing props and arranging them along with lights on a virtual theater set. Rather than creating solid-surfaced objects, however, and separating them in empty Cartesian space, I worked with transparency (as I had while painting) and textured shadow casting to create spatial ambiguity, merging objects and space, figure and ground. These three-dimensional constructions were composed much like a painting, from a fixed point of view with great consideration given to the framing. The resulting compositions were rendered and output to film as transparencies and exhibited as large-scale light boxes. (Along the way, I also collaborated on the 3D computer-animated film *West of Eden* produced at Softimage.) While these images were created in virtual 3D working space, they were output through photographic media as two-dimensional Duratrans stills, thus defeating my original intent. And so, seeking a more effective means with which to communicate a subjective sense of enveloping spatiality, I began to work with immersive VR, or what I prefer to call *immersive virtual space.*

In my experience of constructing virtual environments, the medium of immersive VR offers a unique means of expressing this particular sensibility. This is primarily because of the medium's enveloping spatiality, a spatiality that seemingly allows viewers to enter it, and because of its kinesthetic and interactive properties. As a means of distilling and amplifying the sensations and emotions of being conscious, embodied, and mortal, of heightening sensations "of the body as the site of consciousness occupying space,"[3] immersive VR is far more effective than any other artistic medium I have used. Just as the invention of film—through the technology and craft of photography—extended the stillness of painting into the flow of time, the technology associated with immersive VR extends beyond the two-dimensionality of painting and film into enveloping "circumferal" space. In virtual space, the artist designer can construct three-dimensional, animated, conceptual models of the world, manifesting them within a virtual spatiotemporal arena where they can be kinesthetically explored by others through real-time interaction and full-body immersion. The viewer thus becomes a participant within the artist's world. This is particularly so when approached through an embodying (rather than disembodying) user interface such as that of *Osmose* and *Ephémère.* In works such as these, perceptual boundaries between inside and out may be experienced as permeable because the virtual and immaterial are confused with the bodily felt, experienced as strangely real. This is the paradox of immersive VR and its singular power.

The origins of the technology associated with virtual reality lie in the military and Western scientific-industrial complex: VR is not neutral but by default carries Cartesian values.[4] It should not be surprising, then, if most of the metaphors—spatial, visual, interactive—used in conventional VR design reinforce what Henri Lefebvre called "the reign of King Logos."[5] In this context, virtual reality can be read as a "literal enactment of Cartesian ontology,"[6] the product of the collective consciousness of Western culture issuing from "a techno-utopian ideology ripe with subconscious perceptions and prejudices in which liberation is sought from the body [and earth] by dissolving into the machine."[7] When constructed by artist designers who are aware of the technology's debt to King Logos and who deliberately choose to circumvent its conventions, this medium can effectively be used to convey alternative worldviews, acting as a countering philosophical tool. In *Through the Vanishing Point: Space in Painting and Poetry,* Marshall McLuhan wrote that the role of artists was to create "counterenvironments" to open the doors of perception by correcting the unconscious bias of a given culture. He went on to say that in an age of accelerated change, the need to perceive the environment becomes ever more urgent.[8] Similarly, Lefebvre, in his seminal book *The Production of Space,* called for the production of antienvironments and "counterspace" in the face of the homogenizing absolute space of Western metaphysics.[9] Since these words were written, the paradoxical medium of VR, with all its implications, has emerged.

In my work, I have attempted to push VR beyond its conventions to present a different interpretation of being in the world. In terms of content and sensibility, *Osmose* and *Ephémère* are a far cry from the adrenaline-pumping techno fantasies common in VR. *Osmose* and *Ephémère* shun conventional hand-based modes of user interaction, which tend to reduce the body to that of disembodied eye and probing hand, in favor of an embodying interface that tracks breath and shifting balance, grounding the immersive experience in the participant's own body. *Osmose* and *Ephémère* avoid the hard-edged mimetic realism toward which most VR aspires, instead relying on semiabstract semitransparent figuration to create an ambiguous, evocative painterly aesthetic that actively engages the participant's imagination in the work.

Osmose and *Ephémère* were constructed with a team at Softimage Inc. in Montreal from 1994 to 1995 and 1996 to 1998, respectively. The graphics and animations were created by Georges Mauro using Softimage's 3D animation software; these were adapted to real-time VR through custom programming by John Harrison. Rick Bidlack composed and programmed the sound, and Dorota Blazsczak designed and programmed the sonic architecture. *Osmose* was produced by Softimage. *Ephémère* was coproduced by Softimage and my company Immersence Inc. Both works originally ran on a Silicon Graphics Infinite Reality parallel-processing computer, with a stereoscopic and stereo-sound head-mounted display and now have been successfully ported onto a PC, with a stereoscopic and stereo sound head-mounted display. We designed a user interface that motion-tracks the participant's breath and balance (breathing in to ascend, out to descend, leaning to change direction. The use of conventional hand-based interaction was deliberately avoided. The sound uses a PC (or a Mac in the case of *Osmose*), a Kurzweil sound synthesizer and processor, and an Acoustetron for localizing the sounds in real time in three-dimensional space.

The central experience is that of the immersed participant—the "immersant." During public exhibitions, this rather intimate experience takes place in the company of an attendant in a small private chamber facing a larger audience space of relative darkness with two luminous screens. This public space is filled with sound, as it is generated in real time by the immersant's behavior in the virtual space. One of the screens is a stereoscopic video projection of the three-dimensional world as it is experienced by the immersant, enabling museum visitors to vicariously witness each immersive journey as it takes place in real time. The other bears the projected shadow of the immersant's silhouette as he or she moves and gestures in response to the work. The use of this shadow silhouette alongside with the real-time video projection serves to poeticize the relationship between the immersant body and the work, drawing attention to the body's role as ground and medium for the experience.

To experience *Osmose* or *Ephémère,* the participant dons a stereoscopic helmet through which the computer-generated 3D graphics and 3D sound are displayed in real time according to their breathing in and out and to their shifting center of balance, both of which are tracked by sensors mounted on a vest. There are no gloves, and there is no phallic joystick.

SPATIAL STRUCTURE OF *OSMOSE*

The first virtual realm encountered by the immersant in *Osmose* is a three-dimensional Cartesian grid that functions as an orientation space and makes reference to the technology's origins. With the immersant's first breaths, the grid gives way to a clearing. In the center of the clearing is a tree, into whose leaves it is possible to enter. Surrounding the clearing is a forest, which when entered is never-ending in all directions including up or down, except by following a stream or by becoming still and waiting for time to pass. In the clearing there is also a pond into which one can sink (by breathing out) and then descend deeper into an oceanic abyss

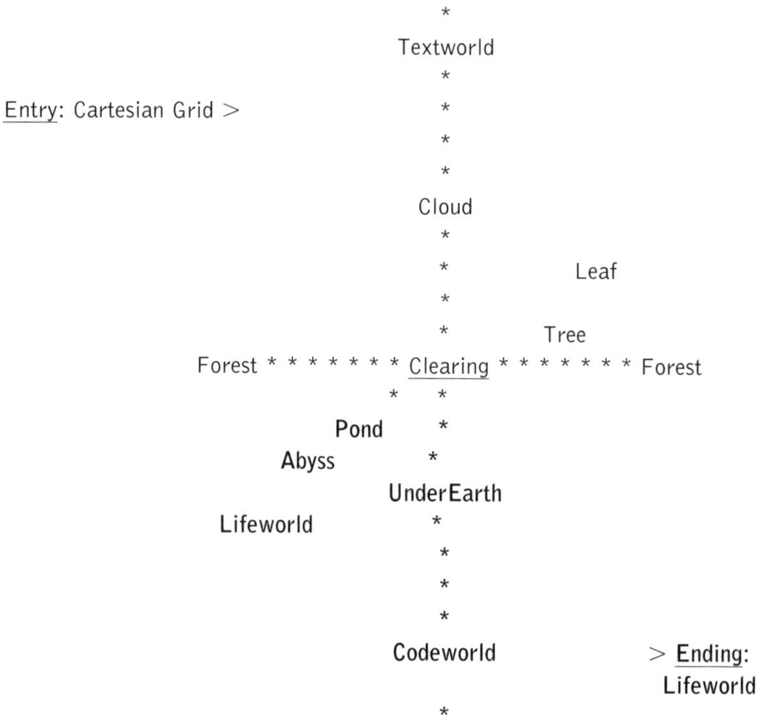

in which a symbolic life-world appears through which one can return to the clearing with its pond, stream, and tree. It is also possible (by breathing in) to ascend into white cloud—or, by breathing out again, to descend into subterranean earth, passing roots and rocks and underground streams. Two other realms—above and below, of text on nature, technology, and the body and of software code—function as the conceptual substratum and superstratum parenthesizing the work. The sounds within *Osmose* were sampled from a male and female voice uttering phonetics and digitally processed to create a range of effects and localized in three-dimensional space. Sound is generated on the fly, in real time, responding, like the visuals, to changes in the immersant's head position, spatial location, direction, and speed. Using breath and balance, immersants are able to float or hover through all the virtual realms and in the overlapping areas between. After fifteen minutes of immersion in *Osmose* (during public installations), the life-world reappears and then irretrievably recedes, bringing the session to an end.[10]

SPATIOTEMPORAL STRUCTURE OF *EPHÉMÈRE*

In *Ephémère,* the iconographic repertoire is extended beyond the trees and rocks and streams of *Osmose* to include body organs, blood vessels, and bones, suggesting a symbolic correspondence between the chthonic presences of the interior body and the subterranean earth. While *Osmose* consisted of nearly a dozen realms situated around a central clearing, *Ephémère* is structured spatially into three levels—landscape, earth, and interior body. The body functions as the metaphoric substratum under the fecund earth and the lush bloomings and witherings of the land. Unlike *Osmose, Ephémère* is also structured temporally. Even as the immersant roams among all three realms, no realm remains the same. The landscape changes continually, passing through cycles of dawn, day, evening, and night, from the pale of winter through spring and summer to the climatic decay of autumn. While participants may spend an entire session in one realm, it is more likely that they will pass constantly between them, immersed in transformation. Throughout the work, the various rocks, roots, seeds, and so on come into being, linger, and pass away. Their appearings depend on the immersant's vertical level, proximity, slowness of movement, and steadiness and duration of gaze.

All the transformations and interactions in *Ephémère* are aural as well as visual. While the visual elements pass through varying phases visibility and nonvisibility, light and darkness—and in the case of the landscape, progress from the more literal to the abstract—the sound is also in a state of flux. Localized in three dimensions and fully interactive as in *Osmose,* sound oscillates between melodic form and mimetic effect in a state some-

Entry: temporal >>> Winter > Spring > Autumn > **Ending**

spatial
level one:
Forest Landscape: swamp > forest, river, boulders > falling trees > falling leaves
*
level two:
Subterranean Earth: earth, roots, rocks, underground stream, seeds > falling embers
*
level three:
Interior Body: interior flesh, organs, bloodstream, eggs, bones > falling ash + dust

where between structure and chaos, adapting moment by moment to the spatiotemporal context of the immersant within the work. *Ephémère* is more interactive than *Osmose:* when gazed on, its landscape rocks transform into other landscapes; seeds activate when gazed on for an extended length of time, rewarding patient observation with germination, inviting entry into the luminous interior space of their bloom. The only constancy is the ever-changing river: when the immersant surrenders to its gravitational flow, it metamorphoses from river to underground stream or artery/vein and vice versa, summoning in the corresponding visual and aural elements of each realm. Deep within the earth, rocks transform into pulsing body organs, eggs appear, and aging organs give way to bone. Depending on the immersant's behavior within the work, there are several endings, of falling leaves, of ashes, and of dust.

The visuals in these works are soft, luminous, and translucent, consisting of semitransparent textured 3D forms and flowing particles: the 3D forms have been designed to be neither wholly representational (that is, recognizable) nor wholly abstract but to hover in between, creating perceptual ambiguity. By animating these forms and by enabling the participant not only to see through them but to float through them as well, it is possible—because of their varying degrees of transparency—to create spatially ambiguous figure and ground relationships. The resulting constant variability of the perceptual field causes semiotic and sensory fluctuations, or what I call "perceptual buzz," in which multiple poetic associations may be evoked—for a single literal meaning closes the work, whereas ambiguity invites further imaginative play. In my work, ambiguity is key to softening, lessening, the distinctions between things. This strategy, developed through years of painting, serves to offset the cultural bias of dualism, of maintaining rigid boundaries between subject-object, I-it, which finds expression in the aesthetic of "hard-edged objects in empty space" so common in three-dimensional computer graphics (an aesthetic clearly embodied in the dinosaurs of *Jurassic Park*).

The interactive methodology of *Osmose* and *Ephémère* does not rely on conventional hand-based VR interface methods such as joystick, wand, trackball, or glove—which tend to support a disembodied, distanced, and controlling stance toward the world. Instead, we developed an interface that is body-centered and that relies on the intuitive, instinctual, visceral processes of breathing and balance. Through breath, the immersant is able to rise and fall in space with ease and precision. By subtly altering the body's center of balance, the immersant is able to change direction. This reliance of breath and balance is intended to reaffirm the role of the living physical body in immersive virtual space, as subjective experiential ground. It is also intended to act as a channel of communion rather than as a tool of control. As in meditation, the practice of following one's breath and being centered in balance opens up a profound way of relating to the world.[11] This strategy has been informed by my own experiences of scuba diving and has many implications for the work as a whole, both on an instrumental level and in terms of metaphor. The experience of immersion in the deep sea, via scuba diving, has significantly shaped my conceptual approach to VR. While diving, the hands are rarely used because touching often means doing or receiving harm; the vertical axis of movement is more important than the horizontal, and one's buoyancy is dependent on skillful use of breath and balance to rise or fall or turn. Most of all, one feels the exquisite sensation of floating instead of being gravity-bound, of being sensuously immersed in the big blue. Here, space is not perceived as empty or passive but sensually embraces, envelops the whole body, inviting reverie and surrender of the self in rapture of the deep.

Since 1995, nearly twenty thousand people have been individually immersed in *Osmose* and in *Ephémère*. The responses of many participants initially caught us by surprise, in terms of their emotionality, including euphoria and tears of loss. Not only do the works' reliance on breath and balance facilitate relaxation, but they also tend to facilitate a mental state in which logical, rational, goal-oriented behavior is abandoned for perceptual free-fall. In comparing participant responses with psychological research into traditionally induced altered states of consciousness such as meditation, it appears that full-body immersion in an unusual virtual environment (that is, one that does not seek to reproduce literal appearances and habitual behavior in the real world) can facilitate shifts in mental awareness.[12] In this state, it appears that perceptual "dehabituation" allows perceived boundaries between interior and exterior, mind and body, self and world, to become permeable. These parallels have been explored in a previous paper.[13]

The intertwinings of the imaginative and the physical, the immaterial and the material, lie at the heart of *Osmose* and *Ephémère.* These works have been nourished by my experience of an actual space, a place partly cultivated, partly wild, on the slope of a mountain in southern Quebec. Though this land is not pristine by any means—having been cleared by settlers in the eighteenth century, mined for copper, logged for furniture, grazed by cattle, farmed for apples, its meadows rooted with European "aliens," and its nonhuman predators, cougars, wolves, and bears, hunted out long ago—the elements of *Ephémère* and *Osmose* have their source here. This land's trees and roots and rocks, its ponds and mountain streams, its bloomings and witherings through time, have become numinous, as present in my imagination as in actuality. As I ramble among their physical manifestations throughout the seasons and flowing light, they in turn appear in my work like apparitions in a haunting reciprocity between the virtual and the real.

On this land, however, fewer birds arrive each spring; the frogs and toads have fewer young; and the maple trees are dying of acid rain from smelters in the American Midwest. This land is but a microcosm: worldwide, wild places of the earth are being dramatically altered due to a litany of human attitudes and actions.[14] Meanwhile, public attention is being directed to the virgin, untrampled territory of cyberspace. And what of virtual reality? Can virtual representations of nature return our attention to the nonhuman living world—conversely increasing our appreciation of the complexities of the natural environment? Or will virtual environments proliferate at the inverse rate of the disappearance of the real—as some sort of psychic compensation? Perhaps the very act of creating virtual environments such as *Osmose* and *Ephémère* point out the danger that soon computer-generated simulations may be all we have left.[15]

When I first began writing about *Ephémère*—before its completion—it was springtime, and apple blossoms were drifting gently on their boughs. As I write these words now, the apples have already ripened, been harvested, and fallen and rotted, and the withering leaves of Canadian autumn are trickling through the October sky. I write on a laptop electronically connected to the human world, and yet I am alone with an encircling horizon of woods and fields surrounding me. Voices from digital recordings of the seventeenth-century compositions of Pergolesi ring out, interlaced with the rustlings of wind, the flowings of water over mountain stone, and the occasional sound of a hunter's rifle shot. As much as I try to focus, the land keeps calling me—away from keyboard and mouse pad, virtual reality, and the abstraction of words—out into the sensations of nature, winds on my skin, scents of decaying vegetation, and the presences of its inhabitants going about the business of their lives. Waking up the other morning, I looked out the cabin window and thought I was inside the autumnal flux of *Ephémère*—the external internalized and reexternalized as art.

As I reread these words, I sit at the roots of a solitary maple tree among its crumpled ochre leaves, in the gathering violet light and tranquillity of dusk broken only by the sound of the international flight path of transatlantic jets—watching for the deer to venture from the safe shadows of the woods into the soft evening meadows of the orchard. I wait for the deer and all the other creatures who pass through here, strands of multichanneled life, life as a river with infinite rivulets pouring through time. These are the living presences who are so absent in the human-made virtual environments of *Osmose* and *Ephémère.*

NOTES

For more information on *Osmose* and *Ephémère,* including an extensive bibliography, go to http://www.immersence.com

1. By *immersion* or *immersive virtual space,* I mean immersion in a 360-degree spherically enveloping virtual environment, in my opinion possible at the present time only through use of head-mounted displays (HMDs) with wide fields of view. While less cumbersome techniques are under development, current alternatives such as wrap-around screens or domes (now known as spatially immersive displays) are not as effective in achieving a sense of envelopment. Neither are the cubed-shaped display rooms known as CAVES. My comments in this chapter refer to full-body immersion through use of HMDs.

2. This experience bears resemblance to Maurice Merleau-Ponty's description of night in *The Phenomenology of Perception* (1945) (London: Routledge & Kegan Paul, 1995), 283: "When, for example, the world of clear and articulate objects is abolished, our perceptual being, cut off from its world, evolves a spatiality without things. This is what happens in the night. Night is not an object before me; it enwraps me and infiltrates through my senses, stifling my recollections, and almost destroying my personal identity. I am no longer withdrawn into my perceptual outlook from which I watch the outlines of objects moving by at a distance. . . . it is pure depth without foreground or background, without surfaces, and without any distance separating it from me."

3. Yasmin Kharim, private correspondence, 1995.

4. Numerous writers have pointed out technology's tendency toward reinforcement of the Western worldview, including Katherine Hayles, "The Seductions of Cyberspace," in Verona Conley, ed., *Rethinking Technologies* (Minneapolis: University of Minnesota Press, 1993), and Katherine Hayles, "Narratives of Artificial Life," in George Robertson et al., eds., *Future Natural: Nature, Science, Culture* (New York: Routledge, 1996). Also see Margaret Morse, "Landscape and Narrative in Virtual Environments," in Mary Anne Moser and Douglas MacLeod, eds., *Immersed in Technology: Art and Virtual Environments* (Cambridge, MA: MIT Press, 1996). My experience in the late 1980s of building a software company into the world's leading developer of 3D computer animation software led me to understand the potency of this technology in reinforcing traditional Western scientific values.

5. Henri Lefebvre, *The Production of Space* (Oxford: Blackwell, 1991), 407.

6. Richard Coyne, "Heidegger and Virtual Reality: The Implications of Heidegger's Thinking for Computer Representations," *Leonardo,* 27, no. 1 (1994): 68.

7. Ziauddin Sardar, "alt.civilizations.faq: Cyberspace as the Darker Side of the West," in Z. Sardar and J. Ravetz, eds., *Cyberfutures: Culture and Politics on the Information Superhighway* (London: Pluto Press, 1996), 34.

8. Marshall McLuhan and Harley Parker, *Through the Vanishing Point: Space in Poetry and Painting* (New York: Harper & Row, 1969), 241, 252. The words quoted here originally were addressed to the creators of psychedelic light shows.

9. Lefebvre, *The Production of Space,* 407.

10. For more information on *Osmose,* see Char Davies, "*Osmose:* Notes on Being in Immersive Virtual Space," *Digital Creativity,* 9, no. 2 (1998), first published in *Sixth International Symposium on Electronic Arts Conference Proceedings, Montreal: ISEA95* (Montreal: University of Montreal, 1995).

11. Drew Leder, *The Absent Body* (Chicago: University of Chicago Press, 1990), 178, 274: "Breath is a potent tool of overcoming dualism. Physiologically, respiration stands at the very threshold of the ecstatic and visceral, the voluntary and the involuntary . . . inside and outside, self and Other are relativized, porous, each time one takes a breath. The air is constantly transgressing boundaries, sustaining life through interconnection. One may have spent years studying the mystics on the unreality of dualism and have this

remain an abstract idea. But in following breath, one begins to embody this truth." "Balance is a question of centering. When we are properly centered, our experience of Being is in equilibrium. Being well-centered, we can encounter other beings in a more open, receptive way. Finding our center is a necessary step in the development of our ontological capacity to open ourselves to the larger measure of being and to encounter other beings with a presence that is deeply responsive. Coming home to our true center of being, we can begin to relax our egological defenses, and begin to experience things outside the subject/object polarization. Being well-centered in Being is therefore at the very root of *Gelassenheil,* that 'way of being' in virtue of which according to Heidegger, we are going to be most favored with a deeper experience of beings, and the presencing of Being as such."

12. Arthur Deikman, *De-automatization and Mystical Experience in Altered States of Consciousness* (New York: Harper Collins, 1990).

13. Char Davies, "Changing Space: VR as an Arena of Being," in John Beckman, ed., *The Virtual Dimension: Architecture, Representation, and Crash Culture* (Boston: Princeton Architectural Press, 1998), and in abbreviated form, in R. Ascott, ed., *Consciousness Reframed: Art and Consciousness in the Post-biological Era. Proceedings of the First International CAiiA Research Conference* (Newport: University of Wales College, 1997).

14. In October 1998, newspaper headlines announced that PCBs and other persistent toxic chemicals—from as far away as South America—have been found in extremely high concentrations in the uppermost altitudes of the Rocky Mountains, imagined until now to be untouched by man. *Globe and Mail* (Toronto), 17 October 1998.

15. Laurie McRobert, "Immersive Art and the Essence of Technology," in *Explorations: Journal for Adventurous Thought* (Fall 1996).

23

Sound Installations and Spatialization

Cécile Le Prado

As an electroacoustic composer working on sound installations, some preoccupations have been recurrent for the last ten years. One of them, spatialization—controlling the position and projection of sound in a virtual space and simulating the specific acoustics of that space—has become predominant.

Since 1986, I have mainly worked in sound installations for outside or indoor sites. For these compositions, I use mainly environmental sounds and some instrumental sounds. A series of collaborations with the INA-GRM (Institut National de l'Audiovisuel Groupe de Recherche Musicale) and now with IRCAM (Institut de Recherche et de Coordination Acoustique et Musique) led me to integrate the computer as a tool for composition both in the studio (for sampling, sound processing, analyzing, and resynthesizing the digital recorded sounds) and also at the installation site (for sound spatialization, especially evolutions in time of the source positions and of acoustic reactions).

SPACE AND COMPOSITION FOR SOUND INSTALLATIONS

I could take a ball of wool and try to unwind it to explain a logical work process, but things often reveal themselves as being the sum of different preoccupations. For ten years, I have done the following kinds of work:

- Recording environmental sounds,
- Making sound installations that include the public,
- Using almost constantly the sound of water and voices,
- Transforming recorded spaces into imaginary spaces,
- Working on hybridization between several elements (such as designing a sound to integrate characteristics of both an instrumental sound and an environmental sound),
- Playing between what is identifiable in a sound, allowing the listener to recognize where the sound comes from, what meaning the sound carries, and what becomes abstract in this same original sound after it has been transformed in the studio,
- Putting sounds out of their original context by editing, making fragments to very short sounds, like microelements, and
- Working on landscapes, memories of sites, and accumulations of stratum.

Among those focuses, spatialization has become predominant. What I mean by *spatialization* is the localization of the position of the sound according to the position of the listener—in front of, behind, on one side, above, below, and nearer or farther from him or her,

- Following the trajectory of the sound,
- Defining the acoustic of the virtual room in which the sound is supposed to be moving (for example, a cathedral with a long time reverberation or a tiny room with a very absorbent acoustic), and
- Correlating the sound movements to the effects of reverberation.

This parameter, spatialization, is taken into account right from the beginning of a new work when recording sounds, in the middle when composing, and at the end when the sound installation settles in a new site with a particular situation for the public.

The works discussed here—*Le Passeur, Vocatifs, The Triangle of Uncertainty,* and *Lignes d'eau, esquisse n°1* all show aspects of spatialization. The structure of the first example, *Le Passeur* (1992)—in which different depths of fields, where sounds are situated in the space, determine reverberation levels—is fixed in the studio when composing. The trajectories of the sound objects are fixed on the chosen site of the sound installation (Jardin de la Treille) when the technical settings such as the position of the loudspeakers is defined. Since this completely depends on the site, the movements of the sounds cannot be reproduced in another place. The "natural" sounds of the site—like car traffic, bird calls, voices coming from the park, rain, and wind—can reveal or mask different aspects of the composition. It is the same in any outdoor sound installation as long as we are playing with the site and making a soundscape rather sonorizing the space.

For *Le Passeur*, Sysdiff software that molds amplification levels has been used. There is in this case no programmed relation between the sound displacement and the acoustic quality of the virtual room (reverberation simulated in studio) in which they move. The term *room* is a generic term including, for example, an open acoustic field with no walls.

In the second example, *Vocatifs* (1994), the sound objects are not moving. The different reverberation effects cannot be reproduced because they are triggered by the movement of the visitors. The purpose is to provoke an opposite effect between the visual and the sound experiences. When one arrives close enough to read the list of names, the speech recognition becomes impossible because of the reverberation level. This simulates a huge virtual room.

In the third example, *The Triangle of Uncertainty* (1996), the spatial behavior of the sound objects is completely determined in the studio. This spatialization, using Spat software, includes the trajectories with a permanent relation between the localization of the sound source in the chosen virtual room, the settings of acoustic reactions of this room, and the duration of the late reverberation. Spatialization becomes a parameter of composition. It takes the global space into account, as a scene in which we place the sound objects and make them come alive. This can be reproduced and heard through headphones (binaural) and two loudspeakers (transural) with up to eight speakers. Each of these configurations can be produced either in a *panpot* of intensity format or in an *ambisonic* format. The composition and particularly the spatialization do not depend on the site of installation. Nevertheless, the restitution of what has been fixed in the studio depends on the ambient listening conditions of the site.

For the last example, *Lignes d'eau, esquisse n°1* (1992), the Spat is also used. As in *The Triangle of Uncertainty,* everything is determined in studio, and the listening of the binaural version through headphones allows a perfect restitution of what had been designed earlier.

The choice of the listening context that we invite the public to enter is important. In all these examples, we are no longer in a concert situation or in a frontal relation to the music. Most of the time, I try to immerse the listener in the middle of the sound, using speakers or headphones. The presence of a certain landscape in outdoor installations or the choice of darkness in a gallery site, for example, influences listening and changes the interpretation of the work. Likewise, inviting the public to sit is an invitation to stay for a while. This concentration is increased with headphones.

LE PASSEUR

The soundscape[1] in the garden in *Le Passeur* was initially very similar to geologic research in that it integrated sound signs, went deeply into the memory of the site, and uncovered strata.

All the sounds used for this installation are connected with the site—voices, wind in the metallic structures of the Parc de la Villette, and sounds coming from a large area surrounding the park in the northeast of Paris, such as industrial sounds or transport sounds. The transformation of those materials in the studio and the combination of different elements lead those sounds to express their own meaning and their own interpretation of the memory of the garden.

On site, loudspeakers and cables are invisible for the viewer. Low frequencies come from a underground gallery. The structure of the garden is organized in parallel espaliers full of holes from where the water runs like little fountains. At night, little spotlights shining from the same holes look like a multitude of eyes on the skin of the garden. Loudspeakers also utilize the specific acoustic of the holes. The elements of composition make a trajectory between seven horizontal and five vertical lines.

VOCATIFS

The beginning of *Vocatifs*[2] is a list of surnames issued from a databank related to children who disappeared in the former Yugoslavia during the 1998 war. This information was collected for the United Nations High Commission for Refugees (UNHCR). *Vocatifs* is a composition based on the reading of these surnames. The voices and the breathing are the only sound materials for the twelve short sequences. The composition is a succession of variations around processes of sampling, fragmentation, classification, and transformations leading to unrecognizable voices.

At the entrance of the installation, a sensor detects the pressure of the visitor passing by and starts the diffusion of the original list. In the room, visitor movements interact with the reverberation parameters. The deeper the visitor comes into the room, the more the direct sound of the sequences is lost in the reverberation.

This sound installation focuses on the ideas of relocations, unsettled positions, loss of spatial bearings, and loss of identity.

THE TRIANGLE OF UNCERTAINTY

Sailors and seafarers find their bearings at sea by means of natural points of reference located along the coast. These points—for example, church spires, hills, water towers, or lighthouses—generally stand out from the rest of the coastline and are called *amers* (seamarks or landmarks). By identifying three such landmarks in complementary directions, the sailor is able to construct a triangle that inevitably includes his ship. This triangle is called the *triangle of uncertainty.*

The sound installation for *The Triangle of Uncertainty*[3] (figure 23.1) takes up the principle of triangular navigation and substitutes acoustic landmarks (information elements used in navigation that can be recognized by listening carefully) for visual seamarks. Lighthouses, buoys, ship radios, and many other technical facilities warn the sailor of hazards or obstacles.

This installation project is concerned with constructing a triangle of uncertainty in a fictive, virtual space on the basis of sound recordings made at the following locations—the southern tip of Ireland (Fastnet Rock), the western edge of France (Brittany), and the westernmost point of Spain (Cap Finisterre, Galicia). In essence, the installation refers to the position of sound in space, constantly chopping and changing between orientation and uncertainty (figure 23.2).

Recording plays an important role for the conception of the work. Both preselected sounds (such as a lighthouse foghorn or the whistle of a buoy out at sea) as well as sounds discovered coincidentally during the recording are used. The sounds were selected because of their extraordinary, innate acoustic and musical qualities, which give the recording a specific, irreversible three-dimensional image. One example is water gushing through a hole in the pier of Malpica in Spain. Owing to its musical aspects, this sound resembles a reworked studio sound, although maintaining its own typical harbor quality. The history of the recording also plays an essential role during the composition of the work in the studio. When recording, the ear is indeed stimulated, but certain visual and other recollections of the circumstances will actually influence the composition and the installation.

In the studio, the recorded sounds are digitally sorted and cut, and some of them are transformed before taking their place in a kind of story board. The choice of a reworking process is often influenced by a harmonic or temporal quality (to widen the concept of rhythm) that already existed when the original recording was made. By simulating the sound sources paths in this space, we write the spatial behavior as an element of composition. It is gradually refined by repeated listening and modification.

This work was created with the help of Spat®spatialisator. The Spat®spatialisator, developed by Ircam and the Espaces nouveaux work-

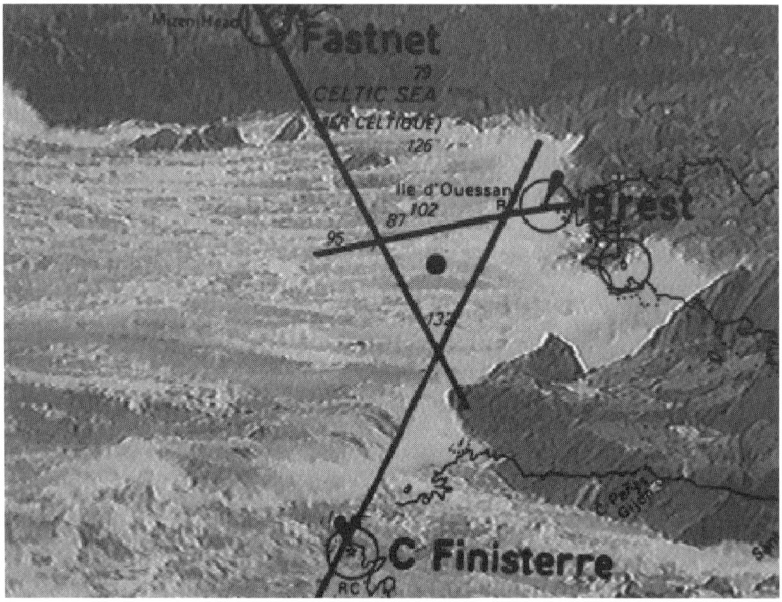

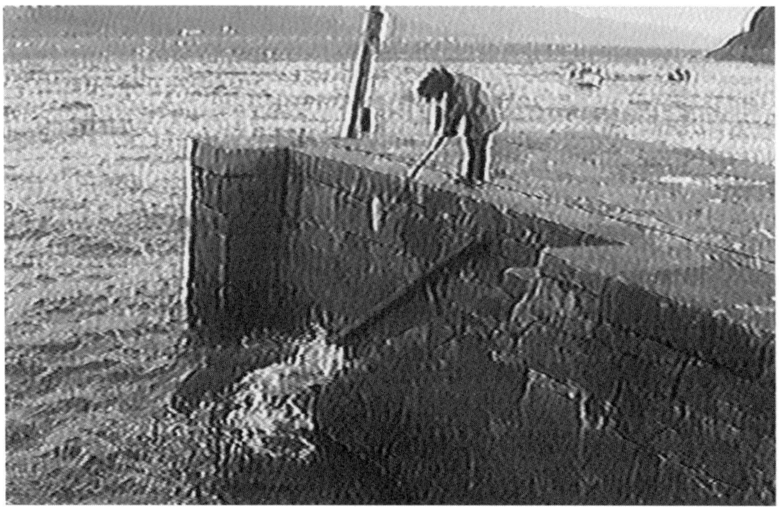

Figure 23.1
Cécile Le Prado, *The Triangle of Uncertainty,* 1996, sound installation. Virtual triangulation between Ireland, France, and Spain.

Figure 23.2
Cécile Le Prado, *The Triangle of Uncertainty,* 1996, sound installation. Recording sounds on the pier of Mapica, Spain.

shop, is a virtual acoustic processor that enables the musician or sound engineer to control the positioning of sounds and their dissemination in a real or simulated listening space. Spat is essentially based on a perceptive analysis of space. It allows this construction (which has been simulated in the studio) to be adapted to meet the requirements of various playback sites, inside or outdoors, over a system of two to eight sounds sources and also through headphones.

The final stage of realization consists in integrating a composition made in the studio into an outdoor or indoor setting. The interplay between the composition, the "natural" acoustic environment, and the acoustic of the installation site is always interesting. Some aspects of the studio composition vanish, while others become more important. The site develops and gradually takes on the form of the suggestion made in the studio, particularly depending on the time of the day, the weather conditions, and so on. The interaction between a given site and the installation determines a specific combination of restrictions and options. In this sense, *The Triangle of Uncertainty* is somewhat constraining. It must be heard under precise conditions. The speakers must be equidistant from each other and set up in a circle. The ideal space of perception, in which the spatial effects can actually be perceived, is limited to a specific area within the circle.

LIGNES D'EAU, ESQUISSE N°1

Lignes d'eau, esquisse n°1 is based on recordings made along the river La Sèvre Niortaise between the town of Niort and the sea, near La Baie de l'aiguillon. The recording evokes thoughts of water, the sea, or water's more secretive aspects, as in the marsh. In the past, the sea reached the town of Niort.

The composition goes from reality (with the sound of water on the beach, occasional footsteps, and no processing), through some intermediate conditions (where identified sounds are mixed with some resynthesized and spatialized sounds), and finally to abstract sequences (creating a rhythm with the original material). Visitors are invited to enter a dark environment lighted only by two lines on the floor. A sound line is parallel to these virtual lines. It is made of a smooth water sound with the steps on the beach. Through headphones fastened on the back of beach chairs, a four-minute sequence runs in a loop.

Displacement is inherent in works where movement is inscribed in the material. . . . In Cécile Le Prado's *Lignes d'eau, esquisse n°1* (1998), the sound emerges and spreads along a well-defined trajectory, underlined by a swath of light. The viewer is projected into the acknowledgment of the passage of time essential to the sound of music and of movement. The consciousness of the "duration" of time, on which Bergen so instead, is reinvested in hearing at the moment when one sits down in a chair and puts on the headphones. This shaping by means of music is exemplary of the plastic quest of the twentieth century: the quest to materialize the abstract relation to the passage of time.
—LILANA ALBERTAZZI, CURATOR OF THE EXHIBITION *REMISE EN FORMES*

NOTES

1. *Le Passeur,* 1992, sound installation created for the exhibition *Parcours sonores/Parc de la Villete/Jardin de la Treille.* Equipment: one eight-track analogic player, one PC computer, twenty loudspeakers and amplification; software: Sysdiff Cidma); scientific consultant: J/M/Gourden; composition realized in the studio Lygis; technical collaboration: Cidma; production: Cidma, Parc de la Villete, with the help of the Ministère de la Culture.

2. *Vocatifs,* 1994, interactive sound installation created for the exhibition *Artifices 3/ Saint-Denis France.* Voices: D. Micic, E. Skralovic (Association Sarajevo); equipment: one Macintosh, one PC, six sensors, seven loudspeakers, two compact disk players, one printed list on a music stand; software: max©Ircam, Sysdiff©Cidma; composition realized at the Ircam studios; production: Cidma, Clameurs, Ircam, Observatoire de l'Image, with the help of Artifices 3; scientific consultant: M. H. Serra.

3. *The Triangle of Uncertainty,* 1996, sound installation created in the Parc Del La Villette/Paris as a suite for maritime landscapes; musique concrète for the sound installation produced as part of *l'Imaginaire Irlandais 1996* by invitation of the French government; composed at the Ircam studios; scientific advisor (Spat®): Jean-Marc Jot; spatial processing carried out using Spat®; musical assistant: Gilbert Nouno; mixing/mastering assistant: Frédéric Prin; assistant sound recordists: Mar Pazos Oviedo, Christian Dubet; software: protools©Digidesign, Spat©Ircam, Audiosculpt©Ircam, Studiovision©Opcode, Max©Ircam; sound installation produced by Clameurs; executive production and distribution: "l'autre rive"; coproduced by the Centre National Dramatique et Chorégraphique Le Quartz de Brest and the Établissment Public du Parc et de la Grande Halle

de la Villete, with the help of the Association française d'action artistique (AFAA), Ministère des Affaires Étrangers, and Audio 33—Amadeus Concept.

REFERENCES

Albertazzi, L., ed. *Différentes Natures.* LINDAU, EPAD, Ministère de la Culture et de la Francophonie, Delegation aux arts plastiques, 1993.

Albertazzi, L. *Remise en Forme.* Paris: Galerie Xippas, 1998.

Billaudeau, B., and C. Le Prado. "Recherche sur la perspective sonore." In Francis Dhoment, ed., *L'espace du son* (80). Ohain, Belgium: Musique et recherches, 1988.

Bossier, J. L., A. M. Duguet, and L. Terrier. *Artifices 3.* Saint-Denis, France: Direction des Affaires culturelles, 1994.

Clostre, A. "Musique en France." *ClingKlong, Frauen musik Forms Schweiz,* 42 (1998): 32.

Delacriox, P. *Sites Choisis* (Chosen sites). In L. Albertazzi, ed., Niort, France: Ville de Niort, 1991.

J. M. Jot, "Real-Time Spatial Processing of Sound for Music, Multimedia, and Interactive Human-Computer Interfaces," *ACM Multimedia Systems* 7, 1 (1999): 55–69.

Jot, J.-M., and O. Warusfel. *Techniques, algorithmes et modèles de représentation pour la spatialization des sons appliquée aux services multimedia.* IRCAM research report, 1996. Available at ⟨http://varese.ircam.fr/articles/textesJot97a⟩.

Julien, P., and O. Warusfel. *Technologies et perception auditive de l'espace.* Paris: Cahiers de l'Ircam, 1994. Available at ⟨http://varese.ircam.fr/articles/textes/Julien94/⟩.

Le Prado, C. *Arts Electronica Festival 97: Fleshfactor* (346–349). Linz, Austria: Springer Verlag, 1997.

Le Prado, C. *Paris ville lumière projets d'artistes pour l'espace public parisien* (93). Paris: Paris Musées, 1994.

Le Prado, C. "The Triangle of Uncertainty." *Proceedings of the Fifth Simposio Brasileiro de Computacao & Musica* (3: 253). Belo Horizonte, Brasil, 1998.

24

A Tool Is a Tool

Pamela Z

I am often asked how recent changes in technology have effected my art. It is difficult for anyone to be alive today in this culture and not be in some way touched by the sudden upsurge of computers and digital technology, and in this regard I am no exception. Not only has this technology had a major effect on my work as composer and performer, but it has infiltrated practically every aspect of my life. I compose music on the computer. I use digital sound processors and MIDI (musical instrument digital interface) equipment in live performance. I record, edit, and construct sound works using digital sound editing software and hardware in the computer and then use peripheral devices with my computer to burn CDs of those works so they can be heard by others. I use musical notation software to create scores when I compose works intended to be performed by other musicians. When I'm not working directly on my art, I'm using the computer to communicate with others through e-mail and to make others aware of my work through maintaining a Web site. And I use the computer to teach others to use digital audio and other software. I use the computer to do my books, manage my promotional materials, and keep track of colleagues, venues, services, and members of the press. The list of computer-centered activity in my life seems endless. I haven't yet found a way to cook in the computer, but if left to my own devices, I'm afraid that I might!

I am aware of the ironies around the effects of the use of computers on productivity. For every task that is made more efficient by them, there is at least as much new busy work around crashes, upgrades, incompatibilities, et cetera to counterbalance that new efficiency with a new brand of inefficiency. Even so, I chose to succumb to all of that and continue to work in the way that I do. In fact, I have to admit that I have been thoroughly seduced by computers ever since my introduction to the Macintosh 128k machine that I started on in the mid-1980s and that I enjoy the actual using of the computer beyond the practicality of what I'm able to accomplish with it. The computer is a tool, and I have a very strong relationship with my tools.

I have made some of my greatest strides and artistic discoveries whenever I have begun to employ a new tool to make my work. I have learned over the years that one of the best ways to stimulate growth or new direction in my work is to introduce a new instrument into my arsenal. I can, in fact, chart major changes in my work throughout my life as coinciding with the introduction of these instruments.

Growing up, I certainly saw a steady stream of new tools for making sound. My first instrument was my voice, and, as children, my sister and I used pods from a tree that grew in our Denver neighborhood as found percussion. As an elementary school student, I began to play the viola in orchestra and in private lessons. While still in elementary school, I was given a classical guitar, and my family shared a set of German block flutes (recorders). An old upright piano entered our home when I was beginning junior high school, and of course I eventually "prepared" this instrument by placing thumbtacks in the felts of its hammers. Also, around the same time, my father sent us one of those newfangled Craig cassette tape recorders, which was followed soon after by a second one. This prompted me to start creating imaginary radio programs—bouncing back and forth between the two recorders to make multiple layers of my voice and other sounds. The introduction of each of these new sound-making tools had definite effects on my development as an artist—shifting my way of listening to and creating music and audio works.

As an adult, perhaps the most profound shift in my work that I can recall occurred when I acquired my first digital sound processor. I bought an Ibanez DM1000 Digital Delay in the early 1980s. Right around that time, I had become dissatisfied with the comparatively conventional music I was playing professionally and began looking for ways to start creating more experimental types of work. My previous influences had largely been rock music of the 1960s, singer-songwriters of the 1970s, and classical opera from the 1800s. But now I was finding myself far more interested in punk and new-wave artists, minimalist composers, and a whole slew of experimental electronic composers ranging from John Cage, Pauline Oliveros, and Alvin Lucier to Laurie Anderson and Brian Eno. All my attempts to jump-start a new direction in my compositional style had been relatively fruitless, and I think this had a lot to do with the tools I was using at the time. For the work I was performing live, I was mainly using my voice and a hollow-body electric guitar. I was working in a kind of standard singer-songwriter way and trying to integrate my classically trained voice into the picture somehow. My recorded work (for which I was using a Fostex portable cassette four-track) started evolving much more quickly than my live work. I was working with layers of sustained vocal sounds combined with sounds from another new tool—a cheesy little Casio synthesizer.

But I was unable to translate any of that growth to my live performance until I brought home the Ibanez Delay. For me, this was a perfect performance instrument! It allowed me to make layers of sound in real time so that I could create fairly complex and dense pieces using just that processor and my voice. Literally, the moment I began working with it, my whole compositional style began to transform. I began to listen to sound differently. Because of the looping (infinite hold) feature on the delay, I became very interested in repetition and gained a keen awareness of the kinds of things that happen to our perception of small pieces of sound when we hear them repeated at length. And because my voice was constantly coming back at me in the delay loops, I began to play a lot more with timbre and texture. These were to become key elements in my work, equaling and surpassing my previous emphasis on melody and harmony. I was so effected by this new instrument (this device that most musicians were more likely to refer to as an effect than as an instrument), that it wasn't long before I went out and purchased a second and then a third digital delay. Digital delays remain at the core of my performance set-up to this day (although I now create them using an Apple Powerbook with MAX MSP software rather than my original outboard processor units).

With the dramatic changes that came about in my music due to the use of this new tool, my hands and my body were freed up for gesture and movement, and I became more focused on the performance aspect of my work. I came to see the sound I was making and my physical behavior while making it as an integrated whole. I learned during this period of time (the mid-1980s) that performance itself was a discipline and that I was as much a performer as I was a musician. Found objects and toys that were both sonically and visually interesting augmented my complement of instruments. I used things like a Slinky ("the wonderful wonderful toy"), a pair of hammer handles, an empty five-gallon plastic water bottle, and some strips of Plexiglas as performance tools. The effect that the addition of these objects had on my work was, in some ways, as profound as that of the introduction of digital processing. I have a piece called *Bone Music* in which I use an Alhambra water bottle (of the type usually found inverted on water coolers). Although I only physically make a sound with this object three times at the opening of the piece, the bottle is so important to the structure that I cannot do the piece without it. (This poses a small problem when touring outside the United States where water coolers are not common!) The three, quick sounds that I make (by slapping the bottle) linger in digital delay loops for the duration of the piece and form its rhythmic base. And the visual impact of this large object remains as I continue to physically manipulate the object silently while singing.

My lack of skill with or understanding of a new tool has often worked in my favor. One example of this would be a discovery I made after I got my second delay. Through efforts to make the two delays synch up with each other, I inadvertently created my first set of "out-of-phase" loops. At that time, there was no MIDI clock to slave one device to the other, so I was trying to adjust the delay lengths and get the two of them in synch. Obviously, I couldn't make them precisely the same length manually, so the loops would start together and then gradually begin drifting apart. When I heard this sound, I was immediately in love with it, and I wondered whatever possessed me to try to get them in synch in the first place. I began writing pieces that deliberately incorporated out-of-phase loops. I was delighted with my discovery. Ignorant of such things as of Steve Reich's early works, I thought I had invented the idea!

"Mistakes" of this kind have long been an enriching force on my work and continue to be. Brian Eno refers to these unexpected results as "happy accidents." Many composers have deliberately imposed external forces on their work as a way of cultivating unpredicted results. Perhaps most notably, there is John Cage with his chance operations. Numerous others have employed mathematical algorithms or other systems to create the structure of their pieces. For me, the introduction of new tools continues to be a great catalyst, and the accidents that happen because of the learning curve are often far more interesting to me than anything I might have thought up on my own.

Once I begin to develop some facility with a particular tool—regardless of whether it is an acoustic instrument, a piece of electronic hardware or a new version of some software—my work is then effected by my newly found proficiency on the instrument. In the past several years, I have been creating recorded works using digital editing software on my Mac. As I gained skill with nondestructive editing programs such as Digidesign's SoundDesigner and ProTools, I began working in new ways. I began making sculpted sound collages with small bits of layered text and found sounds. My first hard disc recording works were informed by the live work I do with digital delay loops, but I quickly started making more complex and varied structures, working with smaller bits of sound, and altering the samples.

As a composer and performer, my choice of tools (instrumentation) is often determined by my own capabilities as a solo performer. I began using the BodySynth several years ago when I wanted to introduce pre-sampled sounds into my live works (figure 24.1). Up until that point, I had used only samples that I created in real time in my digital delays as I performed. I wanted to be able to use sounds I could not sample on the fly, but I didn't want to use a sequencer to play them. I wanted to trigger them myself but wasn't interested in adding a keyboard or some kind of drum triggers to my set-up. I needed to keep my hands free and not do anything that would limit my ability to gesture. I learned about the BodySynth, which was created by Ed Severinghaus and Chris Van Raalte, when they loaned me one to perform a piece called *Dream Encoding* with Zakros New Music Theatre. This instrument, which uses electrode sensors to measure the electrical impulses generated by the performer's muscles, allowed me to use physical gestures to trigger samples and manipulate various sound parameters, so of course I had to buy one. Once I did, I began creating a lot of work that used a variety of sampled sounds. I was able to introduce traffic noises, text samples, and sounds literally from my kitchen sink into my live performance works. The introduction of the instrument changed the way I was composing (figure 24.2).

Of course, tools alone do not make great art. I like to think that the advances I described above stem from the combination of the effects of using the new tool and my strengths as an artist. An important part of an artist's process is selection, and it takes an intelligent, open, and inventive ear to recognize and select good ingredients and then build them into something viable. In the end, the instrument is really just a tool—like my digital processors and my found percussion objects and like my voice. And, as an artist, it is always important for me to be concerned about what work I am actually making with this tool. It frustrates me to see a world so seduced by new technologies that many have forgotten to be concerned about the output. We suddenly see a superabundance of works being created by people who are clearly more interested in what software they have mastered than they are in the value of what they are making with it. One hears endless jokes about content as sort of an afterthought in a project. (And, worse yet, sometimes they're *not* jokes!) The "multimedia" industry (with terminology pirated from the fine-art world) blurs the line between art making and commercialism, thus attracting many people who are seduced by the combination of becoming professionals in a big-money industry and the cachet of being able to call themselves artists. There have always been people who believe that having a great tool will make them great artists or magically result in the creation of great art,

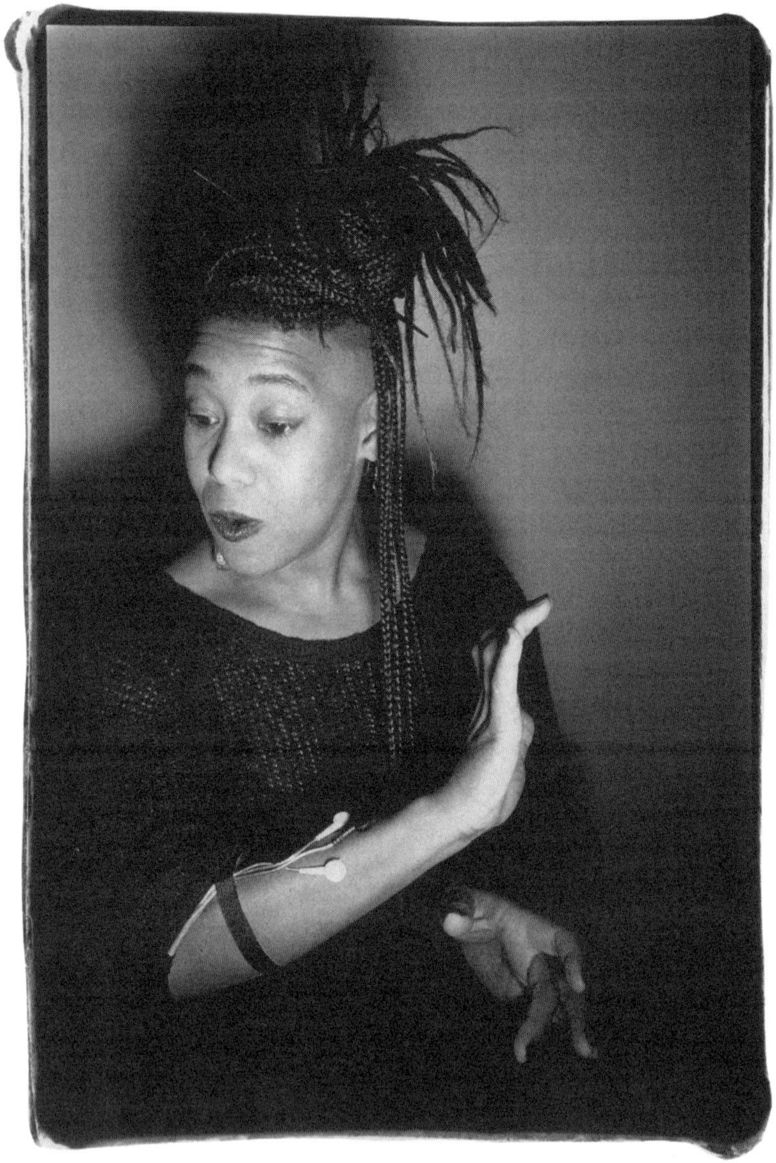

Figure 24.1
Pamela Z wearing the BodySynth. The MIDI controller uses electrosensors to translate a physical movement into sound information. Photo by Lori Eanes.

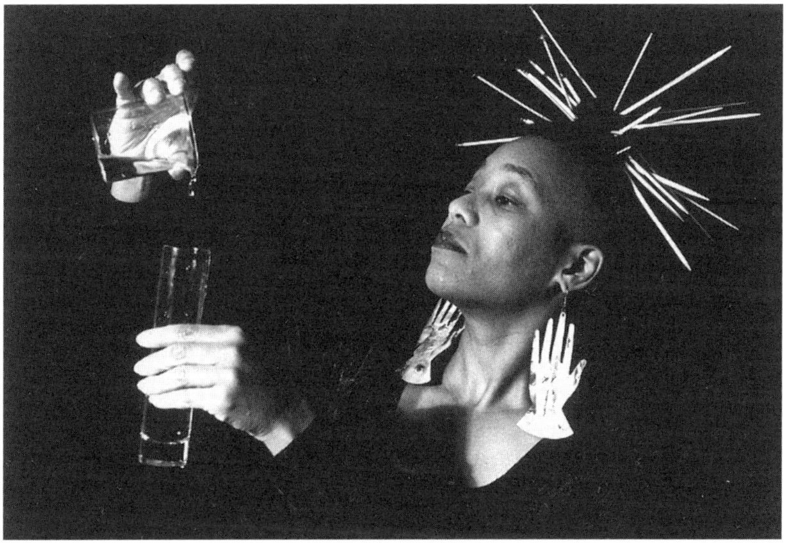

Figure 24.2
Pamela Z manipulating "tools" she used in the 1995 San Francisco Contemporary Music Players' performance of a Louis Andriessen piece. Photo by Marion Gray.

but buying the finest violin or tennis racket does not a great musician or athlete make. However, having a new tool can certainly inspire great work from someone who has the potential to make it. Hopefully, a good side effect of this rush to embrace new technology is the opening up of some new artists who perhaps never realized they had that potential.

In the meantime, it has awakened in me a sort of curiosity about artists' choices of instruments or tools. I find that I am fascinated with artists who work with relatively low-tech tools. and I am also drawn to work by artists who have developed very technically complex tools for making their work. Some of the most exciting work I've seen lately combines very different types of tools—acoustic instruments with electronic ones, mechanical devices with digital devices, machines with flesh and blood instruments. And it is interesting that, in a field that historically has seemed very male-dominated, many of the artists doing this are women.

There have always been a lot of women composers, yet music history books (with a few notable exceptions such as Kyle Gann's *American Music* in the twentieth century) don't tend to reflect that. And since historians of the past generally neglected to acknowledge the contributions of women, I suppose it shouldn't surprise me that in the history of electronic music, few women have been given much notice either. I am aware that there is a disproportionate number of male artists in this field anyway, but there are quite a few more women in the field than one might think from reading most of the books and journals on the subject. I have a feeling that, along with all the other reasons, tools may have something to do with that. It seems that people's expectations of the kinds of tools an artist would use are somewhat separated along gender lines. In fact, when I have remarked about the absence of women's names in various histories or collections of electronic music, I often get responses like, "Well, you know, women aren't as interested in holing themselves up in a lab with a bunch of electronic gear." To which I am inclined to reply, "Actually, I can name for you quite a few who are."

The tool that women seem to be expected to excel in using is the human voice. And when we do excel in that, we do get recognition for it. Cathy Berberian, Diamanda Galas, Joan LaBarbara, Meredith Monk—all these women are very respected and well known for their work with this very technically complex instrument. They are much more celebrated than are any of the men who use extended voice as a main component of their work. But Pauline Oliveros, Laetitia Sonami, Annea Lockwood, Laurie Spiegel, Maryanne Amacher, and the many other women who have done great work in both the designing and using of systems for electronic music are much less likely to be mentioned than their male counterparts. The message seems to be "If you want recognition for what you do, you need to stick with the tools you are expected to use."

At a certain point in time, I became very interested in the male voice. I wanted to hear more men doing modern music that required them to extend their voices. Several years ago when I was doing a regular *Morning Concert* slot on KPFA F.M. (Berkeley), I tried to do a radio program on men doing extended vocal work. I had a hard time finding enough recordings to fill out the program. I saw John Cage at a new music festival (*Composer to Composer* in Telluride, Colorado), and I asked him if he could suggest any recordings of extended vocal pieces for men. After a lengthy pause, he laughed and replied, "I'm afraid there aren't any." I had had such little success in my search that I was inclined to believe him at first, but as I did more searching, I slowly began to discover several (David Moss, Roy Hart, Dimitrio Stratos, and Jaap Blonk to name a few). But it still is easier to find female artists working in challenging ways with the voice, just as it is easier to find male artists doing nonvocal experimental music. And, again, although I don't claim that the actual numbers are balanced, there are quite a few of these artists in existence, and many of them are exceptional in their field. Yet they don't get the amount of recognition that their opposite-gender counterparts seem to get. They are not being rewarded for breaking the norm in their choice of tools. This goes a long way in explaining, at least in part, why it might be that there aren't more women who choose to work with nonvocal electronic music and why there are so few male experimental artists selecting the voice as their chief instrument.

Whether it is because of natural tendencies or because of deeply rooted socialization, men and women can often be very different in their approaches to making art, and these differences are magnified when the art is experimental or avant-garde in nature. I have observed, when teaching performance workshops, that women are often much more comfortable using their voices (and bodies) in untested ways. Sometimes even men who are great improvisers and who feel quite at home making loud, bizarre, even shocking sounds in public when using some kind of external tool (a saxophone, a percussion instrument, a piece of sheet metal) become shy and uncomfortable when asked to experiment with their voices. Perhaps the external instrument is like armor or a shield between them and the audience, so that using it to produce strange or unusual sounds may feel less awkward than making a sound that comes directly out of one's face. When an artist uses his or her own body as an instrument, it is like being naked. Making wild sounds with an external instrument may seem like exerting control over something, while making those sounds with the voice might seem like losing control (i.e., madness, hysteria?). I don't mean to overgeneralize. There are many men who are quite happy using their voices and bodies in performance, and there are many women who are not. But in a general way, women in our culture are not only socialized to feel more comfortable baring themselves in that way, but people are socialized to feel more comfortable accepting it from women.

Likewise, our culture has always socialized women to feel less confident working with mechanical or electronic devices, and people in general continue to have less confidence in women's abilities with them. This perhaps seems like an overly obvious or perhaps archaic observation. But as a female artist working with technology, I get little reminders from time to time (in the first decade of the 21st century!) that this is still so. For example, although it doesn't happen as often as it used to, I still get asked questions like, "Who set this up for you?" or "Who taught you how to do that?" —questions I don't imagine I would be asked if I were a man. The funniest part about it is that, compared to most people I know working with live electronics in performance, my set-up is relatively simple. I tend to use all the devices I have in fairly straightforward ways. I am encouraged that these questions are slowly being replaced with questions like, "Did you design that system yourself, or did you work with a collaborator on it?" I've even been asked that about the BodySynth (an instrument that I use but did not create).

It's interesting that I find myself thinking so much about these issues now. In the past, I have never really been very focused on sociopolitical issues around gender and making art. I always went about tinkering with whatever was needed in order to do the things I wanted to do. I never remember personally having had any concerns about the ability to do something technical having any connection to gender, and I don't remember feeling self-conscious about being the only woman among people doing the kinds of things I was interested in. But then, of course, I was the same person who had to have it pointed out to me by others that I was the only black person at a function or in an organization. I never really thought about these kinds of things much. Naïve as that may have been, I didn't even tend to notice.

A few years back, I was trying to design a program in which I would teach audio workshops at The LAB Gallery in San Francisco. I was applying for a California Arts Council artist residency grant and needed to come up with a strong concept that would be useful to some kind of underserved local community. Laura Brun, the LAB's artistic director, suggested that I do a program for women or girls. I was very hesitant to offer a program that would exclude anyone on the basis of gender. At the time, I was thinking of working with at-risk youth, and I didn't want to go into schools and say, "Sorry, only girls can do this workshop."

The first workshop I offered through that program was open to all high school students. We distributed applications to several San Francisco high schools where I did live performance presentations. Then I met with LAB staff, and we selected applicants who seemed to be most in need of such a program and seemed serious about wanting to do it. Although many girls would approach me after the presentations, in the end I got more boy applicants than girls. In the first meeting of the class, I went around and had people introduce themselves and say why they signed up for the workshop. I was rather taken aback when most of the small number of female participants gave reasons like, "I decided to take this workshop because my boyfriend has a recording studio, and I feel so stupid around him because I don't know how anything works." I began talking to women artists I knew about this, and I began to discover that many of them, as adults, still felt this way. I spoke with a surprising number of women who said that they had taken audio engineering courses and had felt intimidated by the men in the class and even had felt pushed out of the way in situations where many were vying for hands-on time on the equipment. It seemed like a lot of their problems stemmed from a combination of unequal treatment by instructors, classmates, fellow musicians, et cetera, and their own lack of confidence in their own abilities to tackle the tasks at hand. This experience caused me to slowly become convinced that I needed to offer a workshop for women.

For the next two years, I taught several eight-week workshops for low-income women. The workshops focused on both sound and performance. I taught the women about the physics of sound, digital and analog audio techniques, audio art and performance. In groups, we worked on sound performances using found objects, scores, voice, and gesture, and I worked with them one-on-one with digital sound editing on the Mac and how to use devices like a digital sampler, an analog mixing board, and a digital audio tape (DAT) machine. I found this to be a very rewarding experience, and I am delighted that many of my students have gone on to become very deeply involved with sound art.

I don't think that one can really divide tools into categories of feminine and masculine. Throughout history, in various cultures, both men and women have always played a variety of instruments, though some cultures have placed limitations on who could do what (and who could even perform publicly for that matter). But if there is anything to the yin and yang of using the voice versus using an external instrument, then the type of electroacoustic music that combines vocal practice with electronics might be viewed as a way of exerting both feminine and masculine qualities in the performer. This type of work, which was pioneered mainly by women and continues to be female-dominated, could be seen as a modern-day extension of the age-old practice of singing while accompanying oneself on an acoustic instrument (a practice that has historically been done by both males and females). All the various women and men who do this probably arrived at this way of working from a variety of different paths. In my case, the practice of combining my voice with live processing and sampled sounds was a direct descendant of singing while playing a guitar. As I began to develop more facility with using digital processors to create layers of sound in live performance, songs on which I used the guitar began to disappear from my repertoire.

Back in my singer-songwriter days, I always thought of my voice and the guitar as being separate instruments. One was in the role of soloist and the other as accompanist. When I switched over to using processors and found objects combined with the voice, I began to see all of those things as being more integrated. I began seeing the combination of my voice and all my electronic gear as one instrument. One was not there to augment or accompany the other. For me, the digital processors were not "effects"; rather, they were components of a more complex instrument, which included my voice and my physical presence as well. And once again, the choice of tools effects the mode in which I work. I tend to compose works with this entire set of tools in mind. I don't make vocal pieces and then think to myself, "How can I add to this texture using electronics?" All of those components tend to be in there from the start as I create the work. I developed my way of using these tools together throughout the twenty or so years I've been working this way, and the combination feels very natural to me.

In the past, I've had to defend the voice as an instrument. People would speak of vocalists and musicians separately, as if a vocalist was not a musician. For a short while, I was a member of the Musicians' Union when I was living in the Denver/Boulder area. I had to list myself as a guitarist because voice was not included as one of the instruments. I felt that my level of skill and my strength as a musician was much stronger in voice, but I was told that I would have to join an actors' union if I wanted to list voice as my main instrument. I couldn't believe it, but I noticed that there were a lot of musicians listed on the rolls who named things like "tambourine" as their main instrument! A colleague told me that those musicians were probably professional singers who, unlike me, did not also play the guitar. They had to choose an acceptable instrument in order to access of any of the advantages of being a member of the same union to which their bandmates belonged. (I eventually questioned what those advantages might be to the extent that I resigned from that union.)

I also encountered many people in the past who did not consider electronic devices to be instruments. They would say things like, "I don't like those synthesizers. I prefer music that's played on real instruments." "Real instruments" apparently referred to instruments made from wood or brass or silver. These days, I sometimes find myself defending the use of such things as computers, objects not originally created for music making, or samples of text or noises as musical instruments. My definition of *instrument* is a broad one. It basically includes any tool used for making one's work. (For many musicians today, a vinyl disk is a musical instrument!)

There are as many ways of working with tools and as many attitudes toward the tools as there are artists using them. In the end, I think these approaches are more about the artist as an individual than they are about factors such as gender or cultural background. But an artist's relationship with tools, the effects of changes in the tools themselves, and the development of skill and comfort level with those tools have a great impact on the work. What is really required to make good art is a good artist. And a good artist knows how to coax great work out of his or her tools of choice, whether by using lack of familiarity with the tool as an advantage or by perfecting virtuosity with the tool through years of diligent practice. After all, whether the instrument is acoustic, electronic, analog, digital, flesh and blood, or some combination, a tool is a tool is a tool.

25

Production and Reproduction

Nell Tenhaaf

Since I began to seriously pursue an amateur interest in biology in the mid-1980s, I've been intrigued by the idea that organic life and electronic machines would eventually merge in some way. Futurologists are currently hot on the topic of how and when this will happen, and some even play the game of attaching a date to the blending of computers and life, as does Jeremy Rifkin in a recent article where he puts it at 2060.[1] Unless medical science moves ahead in even greater leaps and bounds than it has, I won't be alive then; so I hope that he's late in his calculation by at least several decades.

The disposition that always accompanies this kind of future predicting can vary from gloom and doom to gleeful anticipation; the intense curiosity that I have about what form the merging of life and machines will take falls somewhere in between these two poles. Rifkin situates his thoughts on the matter critically, within an economic analysis of the impending "biotech century" that will be constituted as a bioindustrial environment in which genes spell commerce. His lament for the dearth of public debate to address what is often seen as inevitable because it is technological is something frequently heard in the leftist press these days. But he's right in suggesting that any aspect of the bioengineering revolution has, like science in general, values attached to it. Anything will be threatening or enticing, even human-machine symbiosis, depending on whether we are able to frame it only as manipulation by invisible powers that be or, alternatively, as creative generation of the self.

But I suppose that I describe it in these words because I'm an artist. My work involves technologically based representational practices that can be seen as producing and reproducing subjectivity. I say these things metaphorically, of course, whereas in very real terms, we experience the representations and effects of science as also generative of who we are, physically and psychically. Using metaphoric language allows some slippage between artistic and scientific premises, such that "the age of mechanical reproduction," as Walter Benjamin famously coined it, could now be taken as an appropriate description for the unfolding biotech era. Such slippage points to both the richness and difficulties in a crossover of science issues into art practice, whether transposed metaphorically or deconstructed critically. It's what I've been preoccupied with, both in making art and in writing, for over a decade.

Biotechnology entered my practice in the late 1980s as a perceived solution to problems that arose in my work from a well-established but generalized interest in technology. That is, I wanted a thematic concern to anchor what had become a rather free-floating fascination with computer imaging platforms. Moving from one kind of digital imaging system to another was symptomatic of the broad interest I had in digital technologies throughout the 1980s, such that the technologies themselves were playing too strong a role in determining the content of my work. In retrospect, I could attribute these interests to the Canadian intellectual propensity to identify with the technologically utopic visions of Marshall McLuhan, although he is just the most visible among a historical trinity of McLuhan, Harold Innis, and George Grant. These thinkers outline the links in this country between a tenuously thin population spread across great distances and their interconnection via the early technologies of the railroad and the trading routes before that. Even as a first-generation Canadian, my psyche was infiltrated by this idiosyncratic philosophical brew of geographical location, network mentality, and machinic vision.[2]

Through artists' access to government-run Telidon field trials that had been negotiated by a group of Toronto media artists interested in "communications art" in the early 1980s,[3] I had become interested in networked databases as an alternative to the museum and gallery distribution system. My first online work using Telidon, which is a videotex system based on graphics and text primitives, was called *Us and/or Them* and appeared on a public database in Montreal in 1983. It was a critique of the doublethink of the cold war, a portrait of the strategic similarity of the two superpowers in their standoff even while they persuaded us that their ideological differences ran to the quick. Reflecting back on the random reception of this work by people who logged on, from a dumb terminal in their educational institution, I compare it to the Internet today: it may not be considered counterinstitutional, but it does retain the sense of an expanded venue for art that I was interested in at the time, with its (so far) open access. The proto-Internet work I did in the early 1980s has returned to my practice many years later.

Videotex then became the medium for an installation called *Believable if not always true . . .* (1987), which addressed the illusion of infinite choice in interactive database structures. A touch interface called up an off-line database, located on a computer housed within a set of structures resembling theater flats. The database is structured with a main menu of six choices, each of which is a platitude about knowledge (such as "Too much knowledge can be a dangerous thing"). The imagery reflects the motifs of the sculptures—pop Egyptology and some of its spin-off pseudoscientific theories, mixed in with some equally enticing and fantastical goddess mythology. In the following year, the availability of a three-dimensional animation system elicited a project called *Gramatica* about the language of mathematical perfectibility in relation to the philosopher Ludwig Wittgenstein's foray into architecture in the 1920s.[4]

With the introduction of biological subject matter into my work beginning with *Species Life* (1989) (figure 25.1), my objects and installations continued to use the computer as a way to composite images as well as a device for interactivity. But the focus shifted to a critical engagement with the burgeoning field of biotechnology, with a particular focus on the strong connection between computing and genetics that was appearing in all kinds of popular imagery. The orientation of molecular biology research as reflected in the media was more genetically deterministic then, and I had begun to decry this as a "code fixation," which I defined as attention to genetics at the expense of environment or the supremacy of DNA ruling the somatic matrix in a way that parallels how the (rational) masculine has historically ruled the (irrational) feminine. This position was one aspect of a poststructuralist and specifically feminist critique of science that dominated my point of view. The key conceptual approach of all of my works that involve biology is to bring forward the cultural assumptions hidden in the supposedly neutral representations of science.[5]

In parallel with this strategic interest in biotechnological images and issues, I researched the historical and contemporary politics of science in relation to art and representation in general. Since the Enlightenment, a divergence between the sciences and the humanities (which I think can now more accurately be called *cultural discourses*) has grown into a seemingly unbridgeable gap. In 1959, C. P. Snow named the gap "the two cultures," a split that has not been healed over time. Rather, it has been exacerbated in recent years by the resistance of the establishment on both sides of the fence to any critique from the other.

I emphasize establishment, particularly in relation to the conservative fringe of the sciences that scorns any critical analysis by what they call "academic leftists." They would bunch into this category the many artists

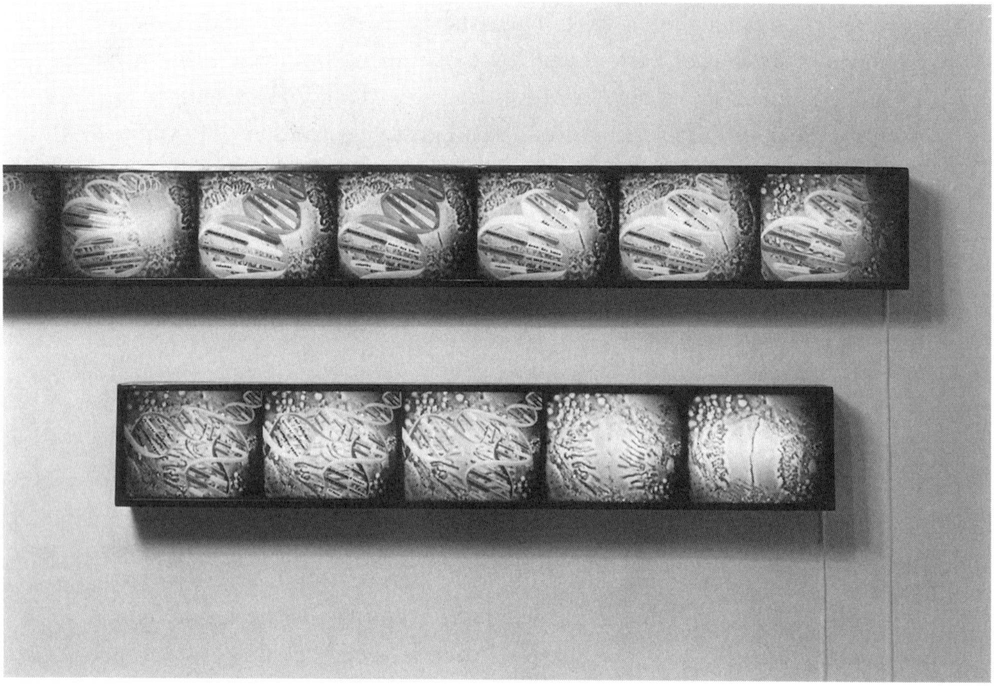

Figure 25.1
Nell Tenhaaf, *Species Life* (detail), 1989, wood, fluorescent lights, duratrans, 367 × 95 centimeters. The monitor-sized images are a digital composite of three layers: video stills of a cell splitting; candy-colored textbook images of DNA; computer-inscribed texts from Luce Irigaray and Friedrich Nietzsche, in their original languages. The texts are deciphered in a smaller lightbox. Both visually and conceptually, this work ironically proposes a binary, gendered inscription of language at a deep level of coding in the organism. Photo by Denis Farley.

who have addressed aspects of science in their work in ways that successfully critique its underlying principles of atomization and objectification and the histories of their negative effects. In recent years, there have been some very public manifestations of the polarization between scientists and cultural practitioners or theorists: most notorious in this respect was the 1996 "Sokal hoax," the publication in the Science Wars issue of the journal *Social Text* of a paradoxically postmodern article by American physicist Alan Sokal. The academic publishers fell for it, proving to the scientists that intellectuals have no grounds for commenting on science, ever.

Nonetheless, I have worked for many years in the vein of science critique, although my core artistic strategies—conceptual layering, aesthetic pleasure, and an ironic tone—usually soften or even mask the critical statement. In a 1993 work called *The Solitary begets herself, keeping all eight cells,* I consider reprogenetic technologies in relation to control over generation of the self. A metaphor for parthenogenesis, this image was shot on video and transferred to computer for processing. This takes the form of a reclining self-portrait that some people say reminds them of Baroque paintings of Christ's entombment. Inscribed on my body are medieval human-beast hybrid figures whose imperfection and even monstrosity were celebrated in the Middle Ages as intrinsic to the array of the world's life forms. Along with these mutant creatures from Conrad von Megenberg's fifteenth-century *Buch der Natur,* I used a readily available image of a dividing cell from a biology textbook to invoke early embryonic development. But it is the work's title that calls attention to a political issue: it refers to embryonic biopsy in the in vitro fertilization process, in which one cell is removed at the eight-cell stage for genetic testing. The idea is that even in fantasized parthenogenesis, here proposed as a process parallel to image production, it may be important to resist genetic perfectibility.

In their imagery and the way the viewer is invited to relate to them, my biology-based works often propose ways to subjectively inhabit objective research data that appear in the form of beautiful but impenetrable digital simulations. In a touch-activated interactive work called *Fit* (1995), a phrase from Rifkin's timeline of the future seems apropos, as he conjectures that in 2012, "*Molecular archaeology* becomes prophecy as a person's genetic past yields a detailed script of the future."[6] In *Fit*, I propose that imagining and opting for an improved genetically altered self would be as complex as deciphering one's genealogy. Pressing on one of the nine images of the interface calls up a small movie showing an art-historical image, with voiceover in one of five languages intoning phrases like "survival of the fittest" and "adaptation." Viewers access the program of videos by pressing on small buttonlike images—as if they were standing at an automatic teller machine. The imagery and languages are all related to my origins, a rather private perception that for the viewer yields to a general impression of cultural complexity. At the opening of all the sequences, a dancing molecular model of DNA and the written ratio "what would be made of me/what made me" appear, to guide viewers in their reading of the piece.

The impasse I've now reached concerning the expression of scientific commentary in my work could be summarized as battle fatigue, a weariness with the negativity of positioning myself critically in relation to science. I'm actually compelled by many facets of the sciences, including developments in digital simulation techniques for molecular modeling and the fact that these are also used for things like cinematic special effects. But to reflect a fascination with such imagery in my art practice can also feel like inadvertently doing advertising for the biotechnology industries because the processes and images associated with them are inseparable nowadays from academic and medical "pure" research.

My sense of frustration with the scope of such issues has led me to focus more on an area that I've been interested in since it first became visible around 1990, the field of artificial life (A-life). A-life is a multiple set of practices based on "evolutionary" or "genetic" computational strategies, set within a backdrop of theories of dynamic systems such as chaos theory. When I first began to inquire about it, I encountered mostly skepticism from my biologist friends and contacts but interest from electronic media artists such as David Rokeby and Norman White in Toronto, artists with established practices based in hacking and robotics. So it took a while to situate myself and my practice in relation to it. A-lifers build models of the natural world in the form of either computer simulations or robots. In my perception, A-life is not research science per se but is parallel to science, a hybrid area with a speculative approach to how nature in a very broad sense works. It brings together biology at both its micro-

and macroscopic levels, organic and inorganic chemistry, computer programming, complexity theory, many aspects of artificial-intelligence (AI) research, semiotics theory, and much metaphorizing.

What distinguishes A-life research from other pursuits of the artificial, in particular classical AI with its basis in preprogrammed rules, is that it relies on the principle of overlapping and interlocked systems built from the bottom up. The idea is that this approach can generate simulations that account for the unexpected, models of systems or artificial agents in which the whole is more than the sum of its parts. At least in theory, new behaviors or entities are evolved that are not preprogrammed and therefore emerge from the system's dynamics.

My first foray into making works related to A-life was still involved with critiquing biopower politics, especially as manifested in the dominance of genetic code over the complex dynamics of whole organic systems. But the imagery I chose suspended these works within a debate: "Should we understand living form—including our bodies—as being dictated by a molecular genetic blueprint or as the self-organized result of the dynamics of life itself?"[7] In *Apparatus for Self-Organization* (1995) (figure 25.2) and *Orphaned Life-Form* (1995), I wanted to develop strategies of representation that evoke the dynamic energies of our biological substrate by using the immediacy of drawing and the direct perceptual effect of light sequences. There is a very specific art-historical citation in these works: the iconography of the cloud shape is from Marcel Duchamp's *Le Grand Verre: La Mariée mise à nu par ces célibataires, Même* (*The Large Glass*). It represents the feminine aspect of the Bachelor Machine, the Bride, who is ethereal and an engine with a feeble dynamo but who is nonetheless matter itself, the female element. The internal light sequences are electronically controlled; *on* and *off* stand for the simplest threshold states of dynamic matter, attractors in the language of complexity theory. The light events are recursive but not repetitive so they seem to be both predetermined and full of potential. This paradox is at the heart of the organization of life and, for the purposes of a cultural analysis, of subjectivity.

A-life programmers and theoreticians are concerned also with this paradox. They use both predetermined genetic rules and environmental randomness to model life phenomena. At the inception of A-life, they were committed to the idea of making synthetic lifeforms that would literally

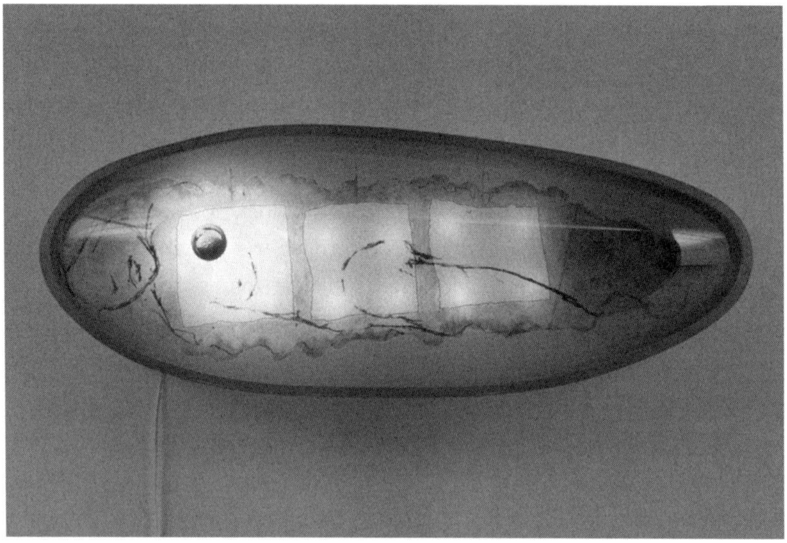

Figure 25.2
Nell Tenhaaf, *Apparatus for Self-Organization,* 1995, duratrans, aluminum, plexiglas, electronically controlled light sequence, 107 × 42.5 × 17 centimeters. The bride has a "reservoir of love gasoline" and is an internal combustion engine, but there is no source for the energy circuit that animates *The Large Glass,* suggesting a playful analogy for the principal of self-organization at the core of A-life thinking. The lights inside this work are visible through the bride's "draft pistons," which are the record of wind-blown squares of gauze hung in a window and represented for Duchamp a portrait of pure chance. Photo by Jean-Jacques Ringuette.

be successors to biological lifeforms. No longer balking at the naiveté of such a proposition, as I did when I first encountered it, I take this notion of "life" in A-life thinking as partly methodological and intensely metaphorical. There is a lot of play in the way that A-life models are interpreted as being lifelike: although a consensus is shared around evolutionary features of organisms such as adaptation to environment and the tendency to reproduce, a vast additional range of simulated computer or robot behaviors are described from a very anthropomorphized perspective. This is in spite of the fact that the behaviors have been built algorithmically—that is, using standardized computer programming. Because of this interpretive factor, A-life is not resistant to cultural narratives in the way that the sciences are; it gives these narratives and their potential meanings much more play.

A-life resonates around implications of dynamic interaction and active interpretation. I would argue that this makes it a creative practice that is very easily linked with art practice. Since A-life is built on creative computational strategies, it certainly enables the art practitioner who is interested in computers and modeling to engage with scientific subject matter in a way not previously available. I have just completed a project that uses A-life methods of evolutionary computation (or "artificial evolution," as it is sometimes called) to set up an interactive situation for viewers that simulates a fitness test. The installation, called *UCBM (you could be me),* asks whether there is an evolutionary advantage in accepting a simulation of intimacy or at least in adapting to a "new" kind of intimacy.[8]

The basis of this work is a computerized system that projects a video persona onto a screen, and this "artificial entity" seems to exhibit some socially adaptive qualities. So one thing that is new about the intimacy in the work is that it hinges on a shared empathy between the human viewer and a surrogate who is driven by a program. In a sense, the surrogate personifies the system that runs the installation. She elicits comparison to some of the now well-known female computer personalities—for example, ELIZA, the interactive psychotherapist program developed by Joseph Weizenbaum at the Massachusetts Institute of Technology in the mid-1960s, or the more recent Julia, a popular online MUD persona programmed by Michael Mauldin at Carnegie Mellon University.

A second simulation factor in this installation is the phenomenon of Internet-based intimacy, or pseudo-intimacy, including CuSeeMe video-conferencing sex exchanges and Webcam displays of real-time self-exposure on Web sites. This content refers back to the online performance *Neonudism*, which I first showed in 1997 and which integrates a live feed from the Net in the form of CuSeeMe two-way video. *Neonudism* is a hypermedia work designed to place viewers into a conceptual and aesthetic environment in which they experience a "live" online sex scene. To participate, the user first engages with a constructed audio and visual environment, which is based on several layers of surrogacy. The top layer of imagery is the point of identification or entry for the viewer—a digital video "self-portrait of the artist," or at least a portrait of a self-involved artistic temperament. This entity does not have a fixed identity; she changes personae through both image and voice several times during the program. The portrait is merged into the next layer of surrogacy, an "eye candy" program of self-generating A-life imagery called *Bomb* that portrays the inner workings of the machine as having a quasi-organic and visually pleasurable life of its own. The composite of these two elements—the talking-head artist overlaid with dreamy pixel flow—interacts with the feeding in of a CuSeeMe image. The "amateur porn" participants who are present via CuSeeMe are transformed into "art nudes," thus pushing the viewer through the boundary of distinction between experience and art and between nudity and sexuality.[9]

What I refer to as the "display phenomena" of the Net are an aspect of the subject matter in *UCBM*—that is, some of the imagery references Webcam displays of people's private lives. The surrogate entity in my installation offers these kinds of images for consideration in her interaction with viewers to find out how they assess the realness and the emotional impact of Internet intimacy display. The surrogate's language is personal and psychological in the kind of information she wants from the viewer, but it is also technical and artificial because it is built on the programming that drives her system. As she probes the relation between her and the viewer, the surrogate "interprets" their body data and assesses their empathy factor. She uses a genetic algorithm to situate a viewer's performance within a population of nine viewers, also displaying the workings of the algorithm as a seductive, dynamic array of bright orange LEDs.

What I want to investigate in this work is the problem of interpretation—that in any process of understanding we use one type of representation analogously or metaphorically with another to explain it. The laws of science have often played the role of ultimate interpreter, but I see A-life modeling as a more sophisticated narrative and interpretive device that foregrounds the interplay between modeler, viewer, and the system itself in generating meanings. In this instance, the A-life element encompasses shifts in key concepts such as fitness and progress (which are currently emerging from complexity theory) and places these within the dynamics of an interaction that is interpersonal, self-reflexive, and attentive to the moment. The question of who comes out on top in the installation is a moot one, since the less fit viewer could be seen as having an adaptive suspicion of artificial relationships with machines. There is a certain voice of scientific and technological authority invoked, but it is counter to that well-established project of science, determining final answers. It calls attention to the integration of any individual entity within complex and coevolving systems.

In my practice, I try to come to grips with the escalating mediation of the natural world through digital media and a parallel sense of increasing artificiality. The voice of institutional critique will always have a place for me. But in my recent work, I want to consider how new hybrid representational techniques tie in to some positive intellectual developments and shifts in perception, in particular how A-life can lift the social burden of hammering away at the monolith that science tends to be for the non-expert. In this phase of my creative life, I'm interested in theories and practices that generate a more optimistic approach to investigating the cultural impact of science and technology.

ACKNOWLEDGMENT

This text is an expanded and revised version of "Nurturing the Artificial," *So, to Speak* (Montreal: Artextes Editions, 1999).

NOTES

1. Jeremy Rifkin, "God in a Labcoat," *Utne Reader* (June 1998): 66.

2. See Arthur Kroker, *Technology and the Canadian Mind: Innis/McLuhan/Grant* (Montreal: New World Perspectives, 1984).

3. See *Art Is Communications*, catalog of a videotex exhibition curated by Paul Petro and Geoffrey Shea, A Space, Toronto, November 1985. Petro, Shea, Bill Perry, and Nina Beveridge were the founders of Toronto Community Videotex.

4. *Gramatica*, 1988, is an installation (not interactive) composed of a slide-dissolve sequence projected on a suspended 2-meter by 3-meter moiré screen, with sound track, showing a tableau of the allegorical figure *Gramatica* and a computer model of the Wittgenstein house, representing the drive to perfectionism and structural beauty.

5. For an in-depth discussion of this work, see Kim Sawchuk, "Biological Not Determinist: Nell Tenhaaf's Technological Mutations," *Parachute* 75 (July, August, September 1994): 10–17.

6. Rifkin, "God in a Labcoat," 71.

7. From the catalog for *LikeLife*, an exhibition curated by Joe Faith, Medeni Fordham, Inman Harvey, and Phil Husbands, held at the Brighton Media Centre in conjunction with the Fourth European Conference on Artificial Life in Brighton, UK, August 1997. I presented a poster on "Aesthetics of A-life" within the conference and showed two works in the exhibition. See Phil Husbands and Inman Harvey, eds., *Proceedings of the Fourth European Conference on Artificial Life* (Cambridge, MA: MIT Press, 1997).

8. The title of the work comes from Kerstin Dautenhahn, "I Could Be You: The Phenomenological Dimension of Social Understanding," *Cybernetics and Systems* 28, no. 5 (July–August 1997): 417–453. A cybernetics researcher at the University of Reading, U.K., Dautenhahn's research involves modeling aspects of social interaction, such as empathy, in robotic artificial agents.

9. *Bomb* by Scott Draves is an interactive program that is based on fractal and genetic algorithms and that responds to users' keyboard input as well as to sound. The original conception of *Neonudism* is by Nell Tenhaaf and Eliot Handelman. For more background, see my "Neonudism," *Parachute* 85 (January, February, March 1997): 12–15.

REFERENCES

DeLanda, Manuel. "Nonorganic Life." In Jonothan Crary and Sanford Kwinter, eds., *Incorporations/Zone 6.* New York: Urzone, 1992.

DeLanda, Manuel. *A Thousand Years of Nonlinear History.* New York: Swerve Editions, 1997.

DeLanda, Manuel. "Virtual Environments and the Concept of Synergy." *Leonardo* 28, no. 5 (1995): 357–360.

Hayles, Katherine. "Narratives of Artificial Life." In *FutureNatural: Nature/Science/Culture.* London: Routledge, 1996.

Helmreich, Stefan. "The Spiritual in Artificial Life." *Science as Culture* 6, pt. 3, no. 28 (1997): 22–31.

Kauffman, Stuart A. *The Origins of Order: Self-Organization and Selection in Evolution.* New York: Oxford University Press, 1993.

Levy, Steven. "Artificial Life: The Quest for a New Creation." *Whole Earth Review* (artificial life issue) (Fall 1992).

Mitchell, Melanie. *An Introduction to Genetic Algorithms.* Cambridge, MA: MIT Press, 1996.

Penny, Simon. "The Darwin Machine: Artificial Life and Interactive Art." *New Formations* no. 29 (Autumn 1996): 59–68.

Tenhaaf, Nell. "As Art Is Lifelike: Evolution, Art, and the Readymade." *Leonardo* (digital salon issue) 31, no. 5 (1998): 397–404.

Tenhaaf, Nell. "Where Surfaces Meet: Interviews with Stuart Kauffman, Claus Emmeche, and Arantza Etxeberria," *Leonardo* 34, no. 2 (2001): 115–120.

Tenhaaf, Nell. "Simorg Culture." *Parachute* (La Bestiaire/Endangered Species issue) 72 (1993): 41–43.

Whitelaw, Mitchell. "Tom Ray's Hammer: Emergence and Excess in A-Life Art." *Leonardo* (digital salon issue) 31, no. 5 (1998): 377–381.

26

Your Words, My Silent Mouth: Trying to Make Narrative Sense out of Nonnarrative Work

(A brief collection of interviews and polemics in the interest of aesthetic quasiclarity)

Allucquère Rosanne Stone

/framegrab from the video/

I was four years old, the family was visiting a friend who lived way out on Long Island, and a hurricane blew in. The power went out, and the big folk lit candles, and with the wind roaring and howling and shaking the house, I wandered away from the warm flickering glow in the kitchen and the conversation around the table and toddled hesitantly into the living room. The darkness in there was attractive and scary in an adventurous way. The living room was furnished in Victorian style, and I was very small, and the furniture loomed hugely over me, and as I passed beyond the kitchen doorway, the area around me became quite dark. There was a musty aroma of long disuse mixed with an indescribable fragrance that my daughter calls the smell of old people.

I was slowly feeling my way across the rug when a terrific flash of lightning lit up an ancient Atwater-Kent radio console right in front of me, the kind with the speaker in a round cloth enclosure on top of the set. In that blue-white shard of time, a stunningly loud clap of thunder slammed into the room, and as it did so, the thing before me was charged with an eerie, malevolent, living presence. It was no longer the passive object of my gaze; instead it glared back, measuring me, testing me, gathering its forces. I was frozen to the spot. I couldn't scream, couldn't run, couldn't even step back. The hair all over my body stood straight up. And in the next instant the room was pitch dark, the spell was broken, and I scrambled shrieking back to the kitchen.

It was a seminal moment for my relation to technology, and over the years I've tried in several works to capture something of that eerie instant in the storm and the dark, heavy Victorian atmosphere when the artifact came alive for the first time. . . .

BACK TO THEORIZING

It's easy to talk about theory, but then you stop theorizing, and you do what you do and it speaks for itself. *Your Words* (1996) was like that. It was an enormous release from the tyranny of theory—my first, really, to leave me feeling so free. But we are so terribly tied to the whole idea of textuality and, its relationship to pedagogy and, even worse, to legitimacy. Folks who still don't get it invite me to perform and then ask for text for the publication part of whatever they're doing, a symposium or whatever. And when I say that text is counter to what I just said during the piece, counter to the current direction of my work, that at this point I'm not really thinking in a textual way, they still refuse to get it even in the face of an evolving body of creative production whose main textual content consists of my saying, "Let's reevaluate the enterprise of textuality and its relation to performance, let's think like New Media/New Studies wonks, let's turn to embodiment and gesture and voice as an experimental arena in which to work out these problems of representation and method." And they look blank. There's a mental switch that turns off. It reminds me of that sweet, ridiculous woman who came into my office and asked me if I'd be willing to support a conference on women's issues that was going to be held at the university later that year. I said sure, show me the poster, and she pulled one out and opened it up. The featured speaker was Joanna Russ, who happens to be on record with some fairly disgusting antitransie remarks. I stopped reading and looked up at this woman, and her smile faltered a bit. "So you really expect me," I said, "to support a conference whose guest of honor halted a major conference on women and writing to demand that a transsexual in the audience be thrown out?" And this sweet, clueless person got her smile back on straight and said, "Well, if you can't support it yourself, would you mind recommending it to your friends?"

That's roughly the level of the discourse. As soon as you move just a bit away from the equilibrium point, the brain switches off, comprehension goes out the window, and you get into areas of inanity that you didn't believe existed. In specific, there is this imagined continuum between performativity and textuality, and it is one of the most tenacious nonthings I've ever encountered. The people who, whether accidentally or on purpose, are acting as gatekeepers for what counts as acceptable academic work have become so fixated on text that everything must be reconstructed as text, no matter how unlikely the topic or circumstance. Which is why, come to think of it, we're sitting here discussing an installation piece and attempting to make sense of it as text.

I think part of our job as new studies practitioners is to focus our attention on that point at which the mental switch turns off and seek out ways of fooling it into flipping back on again.

Let's see if I can open that up a bit. When I entered the History of Consciousness program, the first thing that attracted me was the disciplinary jargon of critical studies. To me it looked dense and complex and impenetrable, which was wonderful. It was wonderful because encompassing it was such a huge challenge, starting from scratch as I had rather than having been trained in the tradition as an undergraduate. And of course as things turned out, my naiveté worked in my favor. It allowed me to develop my own idiosyncratic interpretation of why the language of critical studies had evolved into its particular relationship to other methods of making meaning. I realized it was because in our attempt to grapple with fundamental issues of meaning production, some of us had come to the conclusion that we had finally exhausted the possibilities of formal language as a tool with which to analyze and criticize itself and its relationship to power. And so instead we had turned inward on our own meaning practices, were restructuring our own descriptive language, ripping up *parole* in hopes of getting through to *langue*—using language against itself to tear itself apart. The hope in this was twofold: one, that by using language praxes as a kind of pathognomonic tarot, we might discover something about method, meaning that out of the wreckage of *parole* we might be able to discern a shadow of the lineaments of *langue;* and two, that by reassembling the pieces in new configurations, we might be able to make meaningful statements that were at least partly outside the force field of phallogocentrism, precisely because they were outside language as we know it. Now, this is traditionally seen as more the role of poetry, which, in our still-Enlightenment-soaked institutions, is not an okay thing to employ in, say, the social sciences. But keep in mind that in New Studies the idea of disciplinary specialty doesn't work in easily recognizable ways. So it's not impossible that a part of the work of poetry might be part of the work of understanding method.

I'm sure you've noticed some similarities between the terms I just used and certain things that were going on elsewhere in the cultural enterprise. Of course, there have been many other attempts to break loose from the culturally authorized structure of language while still retaining some of the cultural functions of language. By this I am simply pointing out that since language is always encultured, language always means, whether we happen to want it to or not. If it doesn't mean through a preexisting rigor of symbology, the listener will find some resonances anyway, pleasant or otherwise, for as humans, making meaning is Job Number One, hardwired and always ready. This, by the way, is the point of the Near-Legibility project: near-legibility is about provoking those resonances, though with Near-Legibility we are trying to provoke resonances in a system of symbols that is not the default one but that nevertheless exists in some juxtaposition to it—almost-remembered voices and experiences that flit about the edges of consciousness but don't quite make it into directly capturing our attention.

In our time we've had the Beat movement, and it could be well argued that rap and hip-hop are also attempt to break out of existing meaning structures, but of course the enterprise extends much further back than that. But all of it, really, was part of the general refiguration of symbology itself, which signaled the loss of modernism's grip on meaning. The passive mode I'm using here may be misleading, so let me add that by the passive construction refiguration I don't mean that symbology was changed as a by-product of social forces on other errands. Rather, I am conceiving the rigor of symbology, the rules that symbology gives itself, we may call this *langue,* though it isn't quite—as something that under certain circumstances can come to act as if it possessed agency. Here I'm referring to the larger field within which symbol operates as a kind of potential structure that can spontaneously reorganize under certain gradients of information flow. In other words, a far-from-equilibrium condition, though I think Prigogine would cringe at my appropriation of the term. In this interpretation—and this is my point, really—the refiguration of symbology that began with the inception of the postmodern was not merely a consequence of the play of forces in and around the emergence of early capitalism but rather was an active participant in creating it. In other words—and this is the sticky part—language is more than organic; it's organismic. And it's not an isolated example. The same is true for any complex dynamic structure of symbols that exists in close relationship to dense social systems. Capital has always been the perspicuous example, but it isn't the only example. Okay, so I don't think it's exactly fashionable to talk about, say, Haraway and Rimbaud in the same breath, but whether consciously or unconsciously, they share something of a common method of expression produced by stretching the tools at hand beyond their limits. In another sense, that's merely what poets do, but I'd say it wasn't exactly safe to make such a statement about theory. In fact, I did, to Teresa de Lauretis, and got roundly trounced for it.

But we ignore the tectonic shifts in rigor of symbology at our peril. Ignoring them is what made Alan Sokal possible. Sokal quite correctly perceived what was happening and turned it to his advantage and to Social Text's embarrassment because the very people who were engaged in the enterprise of restructuring method were unable to distinguish between the professional jargon they had hitherto employed and the new symbology that was struggling to speak with their voices and write with their pens. And so they had to pretend that the language they used still meant within the older frame of meaning production, when the meaning structures they produced had already left the older frame and now meant differently. Which indicated that they themselves couldn't understand what the language they were bringing into being was trying to tell them. So when someone arranged that language in random ways, the self-same refusal to see the depth and extent of the transformation that had already occurred made it possible for nonsense phrases to displace meaning. One rarely sees such a moment played out in such a vivid way.

This had a direct relationship to *Your Words, My Silent Mouth*, which should by now be obvious. Its animating force is the old, old idea that the speaker is capable of saying only what language permits. And here I mean *langue*, simply. So that when the structures of meaning production are always already predetermined in the phallogocentric sense, whatever comes out of your mouth is already predetermined in its range of meaning; whatever you say and however you say it, you are always merely uttering the phallus. Because in this episteme that we share, the phallus is the rigor of symbology; and consequently we recreate endlessly the world in which the phallic utterance means. In the specific case of transsexual subjectivity, when the transsexual speaks, those utterances continually erase the signs of the speaker's presence in the world—the language world and the world of situated action. Statements about deconstructing identity and gender, about breaking the deadly grip of binarism, are heard as something else because they are uttered in the language of gender and binarism. So therefore, if the transsexual is to speak and to be heard as a transsexual, it is necessary to find a language that does not continually dissimulate about what is being said, dissimulate about its own nature in the guise of producing meaning.

So how do we get outside meaning just far enough and just long enough to turn back, observe our situation, and act on meaning itself in such a way that we make room for other regimes of utterance? That was, by the way, the original intent of at least some of proto-rave culture beginning in the

1960s—a conscious and deliberate attempt to get at meaning production in a different way. I had dug myself deep into the guts of that movement at the time, happy and stupid and doing things we called psychedelicatessens, which were basically assaults on the senses in an attempt to stun the forebrain long enough to sneak in past it and rearrange the brain's handling of symbology before all those wonderful defense mechanisms snapped into place.

Psychedelicatessens were just a new and funkier electronic version of the Catherine wheel, which was a mechanical-age parlor toy that worked in much the same way. The Catherine wheel consisted of a set of rotating mirrors that created a rapidly changing and confusing visual field, the idea being that someone looking at one would become entranced by going into visual overload. I had an odd experience while trying to explain this. I was being interviewed by the local radio station as part of a promotional blitz for a rock band and light show, and the announcer asked me what light shows were for. I opened my mouth to say, "Well, the purpose of a light show is to stun the forebrain so that we can operate directly upon the preconscious with a carefully orchestrated set of symbols whose purpose is to permanently change the way one processes reality." (Eric Clapton said it much more succinctly: "The object of the game, really, is to send them home feeling they've been struck by lightning.") And I suddenly realized I was about to say this to the audience of a top 40 A.M. radio station in Topeka, Kansas. And my tongue stuck in the back of my throat. The only sound that came out was "Gah." We both stood there, him holding the mike out expectantly because he knew me as a fairly articulate person, and me absolutely stonkered about being able to say a single word of this in such a way that the audience could receive it. Probably not more than ten agonizing seconds went by before he jumped in with something suitably inane, whereupon we changed the subject.

TRANSITION

I originally started the project that includes the video installation piece because for some time, beginning around 1990, I'd had the creepy feeling that nothing I was doing was equal to the task at hand. I was still textual, so to speak, but I'd been dogged by the sense that no matter how eloquent my writing became, I was still addressing the wrong audience and that circumstance had little or nothing to do with the quality or lack thereof in what I was trying to say. Further, and more to the point, the disloyal opposition—by which I mean the transphobes and bigots among us who masquerade as scholars and purport to write interestedly or dispassionately—were peddling their wares at the same old stands. One can write only so many scholarly rebuttals before the topic wears thin.

And then there was the flock of hot-button issues that circulate around the marvelously unstill point that I've been thinking of as New Studies. Hottest among them is what constitutes acceptable academic work, and to a great extent that issue is bound up with what counts as inside a given set of disciplinary boundaries and what counts as outside a single discipline or outside all of them. Consequently, before I can talk about my work, I have to explain where it comes from and where it's going. Hence digressions like this. You know this stuff already, but then again, you may not.

So here's the riff. Likely it hasn't escaped our attention that the really interesting questions these days are not the ones being asked inside the humanities, social sciences, even the arts. Not that what goes on there isn't worthy and important and interesting, but I don't need to belabor the point any more: it's fairly self-evident. Lets call the constellation of technologies, practices, beliefs, and webs of social connection that those questions imply New Media. And let's call the analytical and explanatory practices we use when talking about New Media by the name New Studies.

The first thing we discover is that we can't do New Studies without simultaneously doing New Media—which is to say, there is no privileged, indifferent observer in New Studies. We are simultaneously inside and outside technologies of meaning production, which are the analytical tools and also the object of inquiry. How did we get ourselves into this position?

The term New Media came into existence because some practitioners thereof were upset by having their work referred to as *interactive multimedia.* No surprise here, really. First, *multimedia* is a term coined in the 1940s, referring to phonograph records and slide projectors used together. (For some reason, sound movies don't seem to have been thought of as multimedia.) And *interactive* is a felicity if ever there was one, a word that means everything to everybody and consequently winds up meaning nothing at all. Hence some folks get together with some particularly weird drugs on a particularly large hotel bed during some particularly raucous conference, and presto, name coinage happens and ve have ze New Media. It's a serviceable strategy, much better than referring to new media as "the field formerly known as interactive multimedia" or, worse, "the field that durst not speak its name."

The results were predictable: business as usual. Disciplines of critical study already exist; likewise the social sciences, humanities, film production, performing arts. Once the practitioners therein notice that New Media has arrived, everyone can have their piece of the pie. This was most egregious during the Great Rush to Theorize Cyberspace (circa 1990 to 1997) (and better to say *Colonize,* in the critical rather than anthropological sense of the term): Critical Studies folks did Critical Studies of Cyberspace; Social Science folks did Sociology of Cyberspace; Technology and Policy wonks did Tech-Pol of Cyberspace. *Und so weiter.* It got so that whatever conference one happened to be attending was full of scholars falling all over themselves to get something with their names and the prefix *cyber* into the citation networks.

Except that what each of these groups winds up with is not New Media. What they get instead is a slice of something that looks enough like New Media to serve the purpose at hand something with its critical teeth pulled, something tamed and civilized and reduced to a manageable size and form. When something truly new comes along (and I do insist that the thing we miscall New Media is in fact genuinely new), it's tempting, overwhelmingly tempting, and in fact to a degree right and meet to make a first sally at it from an established position of strength, from a trusted epistemic anchor to which you can return when things get raucous. But once started on that path, it's all but impossible to stray off. And of course, as with all things that bear the name *discipline,* any knowledge production that does happen is always already subverted to a preexisting power structure whose purposes bear little or no relation to simple study of an object of knowledge.

So with full awareness that were doing it, let's grab a useful trope and lay it on. To wit: Once in a lifetime, if you are lucky, you may have the opportunity to be present at the birth of something special, something possessing tremendous power. You know that you yourself don't have that power, but you have been vouchsafed the vision, the knowledge that somewhere very close by, somewhere you can actually get to, it lies, still helpless but huge with promise. And no matter how we manage to reinterpret its purpose later, in the pale illumination that hindsight provides, we know the power of the thing is the force that drives our daily work and us ourselves and the whole damn universe. Whether it expresses itself as speciation or linguistic diversity or the restlessness of fashion, were looking at the same phenomenon: The power of Change. Transformation. Trans.

I used to attend things like the American Anthropological Association's annual do, and the Society for Social Studies of Science, and so forth. But I don't any more. I am part of a different association, a different society, a different community. My community coalesces only in the spaces between disciplinary conferences and, for that matter, in the spaces between interdisciplinary conferences. As with just-in-time funding and drive-by theory, we form an entity at unpredictable moments and in unexpected ways. As a social body we exist mainly in the Net, as a congeries of individual and eclectic discourses, forming a dense Web of information and experience that nourishes current and future practitioners who may never meet us. With few exceptions, we have no colleagues at our home institutions. For that reason, we are used to working alone, and that experience has taught us to cherish our physical meetings to the fullest, to make the moments last, and to be ready to relinquish them instantly. At the same time, it has honed and toughened us for the task ahead. That task, if we choose it, is to create that physical base for our work, to reach deep into the institutions that house us and change them at their deep hearts. Such a change is effected not by rhetoric, not by revolution. In the postmodern prospect, such a change is accomplished by viruslike structures carrying information that changes the way the organism's cells self-replicate, so that over time the organism ineluctably incorporates that new information; the changed becomes the changer. In this method, there is no battle, no confrontation. Instead, change simply is.

In human terms, the way such a scenario plays itself out is when each one of us in New Media acts exactly like our deep selves, as opposed to acting like a version of ourselves cut and trimmed and constrained to fit within one or more of the Disciplines of the Dead. There is little I could ask that could be simpler while at the same time more difficult. We ourselves embody New Media; no one can bring it into being for us. The old academic order falters; failure of nerve and of vision is the order of the day. This is our moment. We have worked hard for it; now let us seize it and live it to the fullest. In the words of a very old joke, be yourself: it's a dirty, thankless job, but somebody's got to do it.

A SIDEBAR ABOUT THE VIDEO INSTALLATION

In my opinion, the installation video *Your Words, My Silent Mouth* explains things better than saying, "Gah." The piece came out of some fooling around that I was doing with a sketchpad and a MIDI composing system. The original idea was an installation designed for a deeply NeoGoth space I'd imagined—fusty, creepy, ultra-Victorian. Actually, I'd directly experienced the space I was trying to evoke. I mentioned it before, when the family was visiting their friend who lived way out on Long Island, and that hurricane blew in, and that quasi-being squatting in the living room shadows came eerily alive.

There was also a connection in this piece to spirit mediums and the atmosphere in which they worked or in which we like to think they worked. I'd been fascinated by a radio program produced by Frances Dyson that was inspired by Avital Ronell's work on haunted writing and her later exegesis of the telephone. Fran's takeoff point was when the prosthetized absence of the telephone's electric voice became tangled with the medium's un-Heimlich production of absent voices, and she signaled that moment in the program with three deep, spooky organ tones that evoked for me that eerie night in the storm when the radio came alive. I took the first two notes of that sequence as my point of departure for the musical foundation of the piece and wrote a brief variation on it by altering its tonality slightly into the first two notes of the *Dies Irae.* This repeats in *moto perpetuo* underneath the instrumental and voice layering.

After the rough mix of the sound sketches was complete, I planned the visuals to accompany the sound. The piece was shot in Hi-8 and largely edited in camera. Its intent is to be pretty scary, but it also deconstructs itself as it goes along. For example, there is a shot that starts off as if a scalpel were cutting into something that might or might not be a crotch but that for just an instant can be seen to be the Vulcan V formed by the third and fourth fingers of a hand. The visuals are about literally deconstructing identity. Most of the facial shots are of me, looped and repeated in various video modes. And there is a lot of scalpel and other types of cutting action looped and mode shifted as well. This visual playfulness is juxtaposed to the somber, welcome-to-the-crypt style of the music. Actually, during the shoot I was conducting a playful conversation with my husband, who was potchkying in the kitchen while I was taping the visuals, so the raw video footage heard with the voice track turned up is ludicrous. The piece didn't begin to make sense until the voice was dropped out and the music inserted.

The soundtrack is in six layers, each of which is of different length, and they are looped independently. The longest track is slightly under six minutes, and the piece repeats—that is, the original layering reappears in roughly four hours, which I felt was plenty long enough to ensure that nobody would hear the same soundtrack twice. The vocal tracks are a canonical variation on a phrase I recorded in an emergency room, scored for three voices, which alternately speak and whisper the libretto.

During the mix we began to pick up weird electronic chirps and wheezes. Holy shit, I thought. It's the spirit medium sending in her part! I finally tracked these down, or thought I did, to an electronic artifact of the disk drives seeking to track zero, which was somehow leaking into the analog track, though I've never heard it happen before or since. But the sound was absolutely perfect for the piece—the timing, everything—so I mixed it into the final version exactly as it appeared on the tracks, and it's still quite audible there. Don't put yourself in the way of fate; just dance with her when she nods at you.

The piece was commissioned for a show in the University of Houston Gallery at Clear Lake. I've always constructed the music for my work first and added images or other elements later. For me, sound design is the primary aspect of almost every work I've done, regardless of the visual complexity that may be involved. But I wasn't able to get that across well enough to save the Clear Lake installation. They used tiny, tinny speakers, and the entire soundtrack and consequently the impact of the piece was lost for that reason. I'd explained the necessity for serious subwoofers if the piece were to be understood properly, but somehow or other that plea got lost in the set-up frenzy. I didn't get to the gallery myself, but friends who did reported that the sound was inaudible, and for me the sound was everything. However, when the piece moved to the *Women and Their Work* show a few months later, Lisa Becker, their curator, provided an excellent sound system, and the piece was finally presented as I'd intended it.

/framegrab from the video/

27

Video Arte Povera: Lo-Fi Rules!

Valerie Soe

When I was a young college student, I wanted to go to UCLA film school. I admired Alfred Hitchcock's narrative stylings, Stanley Kubrick's peculiar characterizations, Martin Scorsese's fluid and expressive camerawork. But as fate would have it, the film school at UCLA was severely impacted, and I instead completed my undergraduate degree in the art department.

What I discovered was that video art's aesthetics appealed to me much more than Hollywood's narrative filmmaking conventions. I found that by working in experimental forms, I could easily access topics such as identity politics, race and ethnicity, and pop culture. This was much more effective using experimental video's direct and expressive poetics than through narrative film's elephantine character and plot machinations. I also found that experimental forms allowed me to utilize all manner of unusual source materials, including found footage. This recycling aesthetic, making silk purses from technology's sow's ears, continues to appear in both my single-channel videos and multiple-channel installations.

In my work I've used video, text, pop culture artifacts, autobiography, and interactive elements, employing technology both high (computer terminals, electronic message boards, video monitors) and low (zoetropes, viewer-generated graffiti, artist's books) to look at issues of culture and identity in the information age. I'm interested in an organic use of technology as a vehicle for the requirements of the creative concept. In other words, I let the artwork's content dictate the choice of media, rather than simply using technology for its own sake.

By working in experimental forms, I continue to readily explore political and social concerns such as racial discrimination and bigotry (*Diversity*), intimate interpersonal relationships (*Mixed Blood*), body image, power, and control (*Binge*), and the representation of Asians in pop culture (*Picturing Oriental Girls: A (Re) Educational Videotape; Beyond Asiaphilia*).

MIXED BLOOD (1992–1994)

Mixed Blood was an interactive video installation that utilized unexpected juxtapositions of found footage, flashing text, talking heads, and scientific footage displayed in a brightly colored graphic installation that challenged potential viewers to stop, look, and listen. Using two discrete channels of video, *Mixed Blood* combined interviews with over thirty concerned individuals, relevant quotes and statistics, and clips from scientific films and classic miscegenation dramas such as *The World of Suzie Wong* and *Sayonara* to examine interracial relationships in the Asian American community.

In addition to the information disgorged via video monitors, *Mixed Blood* also encouraged viewer interaction. A computer terminal and electronic message board output allowed viewers to respond to questions regarding cross-cultural relationships such as, "What do you think when you see an interracial couple on the street?" and "What are three characteristics of Asian American men?" The query changed daily, and each day's responses were strung together and displayed on the electronic message board mounted on the installation.

For those who preferred more direct expression, the piece had a "talking wall"—a blank white panel on which viewers could write their opinions without technological mediation. This element was extremely popular, with comments on a range of topics from politics to sexuality. There were also small stickers printed with phrases including "Identity/Affinity" and "Who Do You Love?" which viewers were free to take and distribute as they wished.

Mixed Blood attempted to present often controversial topics of miscegenation, bigotry, and sexual stereotypes without being didactic and overbearing. By using both high and low technology, I hoped to ease the viewer's interaction with the piece, facilitate participation, contemplation, and debate, and allow the viewer to fill in the blanks and actively contribute to the form and content of the piece.

***BINGE* (1996)**

Binge was a mixed-media installation that was based on Amy Moon's short story of consumption, addiction, and purgatory. *Binge* looked at body-image distortion, power, and control from the perspective of a compulsive eater. The main character is an unnamed woman who dreams idly of escape from her middle-American life and her indolent husband. Her stifling suburban existence eventually leads her to desperate acts, as she first attacks her own body and then strikes outward, with deadly consequences. By telling the story in the first person through the voice of the unidentified narrator, the piece invites the viewer's rapport with the unbalanced and yet sympathetic main character. Her extreme impulses remind us of the frustration and agony of day-to-day existence and the desperate actions sometimes necessary to flee the killing banality of everyday life.

The installation consisted of various altered objects that were integrated with text and that reveal different elements of the narrative, outlining character traits, plot developments, or other aspects of the story line. Manipulated objects included bullet casings, an accordion book, a shooting-range target, a zoetrope made from a KFC bucket, and assorted audio and video loops (figure 27.1) Once seen, read, or heard, these fragments converged into a composite text that explicated both plot and character in a nonlinear, interactive format.

By using consumer-grade Radio Shack electronics, I emphasized readily accessible technology, including battery-powered plastic monitors, a hand-held Watchman television, and a child's portable turntable. I similarly employed other everyday objects, such as fast-food packaging, mirrors, and picture frames. This underscored the quotidian setting of the piece, based on the diary of a mad housewife, and contrasted these items with more sinister objects, such as a shooting-range target and spent bullet casings. *Binge* compared the banality of its source materials to the desperation and violence of the main character's actions.

DIVERSITY (1990)

Diversity was a three-channel video installation featuring footage of Chan Cheong-Toon, a man who stood regularly at a traffic island at the corner of Broadway and Columbus in San Francisco's North Beach neighborhood singing furiously in Chinese to whomever cared to listen (figure 27.2). Through interviews with Chan as well as with his many observers, the piece addressed the projection of individual desire onto a single subject, as each interviewee offered his or her interpretation

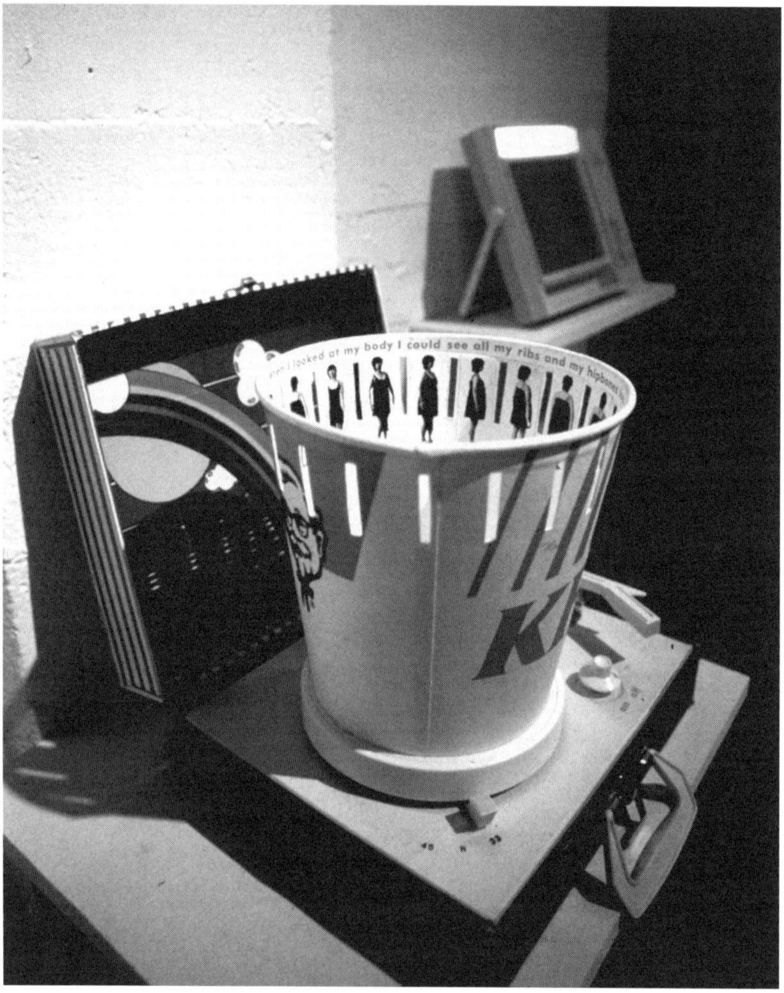

Figure 27.1
Valerie Soe, *Binge,* 1995, text, bullet casings, monitor, altered target.

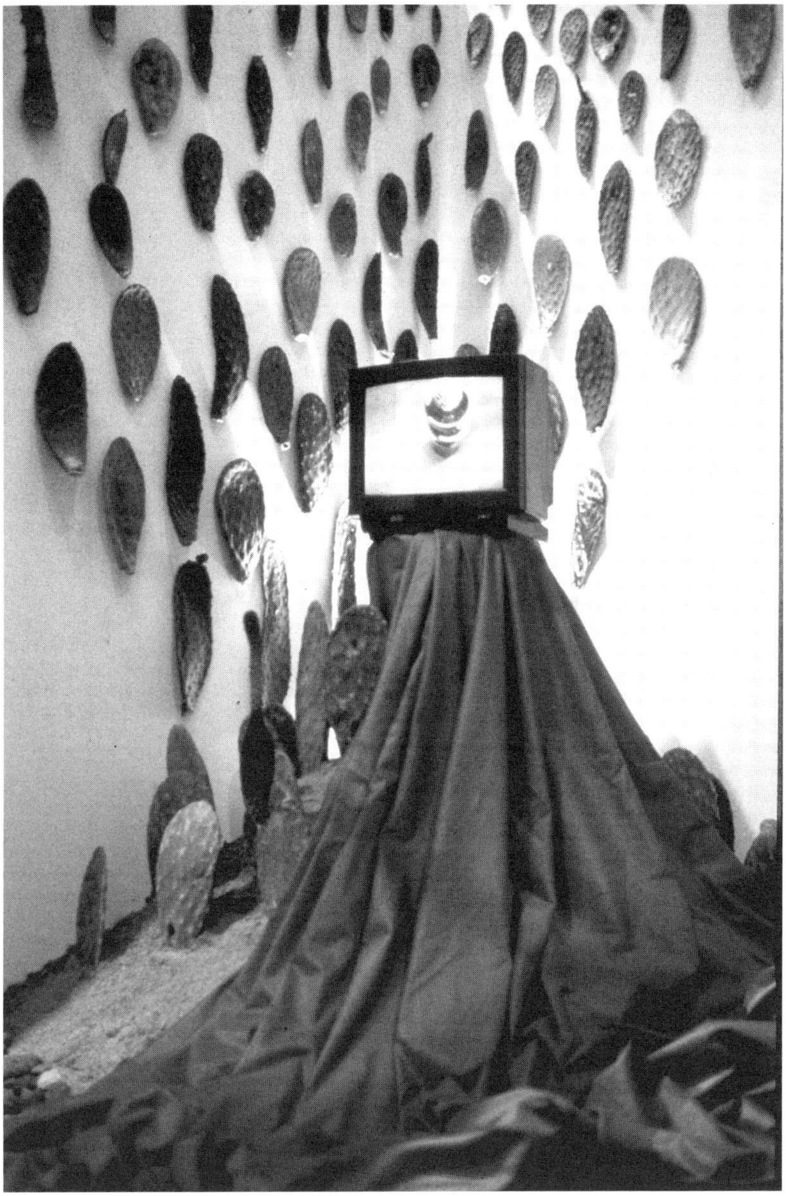

Figure 27.2
Valerie Soe, *Diversity,* 1990, installation view.

of Chan's intentions. In addition, by focusing on this unusual personality, the piece exploded the myth of the model minority, contradicting the fallacy that Asians are quiet, well-behaved, and aligned with social conventions.

The installation consisted of three separate channels of video played simultaneously on individual monitors arranged in a triangular configuration, with screens facing inward. The viewer, positioned at the center of the piece, chose to view one or two monitors at once, although sound from all three was audible at all times. In this way, the viewer sequenced and selected the configuration of the piece, choosing the order in which the images and narrative unfolded. This threefold arrangement reflected the many aspects of the focus of the piece, the singing Chan—as he sees himself, as others see him, and as he objectively appears—suggesting the variety of perception and personality found in a single individual.

On the gallery wall were the names of various Asian Americans who have distinguished themselves in one way or another. Although several, including author Wakako Yamauchi and singer Pat Suzuki, are notable for outstanding achievements, others, such as convicted felons Joe Fong and Wendy Yoshimura, are known for more notorious actions. This strategy pointed out the complexity of Asian American culture, emphasizing again the diversity of a community too often stereotyped as one-dimensional.

In *Diversity,* the simplicity of the technology allowed the complexity of the subject matter to emerge. The stripped-down installation design consisted of three large monitors on black pedestals and a line of names on the gallery wall. Although the simultaneity of the three channels of video and the triangular configuration of the monitors were significant, the installation gave prominence to content rather technology.

WALKING THE MOUNTAIN (1994)

This installation was an *ofrenda* (altar) dedicated to my aunt Lula, who died from a nosebleed at age four in Phoenix, Arizona, in the 1920s. The piece, consisting of sand, cacti, magenta taffeta, video, and text, recounted the fate of my grandparents' cherished second daughter, born into a climate too arid and dry for her genotype. The cacti hanging on the wall and surrounding the video monitor were a metaphor for human tenacity, contrasting with the inability of Lula to adapt successfully to her new homeland. On the right wall, the legend "STAY HYDRATED" reiterated the first rule of human survival, one that Lula was unable to maintain because of her environment, age, and circumstances.

In this installation, I juxtaposed unusual materials—cacti, taffeta, sand, and a video monitor—to draw attention to the unusual subject matter. I hoped to suggest the alien quality of my aunt's new environment, which was a desert full of strange life forms including prickly plants, sand, and Europeans. The piece was a metaphor for the difficulties that new immigrants often face in adapting to their new homeland, especially in the United States, where assimilation is valued over preservation of individual cultural traits.

I also wanted to contrast the natural elements (sand and plant life) with human-made elements (the video monitor and satin fabric) to make Lula seem like a fish out of water in her new environment. The video loop also emphasizes this point, recounting my aunt's tragic story through images of goldfish, faux blood, and a length of scarlet taffeta rippling in the wind. Using a simple combination of video and nontech elements, the piece presents its story plainly and effectively.

LA VIDA POVERA DE SAN PANCHO (1998)

La Vida Povera De San Pancho was an interactive installation made up of melted and made-over Playskool plastic doll houses that had been altered to reflect the ghost stories, histories, and legends of San Francisco. The tag line for the installation read, "Why do you choose to live in San Francisco, a beautiful, fickle, and ever-entrancing city? What is the price of entry into this temperate wonderland? Where does reality end and imagination begin?"[1]

Utilizing tiny TV monitors, sound chips, image, text, and toys, *La Vida Povera* took viewers down Melancholy Alley, through Sound and Fury Avenue, and up Sexolicious Lane, some of the imaginary streets realized in the piece. The installation was an emotional map, a psychic tour of the memories, hopes, and dreams of a city that is constantly in transition.

Each plastic house engaged the viewer in its own way, with points of entry including video and TV images, push-buttons, switches, levers, and other means to play with the structure and meaning of the buildings. The tiny structures became a dreamscape of the imagined and the perceived, the world of hopes and fears drawing the viewer to San Francisco.

I used toy houses with tiny video monitors to get away from the museum-aesthetic definition of *video installation* that often includes huge, beautiful, and expensive video projections by established artists such as Bill Viola and Steina. Instead, I wanted to get back to the cheap-is-beautiful, consumer-electronics look of early video installations, where the ugliness of the monitor was integral to the aesthetic of the piece. This strategy exemplified the do-it-yourself creed, *la arte povera,* the idea that you make what you can out of whatever you can scavenge.

How does this relate to little plastic houses with home electronics in them? It went back to the idea that it is possible to live in an horribly expensive city like San Francisco without being a yuppie or a multimedia professional. *La Vida Povera* celebrated the marginal underbelly of artists and writers and workers who give San Francisco its spice and its texture, as well as the plumbers and painters and UPS drivers who give the city its soul and blood.

UNLEARNING THE LANGUAGE

By working in low-fi, experimental media, I've been able to create work that brings to light untold or neglected stories from outside of the mainstream. Using available and affordable technology frees me from many budgetary constraints and allows me to more readily create work and to engage viewers in exchanges about culture, politics, and representation.

In this way, I'm attempting to reframe and rearticulate an Asian American perspective by countering conventional stereotypes of Asians in American mass media and offering an alternate vision of the community and culture. As Malaysian American experimental videomaker Cheng-Sim Lim states, "These days I'm trying to unlearn the language of Hollywood. I am doing it because I know it's not my language. I am trying to remake my image in myself."[2] Lim understands the vitality and newness of experimentation, which more accurately expresses experiences outside of conventional film and television. Experimental forms are a reflection of the multitude of voices suddenly speaking, each with its own cadence and lexicon, in the new, brilliant cacophony of modern times.

NOTES

1. Valerie Soe, catalog notes, *URL Exhibit,* Southern Exposure Gallery, 1998.

2. Cheng-Sim Lim, "Rojak," *Moving the Image: Independent Asian Pacific American Media Arts,* UCLA Asian American Studies Center, 1991.

28

Face Settings: An International Co-cooking and Communication Project by Eva Wohlgemuth and Kathy Rae Huffman

Kathy Rae Huffman

Face Settings is a female-focused communication project and Internet Web site that reveals the growing Net community of women—⟨http://thing.at/face⟩.

Face Settings, a project initiated by Eva Wohlgemuth and me, was a series of dinner-performance events that took place between spring 1996 and autumn 1998 (figures 28.1 and 28.2). The objective was to join real groups of women in network strategies to expand female connectivity and to engage women in discussions of importance on local, regional, national, and international levels. Through a mailing list called FACES, an online community has developed and expanded beyond the process of preparing meals and eating together. *Face Settings* raised various questions, which were presented at *Face2Face,* an exhibition and meeting of FACES in Graz, Austria, on 8 to 11 July 1998.

Throughout our *Face2Face* events and virtual online communication, women focused on communication practices and how they differ in various cultural settings. Aware of how women are kept out of the technically challenging network, we set an example of what could be done to change that practice, and we constantly mused about how the fact that women speak and communicate differently than men affects our female involvement in online culture.

Face Settings: An international co-cooking and communication project by Eva Wohlgemuth and Kathy Rae Huffman—⟨http://thing.at/face⟩. *Face Settings* is a cross-platform forum for developing and exploring the interests and needs of women online. *FACES* is a female-only mailing list for discussing art, communication, and online policy.

The cooking performances took place between summer 1996 and summer 1998 and evolved directly from our experiences on the (then) recently concluded Internet travel project *Siberian Deal,* which took place in 1995. From Russia, we sent weekly "reports" to Vienna, which were uploaded to the project Web site at ⟨http://www.t0.or.at/~siberian/vrteil.htm⟩. The goal of *Siberian Deal* was exchange: we were traders dealing objects and information to strangers in "the worst place in the world." Our mission was to understand value and learn how our opinions were informed by cold war media and propaganda. In Russia, we established a network of friends we would meet again and again. Our observations of women, their buying and selling techniques, and their status in society, were a

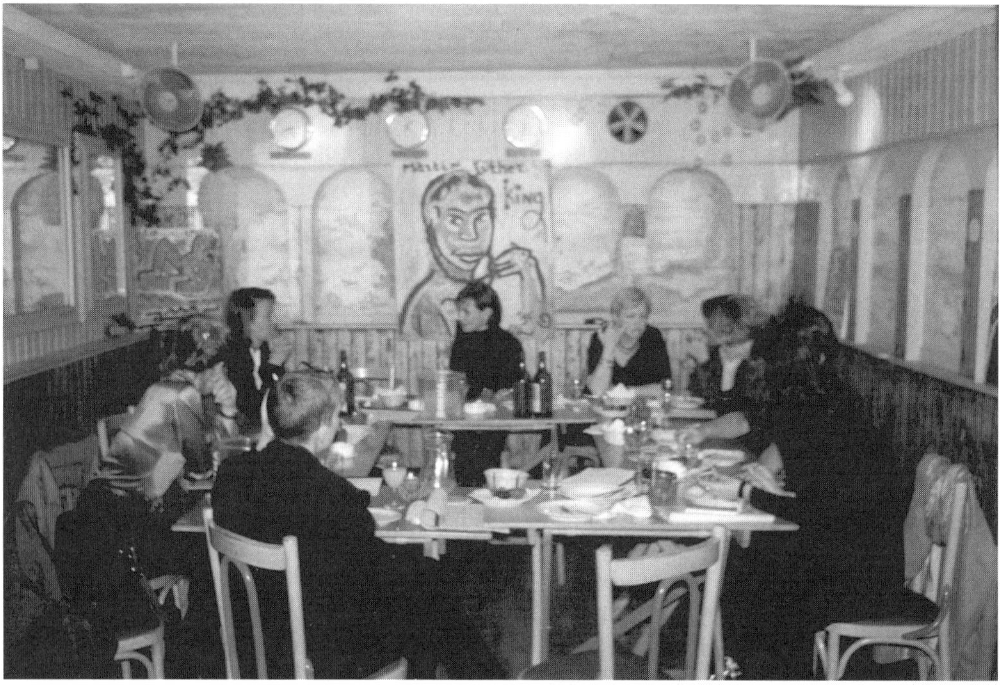

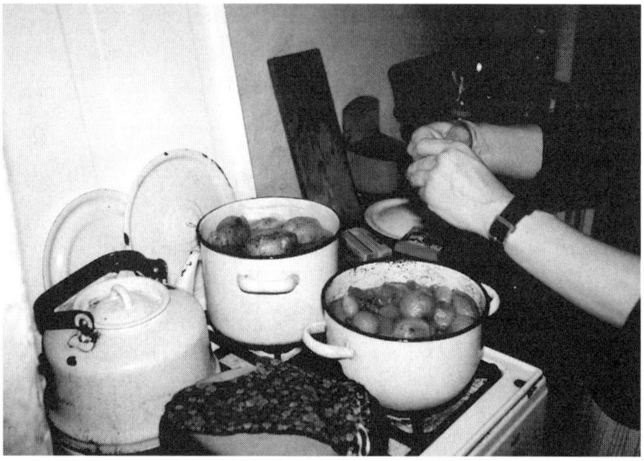

Figure 28.1
Eva Wohlgemuth and Kathy Rae Huffman, *Face Settings,* September 1996, in Rotterdam, the Netherlands.

Figure 28.2
Eva Wohlgemuth and Kathy Rae Huffman, *Face Settings,* October 1996, in St. Petersburg, Russia.

common topic of conversation during our travels and influenced our future work together (see ⟨http://www.t0.or.at/~siberia.vrteil.htm⟩.

Working toward a new project concept, Eva and I realized that our personalities, skills, and backgrounds complemented (and sometimes compensated for) one another. I loved gathering people into social situations for exchanging information and brainstorming ideas. Eva was fascinated by the challenge of translating the process of communication into a visual form that invited participation. She was keen to establish strong conceptual guidelines for a new Internet project. I was immersed in Internet research on a variety of topics, including "cyberintimacy." As we talked, we decided to continue our real and virtual working system and travel to places we found compelling (for various reasons). We also made a conscious decision to focus our energy and dedicate our project to connecting women. We made these choices naturally and without any conflict. In Europe, the lack of structure, encouragement, or reward for women working in the new communication technologies was obvious to us.

As a conceptual artist, Eva had been working on the location sculpture *Systems* since 1989. I had been an active media art curator, producer, and networker (by the old definition) since the early 1980s. In 1994, I was living on my own in Austria, and Eva had become a good enough friend to engage in "girl talk" and relationship problem sharing. During these occasional meetings, Eva and I became aware of the differences in our cultural conditioning, especially how I had been encouraged to develop verbal and social skills in school, while Eva had been encouraged to develop her excellent logistic planning and technical skills. As an American, I had experienced how "citizenship" and "extracurricular participation" were highly regarded during my formative school years. Eva, on the other hand, had suffered a very strict and academic education, without many social functions, in a dogmatic system that did not tolerate deviation from the traditional pedagogic plan.

The basic cultural differences that informed us—which we were aware of—fascinated us. We also noticed that the way we communicated with each other and with women and the ways in which we communicated with men were very different. We began an intensive investigation that included Internet research and the compilation of information on this phenomena. The writings of Dr. Deborah Tannen provided great insights, as did many scholarly studies and linguistic observations about how women and men relate in the online environment. We found corroborative theoretical opinions, and that was enough for us to set out to combine our newly acquired Internet skills, interest in social communication and cultural differences, and art practice. Our new work would evolve with women in five different countries that were located on the borders of Europe.

The events we planned would combine women, cooking, and communication, but we admitted to ourselves that we needed to define what we wanted to achieve as clearly as possible. At this point, we realized it would take time to develop relationships and that it could take several years to build the network we were envisioning. In an effort to proceed logically, we outlined the following list of considerations for *Face Settings:*

- *Communication* It is often difficult to engage strangers in discussions about communication, especially when there are language problems, different social customs, and lack of understanding about the widespread practice of Internet connectivity. Neither of us was the type who readily engaged strangers in conversation, even locally. We would benefit from a defined group in each location.
- *Online representation* The Web site would be a key element of the project, linking the real local events between the regional groups. It would need good stories, provocative photos of the participants, tasty recipes, and theoretical food for thought. We wanted to profile the women we met and give an idea about what their working situation was like. We also decided that the Web site should maintain links to female sites, a suggested reading list, references, our biographies, and personal links.
- *Live performing installation events* With a group of ten to twelve women, we would be able to talk while we cooked and served meals. We wanted to give pleasure to others in these events by serving them and taking care of all the details. This would be a treat, a vacation from the normal work that women know well. This could also be hard work for us. With the understanding that we would encounter various customs, we also knew that dinners "at home" are usually family events, and as strangers cooking the meal, we would be a rare exception.

- *Artifacts* As a Net.art project, there would be little need for physical evidence, but we still wanted to provide some objects that resulted from the collaboration with the groups of women we would work with. Clearly, most artists don't have the time to give to other artists' projects without some real clear understanding of the exchange. We wanted to provide our project partners with specific rewards.
- *Publications* A cookbook catalog that included recipes gathered during our travels was an idea we wanted to pursue. In the beginning, a photo album of each dinner and a collection of memorabilia from each of the events would be created. We would make this book available at each dinner, adding progressive chapters, like courses, as we continued with the project.
- *Goals* *Face Settings* dinner events were to be enjoyable events, celebrations with low stress and lots of personal care (for ourselves and the women we meet). Pleasure connected to food and the sharing of information would be a topic to introduce in our dinners. We hoped to establish real communication that would grow in the digital environment, eventually linking the regional groups in a collaborative cooking performance event.

Our first experiment was to gather a group of women in Vienna for a meal to test our ideas about women and communication. We didn't know what would happen. Because we wanted to explore the practical aspect of this theory, we focused our guest list on communicators—women who were online and who might find the topic a starting point for continued discussions. Like many events held in Vienna, it was a very structured evening, and although a Cuban theme was carried out, a delicious meal was prepared with Maria Pallier (Madrid-Bilbao), and the appetizer was a popular Havana cocktail called Mohito (dark rum, smashed fresh mint, brown sugar, crushed ice, and a little soda water), we found that the women were very reluctant to discuss the topic. We analyzed the situation later and decided that we had not clearly enough stated our mission for the dinner and appeared to have no obvious goals. We had not arranged a seating plan for the table that would pair more experienced communicators with those "newbies" among us. Even so, it was fun, and everyone expressed their interest in our work. The evening conversation ended shortly after Eva's boyfriend came in, around 1:00 a.m. Just naturally, all attention shifted toward him. We didn't yet realize the significance of this reaction.

Reflecting further on the Vienna meal, we decided that dinners in foreign places would need a coordinator, someone who would organize local women, giving us access to people who we would not meet otherwise. We also needed to locate places where women could get Internet access and learn about networking and the potential of electronic communication. Fundraising efforts were begun, and in describing the project, we acknowledged that *Face Settings* would grow and evolve with the participation of women around the world. We were careful to not predetermine the outcome or declare a final conclusion. The work would be continuing research and development and a discovery of how women communicate with each other around the world.

Looking at our travel plans and festival invitations, we realized we could organize two dinners right away. One would take place in Rotterdam, where a number of women we wanted to meet would be attending the Inter-Society for the Electronic Arts (ISEA) meeting and the Dutch Electronic Arts Festival (DEAF). The guest list was a gathering of female artists and representatives of various media institutions from several countries. The Bar Tropical, a small cantina, offered a private space away from the conference center where we were able to gather tables into a circle and talk about our various communication experiences. We wanted to know what it would involve to organize women for dinners in various countries. The conversation was lively, the women (being communicators) were very vocal and expressive. The feasting and fun reached a high level. Many stories were shared and involved the telling of experiences in female groups like literary circles and exercise groups. The tone of the evening was collaborative and supportive. The evening came to an end and the conversation stopped when three ETOY boys came into the dining room and sat down at the table.

The St. Petersburg dinner event was planned because Eva and I were invited to present *Siberian Deal* at the St. Petersburg Biennial. Irina Aktuganova, director of the Gallery 21 (she had also attended our Rotterdam dinner), agreed to coordinate the location and select twelve women to participate. Being a Russian event, the situation was confusing and complicated, but with the additional help of Alla Mitrofanova, we were able to get some background about the various women who were to attend.

At the last minute, we were confronted by Irina and Alla with a new question: they proposed that we invite three men who were also members of their cyberfeminist group. We wanted the project to evolve with local energy, so we said, "Sure, let's try it and see what happens." The men were excited to participate, and it seemed that we could reach some new levels of understanding in St. Petersburg.

Shopping in the local marketplace and preparing the meal in a private Russian home required some flexibility, but everything was ready on time. Customs differ in various parts of the world, but here we waited over an hour for the guests to arrive. We started to understand that there was a different social principle at work and that we were conditioned by our own timetable, not a Russian one. We had prepared an appetizer, a hot meal, had plenty of bread and wine, and had even bought cakes for dessert. Suddenly the table filled up and a discourse began. But, again to our surprise, it was not what we expected. Instead of a sharing of stories and communication experiences as we had requested, we were confronted and told to give a reason for organizing a women-only project. We were not challenged by the women (who remained relatively quiet throughout) but by the men who were members of the St. Petersburg Cyberfeminist Organization. Each, in turn, had prepared a lengthy speech, with specific comments about equality that (in our opinion) misunderstood Western feminism.

Face Settings was not proclaimed as a feminist or cyberfeminist project before this, but we realized that, in this community, we were observed unfavorably for not making a strong position. At this point, we also realized that we would have to rethink the idea of a control group of twelve women for each dinner, our preconceived limit, which we thought would be the maximum number of voices needed to create a situation where everyone could really talk and develop a dialogue. It simply became too difficult to shut out women who really wanted to be involved. We would need to sacrifice real-time communication and hope that online we could make up for the lack of personal attention. Although the St. Petersburg event was a kind of anarchy (in Austrian terms), we enjoyed how it ended up as a dancing party (with boys and girls) in the next room.

Returning to Vienna, we began to prepare the plan for *Digital Care,* the next phase of the project. Eva traveled to Bilbao to participate in the Ciberéa conference, a festival of cyberant, and there, with the help of conference co-founder Maria Pallier, she was able to make contact with Basque feminist artists and to have a discussion about their art, their food, and their communication. At the same time, Kathy traveled to Hull, in northeast England, where she participated in the ROOT festival at Hull Time Based Arts. Here she met with women who hoped to create Internet communication projects in the near future, but she did not meet the women from Site Gallery who had promised to be there (this group would cohost the Lovebytes Festival, which we hoped to participate in). Stefan Korn, with whom we had been communicating for almost a year, brought material from Glasgow for the *Digital Care* gallery installation and furnished a profile of the local Scottish media scene. Our female contacts with Scotland were encouraged by our earlier contacts, Nina Pope and Karen Guthrie, who had made the Internet project *A Hypertext Journal* earlier in the year.

Digital Care was conceived by Eva to highlight the *Face Settings* project. It would focus on three locations: St. Petersburg, Glasgow, and Bilbao. As the Internet research coordinator, I made printouts of online Web site information (there was no computer in the gallery) and created a photo diary of each event, which was added to the project scrapbook. Video recordings of us preparing each entrée were made by Eva, and she created animated gifs from these tapes for the *Face Settings* Web site, which was growing and being accessed extensively from viewers around the world. During the process of *Digital Care,* we made a local promotion for *Face Settings* at various funding institutes and agencies in Austria.

A second *Face Settings* dinner in Vienna was convened in December to review the events of 1996 and to discuss the project's development to date. At this dinner, Diana McCarty, an American living in Budapest, joined us. Diana and I haltingly agreed to moderate a new mailing list that would keep everyone who attended our dinner events informed on a more regular basis. We unanimously agreed on the name FACES. Being complete novices at operating a mailing list, we decided to find a female listserve manager who would work closely with us.

In January, at the *Secret Conference* held at Backspace in London, Eva and I introduced the first version of the *Face Settings* Web site and the FACES mailing list to a group of Net.artists. Josephine Bosma, a journalist from Amsterdam, interviewed us for *MUTE* and especially encouraged us to develop the mailing list. A research trip to California (London was a stopover) was arranged in January 1997 with funds from the Austrian Ministry of Culture. It was Eva's first visit to the American West Coast, and the outdoor lifestyle during the winter was an amazing treat from the cold Austrian climate. During the trip, we interviewed several prominent women artists and composed an overview of the conditions under which feminism evolved in the early 1970s. Eva had fixed an appointment for a 3D digital scan of her body in Monterey, at Cyberware Inc. The second part of the trip was devoted to this event and the beginning of a new Internet work for Eva: *Bodyscan,* which would receive a great deal of attention in the coming year ⟨http://thing.at/bodyscan⟩.

Further funding for *Face Settings* was not forthcoming, so Eva and I agreed that by continuing to accept invitations from institutions and festivals, we could keep the project public and carry on with the actual project plan. The participation at festivals did provide a strong presence and attracted many new women to the FACES mailing list. It did not, however, offer us the focus for and attention to the dinner events that were necessary for a successful project. By summer 1997, the FACES mailing list had developed a regular subscription base and was growing and developing content, especially due to the fact that the Cyberfeminist International, the first public event for The Old Boys Network, joined the list to prepare their workshop as part of the Hybrid Workspace project at *documenta X,* an international art exhibition in Kassel, Germany. *Face Settings* prepared the opening dinner for the all-women Cyberfeminist International, offering support and solidarity to the project. The growth of FACES, in some way, marked the decline of *Face Settings.* In fact, FACES began to self-organize, meet together for dinners, show solidarity at conferences, symposia, and travel for work or pleasure. FACES also became an announcement mechanism for female multimedia projects created by list subscribers and for the discussion of internationally important issues.

During the 1997 Ars Electronica Festival, *Face Settings* was part of the Open Workspace, and we set up an office space and cooking space that offered a daily soup to all. Eva Ursprung, the new curator at the Forum Stadtpark, invited *Face Settings* for an exhibition in Graz the following summer. With a curator in charge, we were guaranteed freedom from the extensive and depressing task of fundraising and planned to concentrate on the Web site and the communication between our regional coordinators.

FACES meetings continued in various cities and were a source of great interest. The women came from all age groups, social backgrounds, education levels, and fields of interest. Their link was an interest in online communication and a curiosity about Net.art. Dialog on these topics was popularized by festivals and symposia convened at contemporary art institutions throughout Europe.

In Graz, from 6 to 11 July 1998, the full circle came together for *Face Settings* at the Forum Stadtpark, Graz, Austria. The exhibition, symposium, and RealAudio workshop was called *Face2Face,* and it provided a reason for about forty women, most of them FACES, to meet, communicate, cook, and exchange ideas. *Face2Face* began and ended around a table. Each woman present (as well as each male guest) was offered the opportunity to speak, contribute opinions, and discuss topics introduced by participants. Artist representatives from each of the five designated *Face Settings* countries (some who had worked with us over the years) presented their work, described projects in their region, and shared their ideas about female communication online.

The *Face Settings* project did not ultimately evolve exactly as Eva and I initially envisioned, but at *Face2Face* it did bring together participants from each country we had originally targeted, and the communication between these women is active and strong on the FACES mailing list. The five regions and their representatives included Beverly Hood and Lindsey Perth, Glasgow; Irina Aktuganova and Alla Mitrofanova, St. Petersburg, Russia; Maria Pallier, Madrid (Bilbao), Spain; Zana Poliakov and Vesna Manojlovic, Belgrade (Yugoslavia) Serbia; Margarete Jahrmann, Doris Weichselbaumer, Sabine Seymour, and Elisabeth Binder, Austria. Active participants in the Graz program included Cornelia Sollfrank (OBN), Hamburg; Katy Deepwell (n.paradoxa), London; Evelyn Teutsch, Leipzig; Birgit Huber, Germany; Betty Spackman, Toronto; Anja Westerfroelke, Linz; Julia Meltzer, New York; lizvlx, Vienna; and Veronika Dreier and Reni Hofmueller, Graz. The topic established by Ursprung—*The Body: Identity and Community in Cyberspace*—was a strategy to define the practice of online presence. She writes, "The FACES community is a mailing list that connects women from areas that border on the fringe territories (and concerns) of the European media centers" (Eva Ursprung, unpublished communication). The body was the central topic of discussion and exhibition.

The *Face2Face* event was an informal gathering of women in a nonhierarchic atmosphere, open to all. It offered enough time for each speaker to develop her presentation without interruption. The topics included Russian cyberfeminist views, the old boys network and cyberfeminist activist strategies, economics and feminism (especially as portrayed in science fiction films), pornography and erotica, and feminist interventions in print. An event report can be read in the *Telepolis Journal,* in pop~TARTS, a column created by Margarete Jahrmann and Kathy Rae Huffman (⟨http://www.heise.de/tp/⟩).

Since 1998, the FACES mailing list has continued to grow. Meetings and projects have been initiated by subscribers, a supportive community of women who thrive in a noncompetitive online environment. A database of women's projects can be found at ⟨http//www.faces-l.org⟩ (also information on joining the list). In 2001, Female Takeover II supported the development of xxero (pronounced shee-ro), a MOO environment built by FACES, at ⟨http//www.xxero.net⟩.

FACE SETTINGS CHRONOLOGY

Summer 1996

Face Settings planning and first dinner event, Vienna

September 1996

Face Settings research dinner, Rotterdam, Netherlands

October 1996

Face Settings in St. Petersburg, Russia

October 1996

Face Settings research in Bilbao, Spain (Eva)
Face Settings research in Hull, UK (Kathy)

November 1996

Digital Care at Galerie Cult, Vienna
11 November—St. Petersburg, Irina Aktuganova (referent)
18 November—Bilbao, Maria Pallier (referent)
25 November—Glasgow, Stefan Korn (referent)

December 1996

Face Settings dinner, Vienna, the FACES mailing list concept developed at dinner

January 1997

Face Settings project announced to the international Net.community, Backspace, London

Research for mailing list and MUTE interview

Face Settings research travel in California (Eva's body scan in Monterey)

Face Settings workshop, *Videofest,* Berlin Film Festival (no dinner event)

Spring 1997

FACES mailing list launched with ten women as an experiment, Diana McCarty and Kathy Rae Huffman comoderate, Vali Djordjevic (Berlin) becomes technical advisor, hosted by Internationale Stadt, Berlin

Face Settings luncheon in Liverpool, during Syndicate meeting at Video Positive, and FACES, an all-female mailing list, announced in a controversial presentation

Jello-Rabbit FACES IRC chat, from ORF Kunstradio, Vienna

Face2Face meeting in Ljubljana (Diana McCarty hosts)

Summer 1997

First Cyberfeminist International—Kassel, at Documenta X, *Face Settings* opens the CI Hybrid Workspace with a dinner event

Face Settings Internet workshop, Kunsthaus Bethanian, Berlin (no dinner event)

Cyber co-cooking event from Bielefeld, Germany, with Radio B-92 Belgrade, IRC chat and recipe exchange online

September 1997

Face Settings, performing installation at Open Workspace at Ars Electronica, a daily soup offered to all; the first FACES meeting, with nineteen women gathering in Linz to discuss the mailing list

Face Settings dinner in Glasgow, as part of the Strut Performance Series, a performing installation and lecture at the Jetai Symposium, University of Glasgow

Grrl Power Posse (with FACES), European Media Art Festival, Osnabrueck, Germany

January 1998

Face2Face meeting (FACES) in New York at Sabine Seymour's, where we meet the Cybergrrl representatives

FACES moved to the Webgrrls Listserv (⟨http://www.webgrrls.com⟩) due to the closing of Internationale Stadt, Berlin; FACES grown to 140 women in more than a dozen countries

February 1998

Face2Face meeting in Toronto (with FACES) at Nina Czegledy's and *Face2Face* in Vienna (with FACES) at Kathy Rae Huffman's

Summer 1998

Face2Face at Forum Stadtpark Graz, Austria, curated by Eva Ursprung (in collaboration with Kathy Rae Huffman), an exhibition, RealAudio workshop, and symposium (with FACES)

A final *Face Settings* event with Evelyn Teutsch (Leipzig) as the performing cook

FACE SETTINGS by Eva Wohlgemuth and Kathy Rae Huffman is now archived at ⟨http://www.t0.or.at/~amazon/FACE/index.htm⟩.

29

Diane Fenster: The Alchemy of Vision

Diane Fenster and Celia Rabinovitch

Mystery cannot be put into words.
—JOSEPH CORNELL

My art works through myth, spirit, science, and technology. At times, I am a modern alchemist who transforms electrical patterns into art. I want to uncover that sense of mystery that guides me to create certain images and recurring forms. Bringing these images together creates a mental sensation—the surrealists would have called it a *frisson*—that becomes a point of departure for both myself and the viewer. I want to create an emotional response and reflective capacity that extends beyond ordinary experience. My recent work, the *Hide and Seek* series, explores this mystery from the themes of intimacy, concealment, and revelation, both of oneself and the other.

The fact that the computer is my chosen tool for creating my art is irrelevant in the face of the art itself, which I use to speak to the earthly, human, and spiritual concerns that have consumed human beings from time immemorial. The computer, like the flint chips and paint brushes before it, liberates a creative part of myself that otherwise might have remained undiscovered.

When I was a small child, I had a persistent interior vision of myself looking out at the world from behind the ordered panes of an attic window. I could see the world below but was somehow removed and disconnected from the life I saw. When I was older, I was told that my last name, Fenster, actually meant "window" and that our family had originally been named Fenstermacher or "window maker." This architectural metaphor of seeing myself looking through a window carried into my artistic life and current explorations of vision using the camera and computer as the technological doubles of the window of my early childhood. Now I realize I am also a "window maker" in that my art creates windows through which others look to see an inner world and to recognize themselves in that invented space.

In the Renaissance, the idea of a frame of reference was begun by artists who, wanting to create a sense of spatial illusion and recession, began to understand painting as a window frame or fixed frame of reference. This idea of a fixed frame of reference, which many believe derives from modern physics, is the origin both of illusionistic art and, ironically, of our ambiguous modern worldview. Modern physics contradicted the Renaissance's fixed frame of reference with proofs that multiple and even contradictory frames of reference coexist in the physical world. Contradiction and ambiguity also apply to my art, which attempts to uncover paradoxes of being, intimacy, and abandonment—or the conflicting emotions of intense human relationships. This is the experience that I generate through contemporary digital tools to depict a new poetry, a multidimensional world of images.

THE ARCHITECTURAL AND THE ARCHAEOLOGICAL METAPHORS

Two encompassing metaphors preoccupy me. These are the architecture of creative work and the archaeology of the soul, which like building out or digging deep are mirror images. Through these fundamental themes, I grapple with internal and personal processes of identity, desire, longing, and the inevitable losses sealed in memory. The metaphor of architecture suggests the processes of constructing, building, and creating a place for the self. The idea of place implies safety and sanctuary, intimacy and warmth, but also isolation and loneliness in the home of one's past, as well as boundaries and limits, both protective and fearsome. The other metaphor in my work is that of the archaeological excavation of memory, which, while it reveals, also conceals through illusion, transformation, and deeply embedded ideas that may obscure the truth. And I am after the truth of myself, the one I can define, as each artist must, with the hope that it extends into a common intimacy with the viewer. Much of Freud's work concentrated on the archaeological metaphor for psychoanalysis, the idea that psychoanalysis is an archaeology of the soul. Freud used archaeology both as a more popular parallel to his own work and as a way to describe the layers of the psyche.[1] Perhaps inadvertently and without direct influence, I stumbled across this metaphor. Ironically, from a technical standpoint, my work in Adobe Photoshop, which is my primary software tool, employs extended multiple layers that break down the crisply autonomous and distinguishable accretions of some digital artists and turn the image into something that appears to mutate and shift as you look at it. In this instance, my work is more akin to the layered and evanescent effects of impressionist painting. The layers represent strata of memory that change and meld into a single image and in this way embody my search for a representation that has an organic history, vitality, and emotion rather than a purely formal or visual "look."

These metaphors—of architecture, boundaries, and transitions—as seen in my earlier series, *Ritual of Abandonment* (1993 to 1994) and *Point of Emergence* (1992), show a kind of spatialization of the psyche and of emotion that I take to be central to most human experience. The metaphor of archaeological layers of meaning runs through my work from the most basic methods of creation to the poetic state of mind that I intend to create. *Hide and Seek* (1995 to 1996) is a culmination of these themes, which have merged to allow me many possibilities.

BEYOND THE ARCHITECTURE OF PHOTOGRAPHY

Juxtaposition—placing two or more seemingly unrelated images together and creating a new relationship, a new reality—is the source of most of my work. In this way, my art reflects the dadaist and surrealist prescription for creating a new art of "the marvelous" or the magical. Similarly, I am drawn to the magical realism of Latin American writers such as Jorge Luis Borges, Vicente Huidobro, Octavio Paz, Pablo Neruda, and others. Juxtaposition allows me to create a sense of surprise and internal meaning. In my use of both digital and photographic media I make these two realities meld and merge, pushing beyond the innovations of the surrealists, who depended on disjunction in the montages of Max Ernst and others. This approach allows me to present a near-cinematic narrative based on the relationship of disparate images in one frame. While in my earlier work I utilized "found" photographs as a starting point for the final image, I now make my own photographs, choosing my models, the setting, and the lighting according to the creative impulse for each work. My use of photography has opened up a realm of possibilities for my work. This pursuit of images with intrinsic meaning and sensual beauty is impelled by intuitive choices that allow me to create metaphors charged with symbolism and emotion.

My work derives technically from two different media—from the computer, which I first learned as a graphic design tool, and from photography, which I initially used as "found" material in the vintage or family photographs that I used in my art. It also derives from the practice of photography itself, which I began to explore in 1992 by taking my own photographs as a source for the images, or photomontages, that I had created with my earlier work. My experiments with photography opened a surprising new realm of meaning for my work, as I was able to find my own voice and create personal landscapes from images that persistently impelled me to photograph them.

Unlike most photographers, who learn photography first, I learned the computer first and photography later. Photography allowed me to stretch and personalize the capabilities of computer software in creating images that overlap the fields of fine art, illustration, and design. Each of these requires knowing computer skills, but additionally for my art I must push myself beyond a given project to creating an image that embodies a feeling I have about the world. My work combining the computer with my own photography allows me to move beyond the limitations of both media—that is, the fixed frame of reference of the photograph and the fractured space of most digital art that uses layers. Instead, I look for an internal resonance between the images, building a final vision that I hope holds together poetically as well as visually. The photograph, for me, has been a means to an end—the starting point of the work. By combining the two tools of camera and computer, my art is more complete.

My involvement with digital imaging began over a decade ago. Its excitement lay in the fact that I could manipulate, edit, and expand the photomontage format. There is a tension or contradiction in my creative practice because I use this digital technology to create psychological narratives that relate both to a "collective unconscious" and to my own process of individuation. These are seemingly nonrational or symbolic aspects of human life that we can capture using a supposedly "rational" machine: the difference is that the computer "thinks" sequentially.

I have moved beyond the current "hard edge" designs of many digital artists and intend to create, as the surrealists did, a world where image, content, and color merge—where borders are not so well defined and meanings can oscillate. Digital technology excels in giving me a way of crafting dreamlike sequences that float into each other, overlap, and merge, giving a sense of inner process and change rather than a defined message. Then, too, the viewer may be unaware that the art was created on the computer, as the images should attract with their own force.

Technically, my tools are a Macintosh computer and Adobe Photoshop software. The images for *Hide and Seek* were taken with an Olympus OM1 camera and then scanned into the computer using an Epson 1200 C flatbed camera; a limited edition of large 34-inch by 47-inch Iris prints on Arches 356 rag paper was printed at Urban Digital Color in San Francisco. I chose to create an Iris edition for this series because the texture and density of the paper allowed a satisfying saturation and movement of color with an organic quality, but the exact color translation from the red, green, blue colorspace of the computer to the cyan, magenta, yellow, black of the Iris inks was a difficult process. I look forward to the time when the technology becomes even simpler, so that it is almost an extension of the artist's mind.

THE *HIDE AND SEEK* SERIES: AN ARCHAEOLOGICAL INVESTIGATION OF MEMORY

> The story I am going to tell comes from a place deep inside of myself, a place that perceives all I have irredeemably lost and, perhaps, what gain there is behind the loss. If some people forget their past as a way to survive, other people remember it for the same reason.[2]
>
> —MALIDOMA PATRICE SOMÉ, *OF WATER AND THE SPIRIT*

I have forgotten and now I am remembering—*re-member,* "to re-call, put back together," from the same root as *religio,* "to bind together," the *re* meaning riches, extra, something innumerable. The state of mind as memory, illusions created by memory and held to, and then lost as awareness enters: thus, *Hide and Seek* represents a search for identity that is not obscured by the illusions of memory or dream any more than by unreflective living. The series is like a game where what is understood plays with what is only suspected, teasing it out through sets of images contained within a frame, as in the window frame of my early childhood. It is a long night that ends in awakening and awareness.

The scanned images are overlaid with fragments of text representing the poetry of the Chilean surrealist poet Vicente Huidobro. His work attracted me with its intensity, romanticism, and ability to confront the complexity of human emotion. I found in his poetry an internal resonance or sympathy with my own creative search, and this became a starting point for the *Hide and Seek* series, which creates an imaginative world of images in its own right. Using the surrealist concept of the "cut-up" or collage as a model, I created a new poem from selected fragments of Huidobro's work. The images in the series are numbered and, when placed in sequence, create a new, autobiographical poem that refers to my own history. Hence, "an archaeological investigation of memory."

This series begins with the house of my childhood and acknowledges the losses that that life sustained:

> She had buried her dreams in a windy closet
> She had come on a dead man wedged in her head.
> —VICENTE HUIDOBRO, "SHE"[3]

In the seemingly simple grid of *Canto One/The Aurora Borealis* (1995) (figure 29.1), I placed images of various parts of the body, vulnerable in their isolation, and a lonely house, like the ones from the area where I was raised. I believe this work suggests my realization of the loneliness and isolation of my childhood, when I buried my dreams of being an artist and shut myself off from the vital reality of myself. It also suggests through the images of hands a way of touching that earlier self, of reaching back to the small girl and pulling her away from the impossible quest for a more perfect childhood. The aurora borealis or the magical northern lights can thus enliven the night with radiance and elusive illumination.

Canto Two/Do Moths Prefer Artificial Light? (1995) (figure 29.2) contains one in a series of seemingly banal light fixtures that I photographed, documenting the interiors of motel rooms I stayed in on a long cross-country journey after a failed family visit. These ordinary objects, a recurring motif throughout the series, became transformed by digital processes of layering and coloration, giving the light and the supine body that they shine on a preternatural glow. The images of the man's hands suggests a tenderness, the magic of connection in even the most mundane of circumstances. Similarly, *Canto Seven/The Path of Light as It Descends from the Sky* (1996) explores the theme of illumination and vulnerability, the man's body passive, suggesting a fragment of a crucifixion or sanctification of the body.

In *Canto Thirteen/The Interior Life of the Dead* (1996), I explored the idea of memory and loss embodied in these lines: "And if my eyes tell you / How much life I have lived and how much death I have died / They will also tell you / How much death I have lived and how much life I have died" (from "The Return Passage" by Vicente Huidobro).[4]

My original plan was to create twelve images that worked with interior resonances called forth by Huidobro's poetry. This thirteenth image was aroused by the death of my close friend and mentor, a woman who was pivotal in my life and with whom I had formed a deep attachment. It

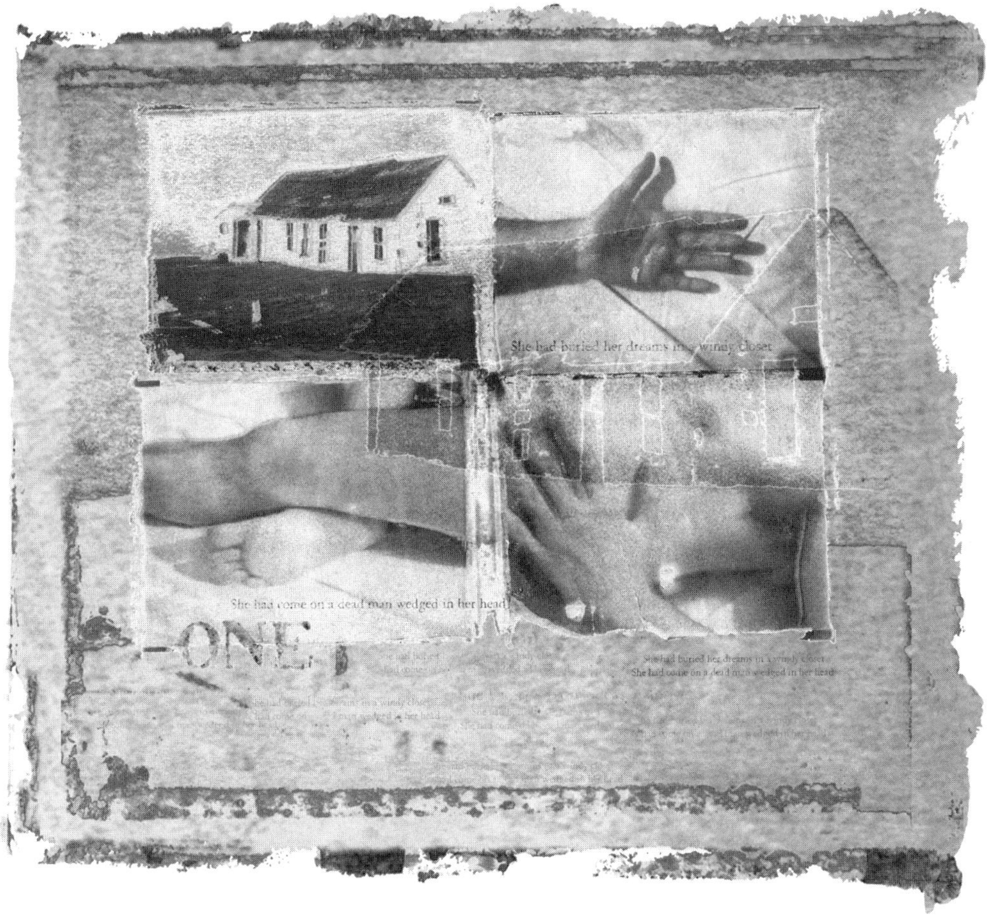

Figure 29.1

Diane Fenster, *Canto One/The Aurora Borealis,* 1995, Iris ink jet print on Arches paper, 47 by 34 inches.

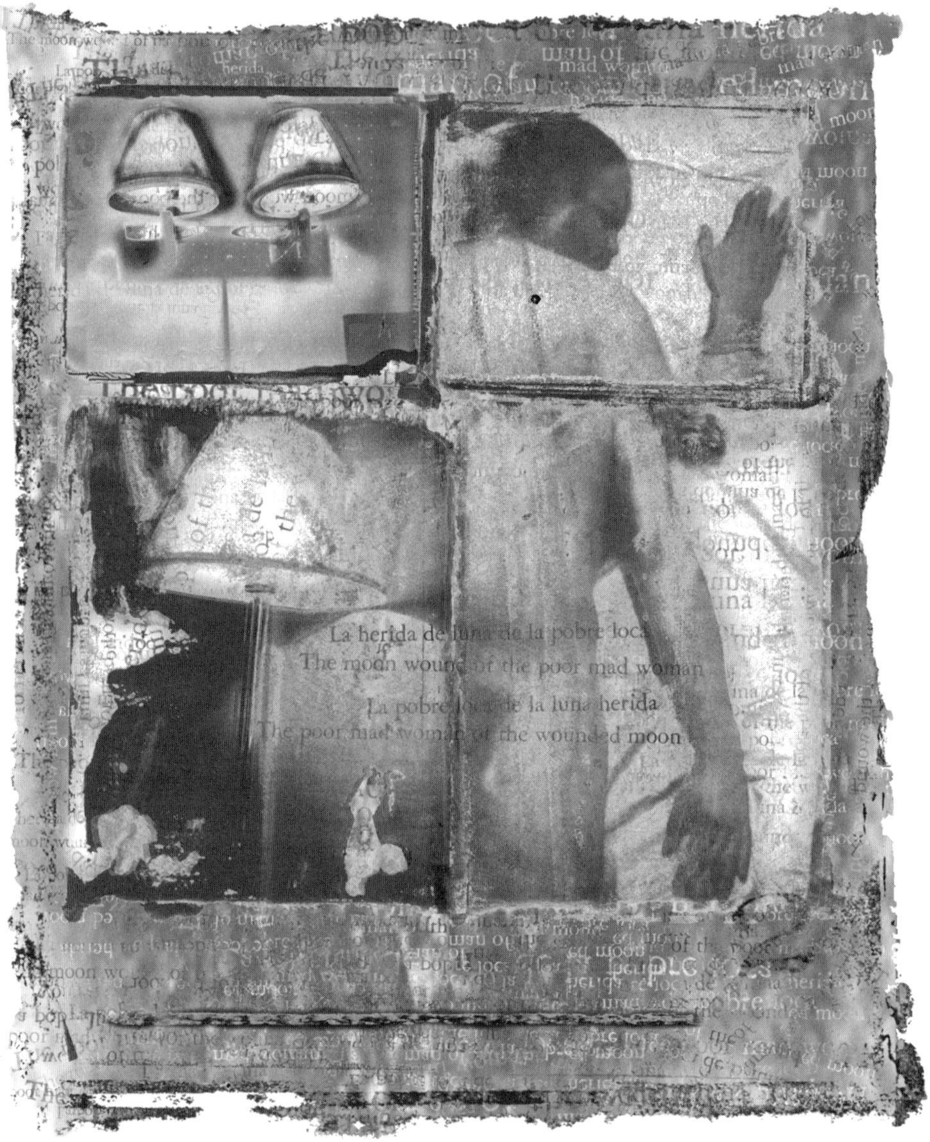

Figure 29.2

Diane Fenster, *Canto Two/Do Moths Prefer Artificial Light?* 1995, Iris ink jet print on Arches paper, 34 by 47 inches.

shows tender images of parts of the body moving as if in sleep. These human images rest in a windowlike arrangement of four quadrants, as the viewer looks on the sleeping figure who is close but removed by the screen of the frame. The light in the upper left quadrant was another in the series of light fixtures referred to previously, while the illumination and urnlike support suggests life after death. The personal aspect of these images, allied with the words of Huidrobro written in multiple layers, allowed me to give shape to my personal loss. This piece creates a layered world of images that expresses how memory illuminates and lingers long after our physical presence and movements become remote and lost.

THE *POINT OF EMERGENCE* SERIES

> I am absent but deep in this absence
> There is the waiting for myself
> And this waiting is another form of presence
> The waiting for my return.
> —VICENTE HUIDROBRO, "POETRY IS A HEAVENLY CRIME"[5]

The *Point of Emergence* pieces are earlier digital works (Fujichrome prints in a 20-inch by 30-inch format) that represent my first long series in which I created a statement that embodies the nuances of the idea of transition. In this series, I explored the frozen moment that is the center of transition, from light to dark, from low tide to high tide, from unconsciousness to consciousness. Each of these states is separated by a veil between the different planes of existence. At the point of emergence, the veil becomes transparent and disappears, allowing a transformation into a new state of being.

To capture a sense of these planes or layers of being, I created a spatial format for my work composed of three horizontal layers. Each of the images is layered with composite photographic and scanned images, which are arranged in a metaphorical sequence from the bottom to the top of the page. Each work was composed of three horizontal sections, the lowest section representing life; the middle section representing death, the absence of life; and the top section representing rebirth, the reemergence of light and energy.

Night Five (1992) is a work from this series that captures the feeling of instability and change. The image of the older woman, sitting placidly self-absorbed in a café, was taken with a Canon Xapshot, as were all the other images in this series. This camera was designed to view images on a television screen and gave me an agreeably low resolution with a jagged and irregular focus that was the effect I wanted. I see myself as creating a new digital superrealism, or surrealism. The second layer, the image representing death, is that of an engraving of hands that I scanned and altered. This image is consistent throughout each of the works in the series. It is significant that you see only the back of the hands, as they are not extended in warmth or welcome. The fingers point upward to the layer of rebirth, which was represented as a flow of lights by the traffic movements that I shot with the Xapshot camera. Its effects allowed me to capture the fluid sensation of the endless movement of traffic in a big-city environment, which is the closest representation of an oceanic state that is available to most urban dwellers.

Night Six (1992) embodies the sense of journey and transition that I intended. Ambiguously, the figure of a woman enters or exits a car, implying change and a journey about to begin or end. Formally, the shape of the car is angled toward the top of the image, toward the hands and toward rebirth. I digitally created a negative, thereby making a less realistic image. This allowed me to imply a sense of strangeness and psychological excitement.

This series embodied my interest in the metaphor of architecture, of building and rebuilding, particularly in that these images try to create a new space, a threshold, between these different states of being. The *Point of Emergence* series thus creates a metaphorical architecture that uses the threshold as the transition between different states of being. However, it has a very modern approach to this theme because each of the layered spaces is discontinuous and must be held together by the viewer's mind.

In the work that followed this series, *Ritual of Abandonment,* I abandoned the formal device of separating the spaces that have different meanings and created a new metaphorical space that merged different images.

RITUAL OF ABANDONMENT: AN EPHEMERAL ARCHIVE OF MEMORY

> Why do I follow the wind through my dreams
> As it stirs my murmuring hair on the roof of this night
> Down deserted roads like sad words
> I could never find you
>
> —VICENTE HUIDROBRO, "THERE IS A CATACLYSM INSIDE US"[6]

The images in this series have to do with loving and losing. I was able to generalize from my own emotional experiences by requesting true contemporary tales of love and abandonment from friends and colleagues across the country. The stories were submitted to me two ways, electronically and in hand-written letters that I later scanned into the image. The text of the story is worked into each piece. All speak of the human situation of falling in love and allowing vulnerability to another person, thereby putting oneself at risk because of the potential loss of the loved one.

Simultaneously with the creation of this series, I became aware of the importance of interactivity and the World Wide Web. After I finished the seven images in this series, I created *Ritual of Abandonment: A Virtual Artist's Notebook* (1994–1995) as an installation on the World Wide Web. I added to this site over the period of a year. This is an interactive site that uses sound, words, visuals, and quick-time movies that allow users to explore their own difficult feelings about abandonment. I received copious e-mails from those who had experienced this kind of alteration in their life, including a striking one from a father in the Midwest, who explored the site with his son, who was recovering from the loss of his first love.

I wanted to find an image that could represent a distillation of all the feelings that the *Ritual of Abandonment* called forth. This image became the icon for the series. I took the symbol for Erzulie, the tragic mistress or goddess of love in the Voudoun religion, who is often symbolized by a heart pierced by a knife. I transformed this image into that of a railroad track that pierces the heart and body. These images are woven throughout this series. It uses the images of a railroad and a reclining nude woman, whose body merges with the tracks, which signify how we run to or away from relationships. This kept with the feeling of this series, that theme of wounding by love, and it connected the female Voudoun icon with a motif of loneliness, a sense of the train tracks and the late-night wail of the train sounds receding in the distance, a cry perhaps of isolation as well as of movement. Each of these images relates a different story of overpowering attraction followed by abandonment; each story called forth an extreme reaction in those who wrote of their sometimes disastrous affairs.

In *Two Running Rails of Mercury* (1993, Fujichrome print), the figure is foreshortened into the one-point perspective of the tracks, while the obscured head and nude body of the woman insist on the corporeality of intense desire. In this image, I also scanned a departing note from the author of the text's lover, which she found attached to her door, which adds yet another layer representing the complexity of relationship. The *Virtual Artist's Notebook* Web installation, in which this image is included, has hyperlinks to fragments of poetry and writing, as well as train sounds, that explore the abandonment theme. The story, written by Kimberly Fisher, is as follows:

TWO RUNNING RAILS OF MERCURY

Two rails. Two lifelines in the night. Two running rails of mercury. They run far out into the bleakness, back to Turlock, fading finally out of sight . . . even the moonlight losing track of them at last. I can't see them anymore. All I can see is your eyes, black, blacker, you black-eyed son of a bitch. I try to see into them, those black eyes, try to penetrate the shielding black lashes, slip in under the black brows, infiltrate the drug-dark centers and arrive inside you. Somewhere.

And I think, that for once, I find you there.

So here we are, you and I; you can't wait to leave, I wait for the train. From your pockets, your belt, and your other secret places come parts of your life: the .45, the .38, the long stainless blade and the treasured titanium . . . the only parts of you that are engineered, planned and precise.

You know, I don't know why we have to do this. I always believe you, but mightn't it be better to make love at this point, and just die the small death . . . holding each other close?

No. I always believe you. Even when I shouldn't.

So you take your pick, and I'm not surprised: you're the only white boy I've ever known who liked to dance with the knife. Not enough drama in the bullet. I just can't place my feelings. I'm here. I'm not here. You're here, then before I'm ready, you're not . . . .
. . . .
. . . .
. . . . Born of the desert, and there you die. When the train comes I'll join you. 'Til then I hold you close to keep you warm.

I Couldn't Stay in Miami (1994, Fujichrome print) was based on the story sent to me by Mark Toal, a friend who relates the painful story of an abrupt abandonment, inexplicable absence, and precipitous return of a lover who, after several years, wants him back and when he refuses because he rebuilt his life, commits suicide several years later. The work is like a dream in which the loneliness of the station in the background and the vulnerable position of the nude in the foreground conspire to create an atmosphere of uncanny isolation, as in the work of De Chirico.

SUMMARY

As an artist, I create nuanced perceptions that physically bring the digital and material worlds together. In my first two series, created in the early 1990s, I believed that digital art should have a different "look," so I took advantage of the intense supersatured colors of the RGB color space, and I chose very bright, almost electronic coloring for these works. Now, in my mind, there is no division. My later work, prints on rag paper, with multiple layers of images and colors, has a decidedly organic and mutable feel.

We are waiting for the rest of the world to understand that the only difference in digitally created work is with the type of tools that we use; the art will express or communicate its reality only as powerfully and precisely as our capabilities as artists allow us. While others have explored the more obvious aspects of digitally created works (hard edges, collage), I use this new medium in a more personal, painterly way. My alchemy, both earthly and spiritual, is most transformative when one forgets about the process and definitions of art and enters instead this new world that I pursue.

NOTES

1. Donald Kuspit, "A Mighty Metaphor: The Analogy of Archaeology and Psychoanalysis," in Lynn Gamwell and Richard Wells, eds., *Sigmund Freud and Art: His Personal Collection of Antiquities* (133–151) (New York: State University of New York and Freud Museum, London, in association with Abrams, 1989).

2. Malidoma Patrice Somé, *Of Water and the Spirit: Ritual, Magic, and Initiation in the Life of an African Shaman* (New York: Penguin, 1995).

3. Vicente Huidobro, "She," *The Selected Poetry of Vicente Huidobro*, ed. and introduction by David M. Guss (New York: New Directions, 1981), 165.

4. Huidobro, "The Return Passage," 219.

5. Huidrobro, "Poetry Is a Heavenly Crime," 209.

6. Huidrobro, "There Is a Cataclysm Inside Us," 175.

30

Pigs, Barrels, and Obstinate Thrummers

Linda Austin and Leslie Ross

[*The authors, a choreographer and dancer and a composer, performer, instrument builder, discuss their respective and collaborative uses of interactive objects and mechanical devices in live performance, focusing on their most recent collaboration. Motivations, techniques, and the give and take of the collaborative process are among the issues explored.*]

LINDA:

The blurring of the boundaries between the mechanical and the human is a traditional science fiction trope that has always fascinated me, most intriguingly the idiosyncratic take on it exemplified in the work of Philip K. Dick. Some of Dick's characters are human and mechanical hybrids, while machines are often animate, seemingly endowed with anima, soul, or personality. At the same time, humans often reveal inexorably machine-like traits. An equivalent fascination in real life is contemplating the play of physical forces that govern machines, objects, and the human body. Anthropomorphizing, to be sure, I find machines with joints and moving parts (such as earth movers or the wind-up devices recently constructed by my collaborator Leslie Ross) lifelike and endowed with suggestions of personality, especially if there is something irregular and unpredictable in their movement. The opposite pleasure, of course, is to look at the body as a mechanical entity. But while Dick often uses the "thingness" or machinelike quality of humans as representative of a sinister reification, a loss of the ability to feel or to respond in other than in a clickety-clack reflex manner, "repeating doomed patterns,"[1] for me as a choreographer, the fascination with the "mechanical" is part of my insistence on the ineluctable materiality of the body—our interface with the physical world it inhabits.

I'm not by any means a Luddite, and yet in an age when more sophisticated technology opens up tremendous possibilities for interactive human and machine relationships, another pleasure lies in primitive bits of engineering where you can see easily how everything works, cooking up jury-rigged solutions to technical problems with cheap, easily accessible materials.

For example, the engineering problem in *Where Fish Drown*[2] was this: I wanted my body's passing to trigger the collapse, without my directly touching them, of a series of precariously balanced 2-foot plastic tubes. The transparent tubes were filled with lime-green colored water that slowly leaked out the bottom, while atop of each was balanced a metal dish containing a mound of earth. My solution was to balance each tube on a strip of waxed paper that was fastened to the floor at the opposite end with tape, and then, while passing each tube/dish structure, dragging a toe, knee, or other body part over the waxed paper strip, pulling at it, at times almost excruciatingly slowly, until the carefully balanced edifice fell with a satisfying clatter of the dish, spilling out the mound of earth. I find the almost absurdly homemade and physically tangible qualities of my solution humorous, especially in an age of technologically sophisticated systems in which human movement can "magically" trigger sound, image, or lighting events by passing in front of a video camera or photoelectric sensor.[3]

LESLIE:

After music school, studying both baroque and modern bassoon, I spent some time traveling to libraries and museums looking for both seventeenth- and eighteenth-century bassoon literature and seventeenth- and eighteenth-century bassoons. This brought me to build replicas of historical instruments, which fourteen years later I continue to do. Most of the conventional orchestra and chamber playing I've done has been on period instruments, and most of the modern bassoon playing has been in new music (including working with MIDI and computers), with a strong emphasis on improvisation and collaboration with choreographers.

From the day when I was first permitted to bring the high school bassoon home and immediately took the instrument entirely apart, my interest and involvement with music making and construction have always worked hand in hand. That I had to stay home the next day to put it back together only meant that for the weeks to follow I was more systematic in the removal and placement of the keys until I could truly throw everything haphazard into a box without any problem.

My approach both to construction and playing remains alternately (or simultaneously in different proportions) impulsive/cavalier and systematic/exacting, and whether we are talking about the making of bassoons or the construction of sound objects, the material technical qualities of musical instruments are in no way felt foreign to the musical qualities of those same instruments, in the same way that using electronics or MIDI or tape are in no way separate from composing and playing.

My involvement in making replicas of seventeenth- to nineteenth-century bassoons has evolved from an initial interest in the relevance of playing baroque repertoire on the instruments it was written for to an interest in better understanding the broad spectrum of tonal characteristics these many different instruments offer, compared to the somewhat homogenous tone concept of a modern bassoon. There is also the continued challenge of trying to understand the workings and acoustical properties of the instrument. Though the bulk of the work is done to spec with many aspects of construction being to .1 millimeter tolerances, there are always individual adjustments that need to be made. These fine adjustments are increasingly made with a specific understanding of how these alterations will effect the playing characteristics of the instrument, but intuition continues to play a huge part in the making and tuning of an instrument.

Despite variations in the playing characteristics of a bassoon, the instrument is built with clear objectives for tuning and scale, flexibility in tone, and dynamics and resonance. There is a continuum to my understanding, as I am always building on past knowledge. With installation and sound sculpture work, I usually start with an image or concept, go out to collect some raw material, and just start building without really knowing what the end result will be. One unifying quality underlying all of my work is simply the pleasure in working with matter, within its fluidity and transformability: feeling the shifting balance and play (the *room*) between what transformation I force and control and what room or level of control the changing material itself exhorts.

With the *Tentacled Bellows* (1989), for instance, I started with the idea that I wanted to control a series of drones but leave my hands free to play the bassoon. This instrument developed into a portable structure where bellows that are attached to shoes and activated by walking pump air into a reservoir made from a tractor inner tube. This tube rests on a frame constructed from PVC tubing where the frame (which also like a doorway reaches over the player) is suspended from a girdlelike harness attached to the hips of the player who stands in the center of the tube. From this inner tube run sixteen to eighteen rubber tubes that are connected to beating reed drones all attached or snaked around the structure (the PVC structure itself comprising two of the drones). Cut-off valves on each rubber tube enable air to pass or not, and variable air pressure reeds also add to the control of whether and how the drones sound (figure 30.1).

That this construction might have a visual impact did not occur to me until after it was built, and in fact I was a little irritated by the attention its visual appeal had when I was so happy with the broad range of groaning, howling, multiphonic, percussive, and pure pitch tone it could generate. This first construction included many elements that have continued to find themselves in my work—notably, drones and ostinatos, portability and self-contained worlds, along with a degree of unpredictability. One visual aspect that I have learned to enjoy cultivating in most sound objects is the impression that the object is about to fall apart and that it is barely holding together—this in sharp contrast to the finished look of a bassoon.

The *Teetering Plucked* (1996) was a structure that came from a purely mechanical concept, where I wanted one single motor to activate the entire sound sculpture. Built out of metal tubing with piano wires strung to the tubes, the structure, round at the bottom and oval at the top, is fan-shaped and about 8 feet high with the spindle up the center. This spindle has a drum with plectra attached that rotates at the very top to pluck the strings. The bottom of the spindle is connected to a series of pulleys and a crank shaft (all constructed out of wood and set to motion by resined rope) and sets the entire fan-shaped structure teetering side to side from a pivot point at the bottom. This is to allow for the strings at the top to come in and out of contact with the plectrums; contact microphones and electronics were used to pick up and modify the sound. The immense strain that the one motor was under in combination with the give of the wood and rope connection caused the movement to be interrupted by constant hiccups from the build and release of tension, adding an unpredictability to the plucking of strings.

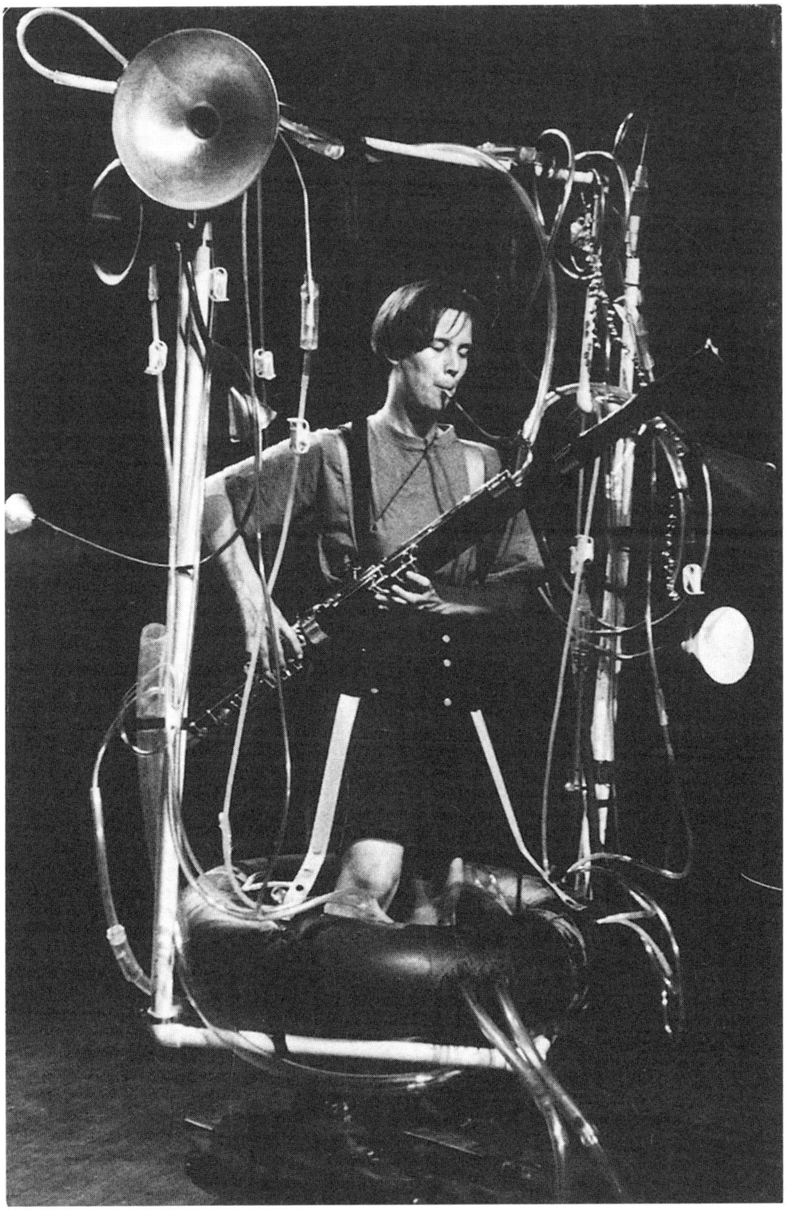

Figure 30.1
Leslie Ross, *Tentacled Bellows,* 1989. Leslie Ross at Performance Space 122, New York City. Photo by Dona Ann McAdams.

LINDA:

PIG (1998)

PIG[4] may seem like an unlikely name for a dance. My obsession with the body's material, mechanical, and kinetic expressivity leads me to look for thematic material that lends itself to structuring this obsession. So when I came across an anecdote in Clarice Lispector's writing of a young girl's ecstasy on the fleeting sight, through the window of a moving train, of what she excitedly exclaimed upon as "a real pig!"[5] I seized on it as an earthy metaphor for the delights and epiphanies yielded by a devotion to the world as perceived by the senses. This metaphor in turn provided a framework for both the movement exploration of the piece and the creation of visual, mechanical, and sound-making devices. It also provided an opportunity for the intersection of human and constructed entities, bringing out the mechanical limitations and possibilities of the human body as well as the expressive and animate qualities of nonhuman objects and gadgets.

PIG has three sections, each with its own title in some way inspired by Lispector's language. In Part I, "we are that thing which must happen," an investigation of the articulatory and gestural possibilities of the limbs and joints generates micronarratives enacted by a trio of dancers. In Part II, "in search of the real pig," a solo dancer (myself) searches for the kinetic pleasures to be found in both stylized set movement and improvisation. In Part III, "avid matter, hungry matter"—the "big group section" (six dancers)—a hungry desire for physical sensation leads to collusions and collisions within the individual body, between body and body, and between body and environment.

My collaborator for this project—composer, performer, instrument maker Leslie Ross—was responsible for all of the real mechanical wizardry involved in the piece. Leslie's creation of sound-making constructs has impressed me for a long time; seeing and hearing her performance in one of her first creations, the *Tentacled Bellows*—Leslie as a "groaning, howling" hybrid human machine thing enveloped in a spaghetti-like tangle of tubes, pumping with her feet on the bellows, opening and closing valves, and puffing on her bassoon—made me laugh and cry. It also influenced me in the creation years later of a self-contained toy theater with its own lighting and sound system that I wear on my body. Leslie and I have collaborated on several previous occasions, and I thought, and rightly so, that her contributions would take *PIG* to another level, not only sonically but also kinetically and visually.

LESLIE:

In *PIG* I found myself using many different modes of construction, composition, and performance, incorporating ideas that I had been wanting to work on for a while, as well as new ideas stemming from visual stimulus in dance rehearsals.

Rolling Barrels (1996)

One idea that I had been carrying around for years was the idea of transforming a large barrel into a rolling musical object—with a person rolling in the barrel. The first idea was to have three large rubber flappers attached to the sides that flopped to the ground in loud smacks as the barrel rolled. I was originally expecting to find a way of playing a musical instrument inside at the same time, which proved to be impossible. What I did not expect was the great delight in rolling in a ball inside of a barrel and the incredibly satisfactory smacks of the rubber hitting the floor.

LINDA:

Barrel/Dancer Duet (1998)

In at least one instance, a sound-making contrivance did double duty as a dance element. Leslie's barrel with rubber flaps gave me a chance to choreograph a direct physical interaction between human and object: a duet between performer Katherine Marx, a skilled contact improviser, who functioned as kind of framing element, fancy stage manager, and object animator in the piece, and Leslie inside the "flapper" barrel—"a barrel that flirts," a reviewer called it.[6] Besides flirtatiousness, the relationship between the round flippered creature and its human partner also displayed a bit of pathos, as when the barrel tried to keep up with Katherine's ease in traversing the space with a series of painfully tiny lurches, inching its way forward, until Katherine finally ran back to push the barrel from behind, stopping just short of the edge of the performing area, where they finished off with a miniature "folk dance" (figure 30.2).

LESLIE:

Barrel Plus Eggshells (1998)

In rehearsal one day, watching Linda go through her solo section (a piece we had previously worked on for which bassoon solo was used), I had the image of a barrel rolling back and forth over a bed of eggshells as accompaniment for the first half, to be overlapped with a sound tape that would continue to the end of the solo. Concerned over whether the sound of crushed eggshells would carry in a hall, I built a 12-foot sound box platform that also served the purpose of keeping the barrel on track; amplification turned out to be unnecessary. The sound tape was comprised mainly of samples and recordings of crushed and rolling eggs, with much of the raw recorded material being electronically manipulated and scored in layers to create the musical direction I was after. Though both parts shared a relentless quality, in contrast to the minimalist score of the barrel moving back and forth over the eggshell (albeit erratically), the sound tape score had a linear build in tempo, density, and dynamics. As different as these two sections were in their use of technology, the musical and artistic development remained the motivating factor for the technology used. Structurally and dynamically, the music score of this piece was the same as when it was just bassoon solo.

LINDA:

The eggshell sound score did indeed resemble the structure and dynamics of the bassoon version. The important difference to me, as the all-too-human dancer interacting with the music, was that the taped version for the second section did not have to rely on the stamina of a human performer for its linear build. It achieved an increasingly relentless momentum that was hard for me to keep up with. How I resolved this is a good example of the back-and-forth influences exerted by collaborators: Leslie was inspired by watching me and made the tape; this taped version then forced me to change my physical approach to the corresponding part of

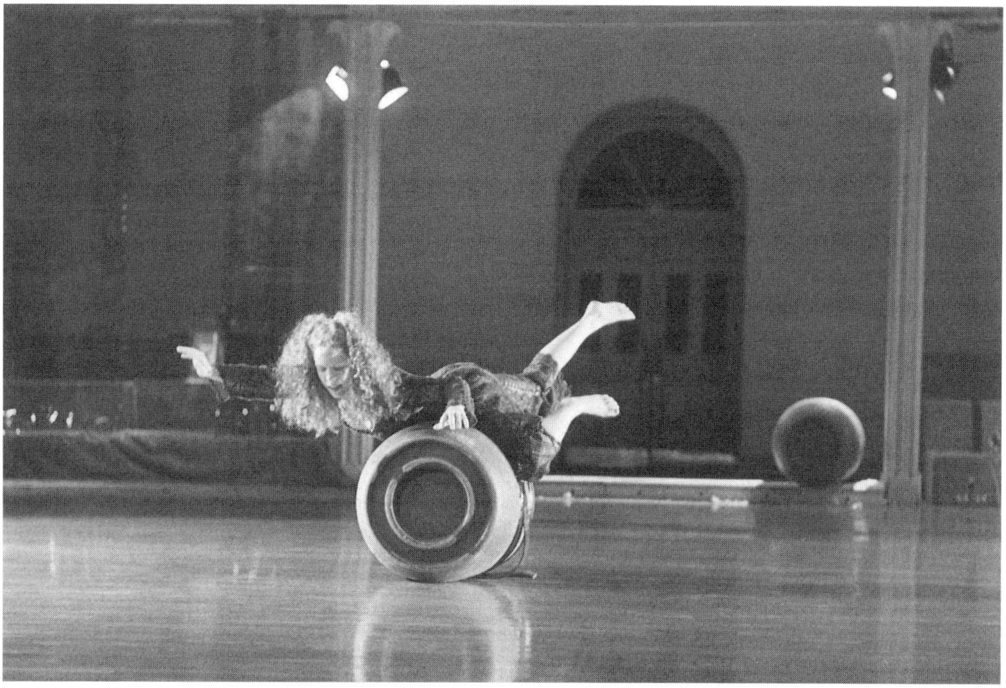

Figure 30.2
Linda Austin and Leslie Ross, *PIG,* 1998. Dance performance at Danspace Project at St. Mark's Church, New York City. Dancer Katherine Marx in a duet with "flippered barrel," Leslie Ross inside the barrel. Photo by Tom Brazil.

the improvisation (something I was resistant to at first). I started to experiment with a very literal image—imagining an egg whirling around inside my pelvis in the same way eggs whirled in a metal bowl to produce one of the effects in Leslie's tape. The resulting movements started out small, contained in the pelvis, and then began to travel down the legs and out the arms; this built until finally there erupted a bout of excited flinging shaking tossing, which took me off the careful diagonal I'd been making and out into the space—a bit of dancing that was a favorite of many audience members.

LESLIE:

Quilled Piano (1998)

When Linda first asked me to collaborate on this piece, she already had the idea of using an old upright piano and asked me if I thought that I would be able to do something with it. How could I possibly pass this by? This instrument was realized gradually as it was being taken apart without having any clear idea of how it was going to be mechanized. All action but the immediate hammers and strings were removed, and a 6-inch-diameter drum cylinder with holes was made that ran the entire length and was rotated by peddling a crankshaft. Pegs that trip the mallets and hit the strings could be inserted or removed as the cylinder rotated, thus allowing the tune to change in this music-box-principle design. The sound board being directly in front of the player was also easily prepared (and unprepared) with alligator clips, cloth spins, and so on as music was being played. This instrument, like the *Bellows,* left my hands free to play bassoon at the same time.

LINDA:

Pig Lamp (1998) and *Pig Lantern* (1998)

When I first offered Leslie the use of my old upright piano for the making of a new instrument, I had envisioned it moving through the space as a counterpart to the trajectory of a trio I was working on. That proved impracticable, so I began to cast about for some other object-being to take on that role. I had previously done an improvisatory "duet" partnered by an old brass floor lamp. In my mind's eye, that same lamp began to be seen making a diagonal gliding path through the space, to sport a slowly revolving pink lampshade, and to emit pig sounds.

First, I looked for a battery-driven small motor that would move very slowly, only to discover that the battery-driven ones all had too high revolutions per minute. I also tried unsuccessfully to rig up "sails" inside the shade to catch the wind from a small battery-operated fan. The ultimate solution was even more primitive but had its own charms.

Two battery-powered florescent lights were fastened inside the lampshade along with a tiny cassette player containing a recording of pig grunts and squeals. The lampshade itself functioned like a spool, sitting loosely on a spindle. Wound around the lampshade was a long pink cord. When the audience entered, the lamp was already illuminated and standing on a little piece of fake grass downstage left. The free end of the lampshade's pink cord was stretched diagonally up and fastened to a spot in the balcony in the downstage left corner. When the lights dimmed to signal the beginning of the piece, performer Katherine Marx entered, turned on the tape of pig sounds, fastened a leash to the lamp, and slowly pulled it in a diagonal path to upstage right, where the bit of grass on which the lamp rested would fit into a gap in a larger patch of grass, like the last piece in a jigsaw puzzle. As Katherine pulled the lamp, the cord unwound, and the spool/lampshade slowly turned. Down the long slanting cord that remained in the space would glide, at the end of the piece, one of Leslie's buzzing or clattering wind-up gadgets.

The only overt "pig" image in the piece was provided by another illuminated and revolving object, a kind of lazy-Susan lantern that sat on the floor and was spun manually, again by Katherine Marx, throwing out irregular swathes of light into the darkened space and providing the audience with here-again, gone-again glimpses of the image of a pig.

LESLIE:

Obstinate Thrummers (1998)

These were a couple of dozen wind-up objects that were also first conceived as a concept a couple of years prior to *PIG.* There was something about our current global corporate world's preoccupation with budget balancing and better efficiency and productivity and my own failed attempts of trying everything from date books to personal information managers to "manage" my life that made me think of elaborate wind-ups that would produce one simple repetitive sound.

The result was a series of objects perched on spindly legs that strummed, thrummed, hit, or stroked strings, rubber bands, shaped tin forms, and other objects. The vibrations set by the rotating arm sent the object dancing across the floor, and the lopsided position of the suspended hit and strummed object set up an erratic arrhythmic cadence. Other *Obstinate Thrummers* are built from metal construction studs with a single string running down the length. The wind-ups with a plectrum attached to the end of the rotating arm repeatedly strum the string, which ranges from 2 to 10 feet in length. These were suspended with contact mikes attached so that the sound was amplified and manipulated.

Not only were these humorous, creaturelike, spastic objects single purpose, but they possibly exemplified one of the most inefficient ways of going about it, embodying, among other things, the hurt credulity of living beings.

For the closing section of the piece, there was an additional tape of recorded bassoon duets over more eggshell crunching and rolling. The meter of the composition was in 5 with constant shifting 2/3 beat patterns. It was a dance with no solid ground, a sensation I further wanted to enhance by having two or three copies of the tape running just out of sync with each other, each inside of a barrel rolling through the space or suspended from ropes swinging through space. First attempts at this failed to produce the results we were after, so I mastered a mix that sent one recording to the left and just a half beat off the other to the right, each on their own fade-in and -out timing. The seasick rollercoaster quality that this produced was exactly what was intended.

LINDA:

I like how leftover ideas from one piece get taken up again in a later piece. Leslie's inefficient wind-up gadgets found their way into *PIG* after being first conceived for an earlier piece. The idea of having tape recorders inside objects rolling through the space is something Leslie and I hope to use within the next couple of years in a piece called *The Use of Rumor*. Big wooden spools will house the tape recorders; their moments will be choreographed just as thoroughly as the dancers' movements; the play between movement decisions based on sound versus those based on visual or kinetic or dancer-interactive motives will be an interesting multidimensional puzzle to work out.

LESLIE:

There is no question for me that a musical instrument is in itself technology and that as an artist I can regard any technology as an instrument or tool to be used. Changes and influences of technology are intrinsically linked to artistic and aesthetic demands and vice versa, as working on many parts of *PIG* have shown.

NOTES

1. Philip Dick, *A Scanner Darkly* (New York: Daw Books, 1977), 55.

2. The final of version of *Where Fish Drown* premiered in 1992 at Performance Space 122 in New York City with choreography and visual design by Linda Austin and music by Paul Hoskin and Linda Austin.

3. See Robb E. Lovell and John D. Mitchell, "Using Human Movement to Control Activities in Theatrical Environments," paper presented at the Third International Conference on Dance and Technology, 18–21 May 1995, York University, Toronto, Canada.

4. *PIG* premiered in 1998 at The Danspace Project at St. Mark's Church, New York, with choreography by Linda Austin; music, including the design and construction of instruments, by Leslie Ross; and visuals by Linda Austin and Leslie Ross.

5. Clarice Lispector, "Sea Bathing," *Selected Crônicas*, trans. Giovanni Pontiero (New York: New Directions, 1992).

6. Jennifer Dunning, "A Strange Little World with a Barrel That Flirts," *New York Times* (9 March 1998): E5.

31

FleshMotor

Dawn Stoppiello with Mark Coniglio

/ (SLASH)

When considering how to describe my work here, I am immediately drawn to use a term that my collaborator Mark Coniglio invented several years ago—*Slash (/) Artist.* The term grew from a time when all of the artists that we met seemed to describe themselves as a "dancer slash performance artist" or "poet slash technologist slash actor" or some other similar string of métiers separated by the all-integrating slash. What these people were trying to say was that they were attempting to hybridize multiple forms into some-other-thing that they could not, as yet, put words to. This is the way I have felt as I have attempted to bring my primary métier of dance together with media and technology over the last ten years.

Through this process, my choreography has changed in response to my close contact with computers and computer-controlled devices. As a choreographer and dancer, my relationship to the world begins with my relationship to my body. As an artist working with computer technology, my relationship to the world is filtered through a hyperriver of bits performing multiple operations in parallel as they flow madly through computer space-time. This duality has infiltrated my choreographic sensibility. It manifests itself as accumulative phrases that are orderly, repetitive, and organized, like a program, but that are interrupted by material that is completely human in its unpredictability and occasional violence. This duality, between what is most human and what is most machine, has become the inspiration for much of my recent work.

This human and machine dialectic is widely apparent in our time. Insulating technologies such as computers and the Internet have become integral parts of our lives. My generation is the first to grow up with the television always on. We have seen the body projected into space, we have seen it explode, we have seen inside it, we have seen it completely altered. We have looked out from TV space and seen that same body slumped motionless, in near decay on the couch. We have been enticed by the perfect TV body—no smells, no secretions, no flaws. We live with the neurosis of wanting to obtain that simulated perfection through intellectual escape into screenal lands where such a body can exist. We try to keep up with that image by absorbing products that relieve us of smell, hair, and fat in some vain attempt at immortality. Technologies that allow us to connect without touching keep us in blissful denial of our bodily imperfection. Yet every cell knows that it will eventually rot, decompose, and return to the earth from which it came. I have concern for this confused and displaced body.

So I am a choreographer slash computer media artist. The slash—the intersection of flesh and silicon, blood and television, body and computer—is the place that I find my work at the end of the twentieth century.

+ **(PLUS)**

Of the several works that I have collaborated on with composer and computer media artist Mark Coniglio over the past several years, perhaps the most important for us both was *In Plane* (1994). This piece was seminal in the development of our thinking about the relationship of the body to technology as an aesthetic idea, as well as the technological innovation required to realize it.

We have used many varieties of homemade and commercially available sensing systems in our performances, but the most sophisticated is MidiDancer. Mark first had the idea for this device while we were both students at California Institute of the Arts. He had been inspired by *Hungers,* (1989) a collaboration between his mentor Mort Subotnick and video artist Ed Emshwiller. In that piece, singer Joan LaBarbara controlled MIDI synthesizers using a small baton that responded to the way that she moved it through the air. Mark was immediately inspired to attach this device to the leg of a dancer but was discouraged by the wires needed to get the information to the computer. So he envisioned, and shortly thereafter implemented, a wireless device that would allow a dancer to make music with the movement of her body.

The first MidiDancer system was built for a collaborative project at CalArts in which Mark and I took part. The original device was quite primitive. It was made from radio-controlled car transmitters. Attached to each transmitter were two sensors in the form of metal levers that we taped (at the loss of much body hair) to our arms and legs. Each sensor measured the flexion of a joint and sent that information via the radio transmitter to a computer, where it could be used to control music synthesizers.

The piece we made was for four performers, each of whom was wearing two sensors, one on the elbow and one on the knee (four individual MidiDancer systems in all). The idea was to give each dancer two sounds to control, one on each sensor, that would stay the same during the course of the piece. Our hope in keeping this fixed relationship was to create a kind of sonic identity for each dancer that the audience could recognize. After creating material separately, we came together to work with the dancers and quickly realized that this one-to-one relationship, one gesture producing one (and only one) sound, did not make for the richest composition musically or choreographically. We came to call this technique the "bleep-bloop" method, as this is all that the first attempt ended up being—a series of bleeps and bloops in conjunction with the robotic choreography required to trigger the system. We were disappointed that the piece lacked the kind of complexity and subtlety that we had envisioned and knew right away that we were going to have to try again.

What we didn't know at the time was that, in that moment, MidiDancer had changed the way we thought about composing. In retrospect, it should have been obvious that we had begun to compose for a new and unfamiliar instrument and that, of course, the artworks that we made with it would be directly influenced by its nature. For one thing, it was clear that we could not work in isolation when creating our materials but instead needed to work collaboratively on both sound and movement. We didn't know the instrument well enough to imagine the outcome, and we needed to really see and hear it happen. Also, we found that the physical gestures required to play the instrument were not as inherently interesting or meaningful as choreography. To understand what I mean, imagine for a moment that you are watching a great violinist play. You may choose to watch her fingers move along the neck of the instrument, but I don't think that you would expect those same finger movements to give you any dramatic information about the piece. We were faced with a challenge: the dancer needed simultaneously to make both meaning and music with the same movements. This is a problem that became even more complicated as we added other media into the mix.

In the summer of 1990, Mark and I first collaborated with Kit Galloway and Sherrie Rabinowitz at the Electronic Café, their performance space and lab in Los Angeles. In this pre-Internet world, Kit and Sherrie pioneered the use of various kinds of telecommunication links to create live artworks between distantly located sites. At this time, one of the most common ways for them to get video between cities was a slow-scan, handheld, black and white video phone. Mark's and my experiences with that device would begin our next series of insights regarding the combination of dance and media.

Tactile Diaries (1990) (figure 31.1), our first collaboration with Kit and Sherrie, had performers at the Electronic Café and the New York University Television Studios in Manhattan performing together using slow-scan video phones and telephone-grade audio connections. One section of the piece was a solo that I performed using the MidiDancer. In this section, Mark programmed the software to trigger the videophone when I made a particular shape with my body. It would capture an image of my performance in Los Angeles and then send that image to New York. At the other end, the still image would arrive on a television monitor, slowly scanning in from top to bottom over a period of five to ten seconds. I carefully chose all of the movements that would trigger the video phone because these would be the only representation of the dance that the New York audience would see. I became very interested in selecting body shapes that, when seen in sequence in New York, would create a different narrative experience from the one that the live audience would have in Los Angeles. It seemed essential to find a way to have the choppy, low-bandwidth video express something different than the full-bandwidth (live) dancer could provide. What was important about this approach was that it emphasized what was distinctive about the technology and provided a different way of seeing the dance.

The use of video in this piece introduced me to a new theatrical element (beyond sound) that could be manipulated with the MIDI data coming from the MidiDancer. *MIDI* was no longer just an acronym for Musical Instrument Digital Interface or simply a word in the name of our device but now represented to me a pathway that would allow my gestures to control basically any media device.

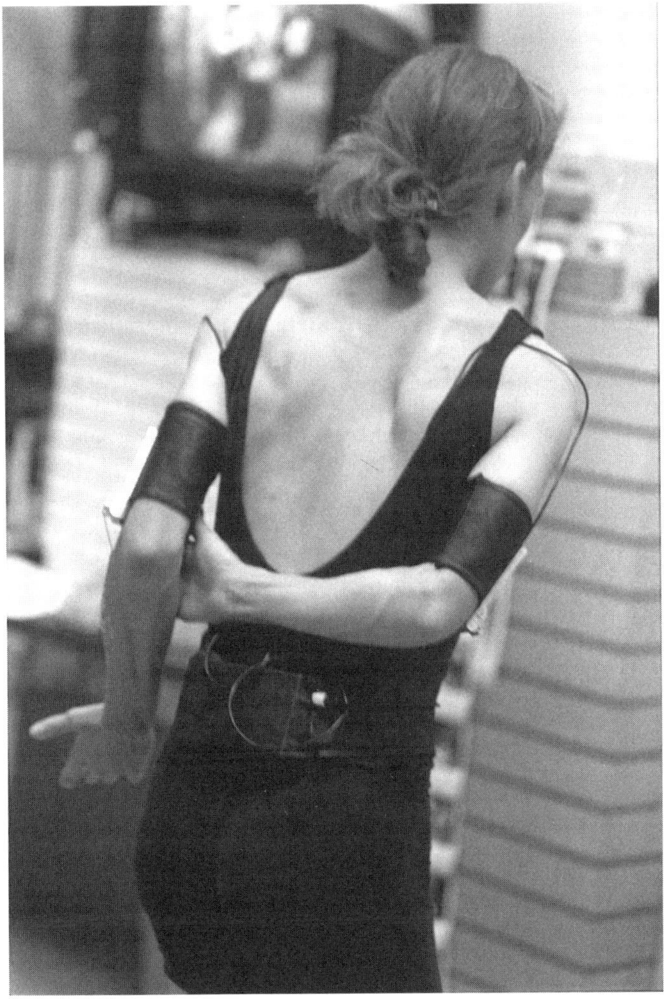

Figure 31.1

Dawn Stoppiello in *Tactile Diaries,* 1990. Performance with the original MidiDancer at The Electronic Café International. Photo by Steve Gunther.

My understanding of how extensive these pathways could become expanded further when I saw Steina during a lecture and demonstration at the Electronic Café some months after *Tactile Diaries.* I was watching Steina use her MIDI violin to "play" a computer-controlled laser disc that contained video images of water, fire, bubbling mud, and other natural environments. The MIDI information was used to randomly access specific frames on the disc, to play forward or backward at varying speeds or to freeze on a frame with no distortion of the image. The flexibility of the laser disc, as demonstrated by Steina, was extraordinary, and Mark and I were instantly taken by its possibilities.

Soon after this demonstration, these influences came together as we developed the initial plan for what would become *In Plane* (1994). Our idea was to make a video tape of my dancing, transfer it to laser disc, and then have me control the playback of that image using the MidiDancer. We wanted to create a duet between me and a "virtual" me stored on the laser disc. This duet was appealing because it emphasized something that was of growing importance to us—the duality between the fleshy body and another body that we didn't have a name for at the time but later came to call the electronic body. The corpus and its electronic Dopplegänger became characters that would find their way into several of our future works.

* (ASTERISK)

As I mentioned earlier, *In Plane* (figure 31.2) was a seminal work for us. It was not only our most technologically complicated piece but it became the cauldron in which we synthesized the theoretical paths that we had been on for the past four years.

The piece was to be a competition between the corpus and its electronic Dopplegänger, a body that bleeds, sweats, gets tired, and feels pain versus a body made of light that is not bound by time, space, or gravity. I became the fleshy presence, while my video image, stored on the laser disc, was my electronic counterpart. Which was the more powerful and beautiful presence? The flesh-and-blood woman exerting herself to an exquisite extreme with the potential of physical failure at any moment? Or the ethereal video body who flies so gracefully through space, can freeze in

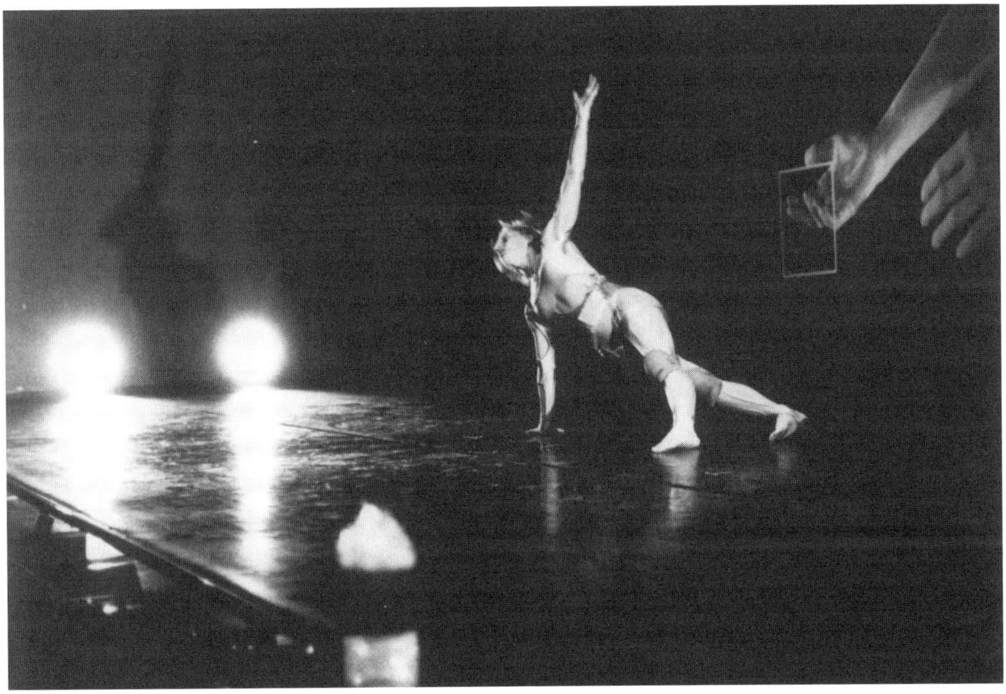

Figure 31.2
Dawn Stoppiello in *In Plane,* 1994. Performance with the newer MidiDancer at the Walker Art center. Photo courtesy of the Walker Art Center.

midair, and never tires? This was to be the essential question posed by the piece.

On a technical level, we wanted my gestures to control the musical score, the playback of images from a laser disc, the movement of a robotic video projector, and the theatrical lighting for the piece. We realized that this was ambitious, but we wanted to see how far we could go. We wanted to find out how much media one performer could play.

We began our work by collaborating on choreographic and musical materials that echoed the traits of the two bodies. The music, representing the electronic, was comprised solely of sampled sounds of machines, while the choreography, clearly representing the corporeal side of the equation, was constructed from a fundamentally human movement vocabulary consisting of running, jumping, falling, and rolling.

These movements were consistent with my stylistic leanings. I am not too concerned with taking a gesture through all of the compositional gymnastics required to expose its many possibilities for interpretation. Instead, I want to guide the audience through the energy of the movement itself. I want to see the relationship of the performers on stage. Of course, *In Plane* is a solo if you only count the number of fleshy bodies on stage. But it is actually an ensemble piece because I consider video, sound, robotic set pieces—whatever—simply to be additional performers. The beauty of using the MidiDancer system was that the notion of a duet with the video was much more than a conceptual idea and was in fact the result of a tangible physical relationship—body → sensor → video.

And, like dancing with a live performer, this was not a one-way street. During the process of creating and rehearsing *In Plane,* I became acutely aware of how information would flow back in the other direction. I would see the video move in response to my gestural control, and my dancing would be influenced by my "playing." Mark prefers the term *reactive* over *interactive* because he claims that it is more true to the actual flow of information, his point being that the computer does not have the intelligence of a human being and cannot interact in the truest sense of that word. As a performer who feels the feedback loop that I describe above, I feel certain that I am interacting with something, even if the modulation of image and sound originates solely with my own gestures.

In setting out to create these kinds of performative relationships, one thing was readily apparent: the radio-control car transmitter anc dual-sensor design were not going to allow us to make the piece we had in our minds. Mark created a new MidiDancer with a significantly smaller transmitter box and eight thin, flexible plastic sensors that could be placed at almost any joint on the body. When I first danced in this new costume, the difference in my movement was immediately obvious. It was less restricted and more fluid because the new design allowed it. We realized how much the sensory device imposes its own limitations on the choreography. Every instrument needs to be played in a particular way to get it to sound, and the MidiDancer was no different.

Traditional instruments respond to gestural input in a consistent way, and the audience can generally come to understand that relationship, even if the instrument is unfamiliar to them. Based on this traditional model, we felt a certain pull to establish a clear relationship between my movement and the media I was controlling. But we both remembered how stifling this fixed relationship was in the first MidiDancer performance we had given at CalArts. Further, this time, I wanted the choreography to serve my aesthetic intention first and the requirements of the sensory device second, something that was already easier to accomplish with the more sensitive MidiDancer. So we chose to allow the possibility of a joint changing its function during the course of the performance. For example, in the first section of the piece, the angle of my elbow directly controlled the volume of a rhythmic musical phrase. In the next section, that same elbow movement would trigger the playback of a video sequence. We chose to sacrifice the audience's clear understanding of the instrument in order to keep our expressive options open.

In the end, there were a myriad number of lessons learned as we made *In Plane*. Each day felt a bit like my first dance class, overwhelming because I was not yet familiar enough with the instrument to keep track of all of its parts. But perhaps the most important experience for us both came late in the creation process, when the elements had begun to coalesce. There was one rehearsal in particular in which I felt that the laser disc images weren't just some external object to which I was weakly linked via some sensory interface. Instead, they started to feel like a hand or a torso or some other part of my body. The medium wasn't separate from me any longer; it was an extension of me. This was curious, in one sense, since my video counterpart, with whom I was supposedly having a fierce competition, was actually under my control all the time. Perhaps this is the hidden message of *In Plane.*

= (EQUALS)

As a dancer, I inherently understand the realm of the body. I had no idea that technology would enter into that understanding until I chose to entwine myself with the machine. I was altered, and so was my body as it expanded to include sound, light, and image. The slashes in my art are inserted between my flesh, the media that moves with it, and the machine that locks the two together. And this puts me at the intersection of flesh and silicon, blood and television, body and computer that our culture is in the midst of splicing together.

III

Concluding Essays

32

Embodiment and Narrative Performance

Jaishree K. Odin

For several decades now, postmodern writers in general have used strategies of disruption and discontinuity to explore the problematics of language that shaped the literary discourse of the second half of the twentieth century. Their self-reflexive fiction embodied the destabilization and deconstruction of traditional concepts of truth, meaning, and knowledge as it spiraled inward into a linguistic labyrinth without any center, foundation, origin, or end. Contemporary narrative reflects yet another mutation as writers step back from their exclusive preoccupation with the bricks and mortar of their fictional worlds and focus once again on the architecture of their world making. In fact, in hypertext fiction, bricks and mortar themselves become a part of narrative making.

Hypertext fiction writers achieve fragmented or dispersed textual surface easily through an elaborate system of linking as well as through intermixing of media. Hypertextual linking fractures the textual surface, turning the otherwise continuous and linear narrative into a discontinuous assemblage of textual fragments that can be folded, unfolded, and refolded in a variety of ways. The resulting space, in Deleuzian terminology, is an intensive space characterized by forces rather than forms, directions rather than dimensions. The hypertextual breaks create temporal and spatial dislocations that mark the points of disruption as well as provide frames for alternative narratives whose relationship to the main narrative is parallel and extrinsic or embedded and intrinsic. The reader of the hypertextual narrative, then, must actively engage the text to discover the complex interrelations among the disrupted narrative threads with one another and in themselves.

In politically informed narratives, hypertextual strategies are indispensable for exploring the issues of embodiment and reinscription of cultural or literary history that are pivotal in the works of women and minority writers. For writers who write out of their own cultural experience, narrative making involves a continuous negotiation of difference with the dominant culture. While nobody would deny that the body is the stuff of concrete physical matter, the materiality of the body is another matter altogether. The materiality of the body that determines how we are oriented toward the world, both in material as well as psychological terms, is very much determined by our location. Katherine Hayles argues that the dominant culture provides abstract models that inscribe cultural practices but that it is in their enactment that incorporating practices materialize that enculturate the body. "Embodiment," she notes, "is akin to articulation in that it is inherently performative, subject to individual enactments, and therefore always to some extent improvisational."[1]

In a further elaboration of the inscribing and incorporating practices, the feminist philosopher Judith Butler argues that the materiality of the body is a construction that emerges out of a field of power that shapes its contours, marking it with sex and gender. Butler points out that we need to rethink the very meaning of construction and the grammatical structures that we use when we talk about construction. For her, it is "neither a single act nor a causal process initiated by a subject and culminating in a set of fixed effects. Construction not only takes place *in* time, but is itself a temporal process which operates through the reiteration of norms."[2] To describe the materiality of the body as a construction in Butler's theorizing, then, is not to resort to linguistic determinism or cultural constructivism. We take it for granted, she notes, that somebody—that is, a human subject or in more recent formulations, something, such as culture, discourse, or power—does the act of constructing. In the first, we resort to metaphysical claims, assuming there is a subject that exists prior to any sociocultural induction, and in the second, we forfeit the agency of the subject and replace it by a surrogate agent in the form of culture, discourse, or power. Rejecting both claims, Butler describes the materiality of the body arising in a matrix of power relations, so that the agency of the subject comes after and not prior to the materiality of the body, emerging through a process of enactment. "To claim that the subject is itself produced in and as a gendered matrix of relations," Butler notes, "is not to do away with the subject, but only to ask after the conditions of its emergence and operation."[3] By reformulating the very meaning of construction, locating it in time, and describing it in terms of a temporal process, Butler reveals the constructed nature of naturalized states of sex and gender. The performance of gender is thus a constant reiteration of the regulatory norms, and it is in the performance of these norms, that the materiality of the body emerges.

Through disrupting the metaphysical claims of the humanist subject as well as displacing the traditional subject-verb grammatical structure, Butler brings to light the importance of the cultural and discursive matrix that shapes subjectivity and agency, without turning them into absolute terms. At the borders where the dominant culture meets the minority cultures, the border subject emerges out of the perpetual encounter of the dominant regulatory norms and the minority experience. Whereas at the dominant site the replication is marked by performance of the same, at the border zone the repetition takes place with a difference. The performance of the same with difference challenges naturalized dominant social and cultural norms.

The perpetual negotiation of difference that the border subject engages in creates a new space that demands its own aesthetic. The new aesthetic represents the need to switch from the linear, univocal, closed, authoritative aesthetic involving passive encounters characterizing the repetition of the same, to that of a nonlinear, multivocal, open, nonhierarchical aesthetic involving active encounters that are marked by performance of the regulatory norms with and in difference. The interactive hypertextual aesthetic is most suited for representing border experiences since it allows both reinscription of cultural history as well as women's literary history in terms of narrative performance, which is empowering both for the writer who steps aside by creating a weave of texts and for readers who step in to weave their own text.

In hypertextual narratives, the meaning does not lie in tracing one narrative trajectory but rather in the relationship that various tracings forge with one other. Artists of electronic media use strategies of disruption and discontinuity to create visual and textual narratives that are multilinear and multivocal. Judy Malloy uses this strategy in *its name was Penelope* (1993) to reflect on questions of women's artistic creativity in both historical as well as contemporary contexts.[4] In a still more complex work, Shelley Jackson's *Patchwork Girl* (1995),[5] the splitting of the narrative into multiple threads through fragmentation becomes the structural as well as the thematic focus. These writers make an attempt to unearth the buried history of women even as they bring to the surface the constitution of women's subjectivity in a patriarchal society. In exploring subjectivity through this process of reinscription of history, they unravel crisscrossing threads of connections that have been historically buried under the homogenizing tendency of the dominant discourse.

Shelley Jackson's hypertext fiction *Patchwork Girl* is capable of countless mutations. The expressive representational surface that the electronic media provides becomes a part of her storytelling. A tree map, a story space map, a chart overview, an outline, and drawings of the female body all contribute to the reading experience. The tree map—a patchwork arrangement of brightly colored horizontal and vertical bars of varying lengths, shapes, and colors—becomes itself the focus of aesthetic attention. Underneath the electronic patchwork arrangement hides the textual patchwork that readers must sort out and reassemble to create a coherent whole. One of the lexias under the section "graveyard" tells readers that if they want to create a coherent whole out of the text, they will have to sew the pieces together.

Jackson's text can be seen as an encounter between Shelley Jackson, an aspiring writer attempting to find her voice, and Mary Shelley, the writer of *Frankenstein.* The title, *Patchwork Girl or A Modern Monster by Mary/Shelly and Herself,* alludes to this encounter as Jackson plays on the name Shelley that she has in common with her foremother. Jackson's confrontation with Mary Shelley reflects her anxiety of authorship: the Miltonic questions "who am I," "where am I" and "whence am I" that the monster in *Frankenstein* asks to understand his place in the world become the thematic focus of her text. In exploring her literary history, Jackson must confront the historical labeling of women's creativity as monstrous, which makes Mary Shelley refer to *Frankenstein* as "my hideous progeny."[6] Whereas Shelley's narrative ends with the exit of the male monster out the window onto an ice raft as he is "borne away by the waves, and lost in darkness and distance,"[7] Jackson's narrative begins with the rebirth and metamorphosis of the female monster, which is made into an empowering symbol of the female artistic subjectivity.

Feminist critics regard the narrative of the monster, a female in disguise, as the heart of Shelley's *Frankenstein.* Gilbert and Gubar see in it a mock rewriting of Milton's *Paradise Lost* and describe it as a narrative about the fall of woman into gender. Thus, "the monster's narrative is a philosophical meditation on what it means to be born without a 'soul' or a history, as well as an exploration of what it feels like to be a 'filthy mass that move[s] and talk[s],' a thing, an other, a creature of the second sex."[8] Jackson's text opens Shelley's text from within and rewrites its innermost core, which is constituted of the female voice struggling to understand both its rejection by its creator as well as its outcast status as the outsider and the other. Jackson subverts the inside-outside distinction of the male discourse, which turns difference into otherness that is assigned a secondary status. Her text exfoliates outward and makes difference and multiplicity the basis of identity and politics. As a result, whereas Shelley's monster feels alone, ugly, and disconnected from society as it is without a (his)story, Jackson's monster experiences her connection to (her)story and revels in her monstrosity, her multiplicity, and her difference. The representation of the female artistic subjectivity as monstrous becomes synonymous with the expression of difference that refuses to disguise itself or be suppressed by the male tradition.

Jackson's narrative, beginning where Shelley's ends, reveals through a fractured surface what Shelley's conceals in an elaborate system of framing and reframing. Shelley disguises the authorship of her text by framing the narrative with letters from Walton to his sister, followed by the narrative of Frankenstein and the innermost narrative of the monster. Jackson's text, on the other hand, contains multiple narrative folds that do not disguise the female creator of the text but rather reveal her connections to women in time and out of time. The drawing of the female body titled "her," the opening screen of *Patchwork Girl,* holds the disparate parts of the text together as an invisible patchwork of relations through a complex system of mirrored reflections as well as refractions. The female body serves as the doorway to the story space map that shows a very symmetrical structure in this very asymmetrical text (figure 32.1).

The map corresponds to the superimposed figure of the female body (mirror image of the reader) with which Jackson's text opens. The top of the map (head) contains the figure of the whole seamed female body. The middle part of the text (trunk) narrates the rebirth and metamorphosis of the female monster from Mary Shelley's *Frankenstein.* The bottom of the map (legs) "crazy quilt" rewrites the male monster's narrative from *Frankenstein* as he is given a new voice that emerges through the gaps created through juxtaposition of fragments from Mary Shelley's *Frankenstein* and Frank Baum's *Patchwork Girl of Oz.* The right side (the reader's left side) of the map is the modern monster's narrative, whereas the left side narrates the constitution of her body out of the fragments of the bodies of the women from the past. Both the body of the text and the text of the body come together in Shelley Jackson's conception of *Patchwork Girl.*

Aside from the structural correspondence of Jackson's text to the female body, the narrative can also be explored in terms of four folds. Each fold has two layers, the figure of the female body as the outside layer and the

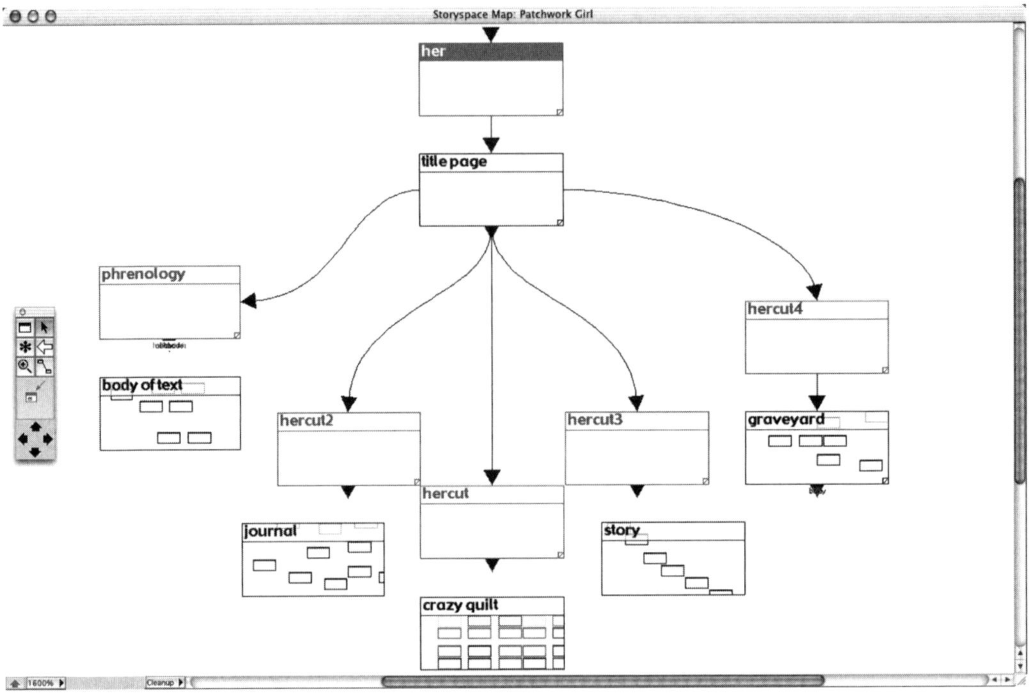

Figure 32.1
Shelley Jackson, *Patchwork Girl,* 1995, story space map. Eastgate Systems, Cambridge, MA.

textual narrative as the inner layer. The text of the body and the body of the text become interchangeable. Starting from the inside, the innermost fold is "hercut," with its textual narrative "crazy quilt." Encircling "hercut" are "hercut2/journal" and "hercut3/story," each constituting one-half of the second fold. Moving outward, the third fold is "phrenology/body of the text" and "hercut4/graveyard," again each making half of the fold. The three narrative folds are held together by the opening screen "her," which displays the figure of the seamed whole female body.

The outermost fold "her" and the innermost "hercut" are reverse images of each other. If "her" is the figure of the seamed whole female body, then "hercut" represents the fragmented female body, and the corresponding textual fragments under it called "crazy quilt" show how each continuous textual fragment is in fact an aggregate of many texts (revealed in "notes" attached to the "crazy quilt") that coalesce to give rise to a new text whose meaning does not lie in any single constituent but in emerging interrelations. The juxtaposed fragments in "crazy quilt" are primarily from Mary Shelley's *Frankenstein* and L. Frank Baum's *Patchwork Girl of Oz.* In Jackson's text, Baum's Patchwork Girl and the male monster from Shelley's *Frankenstein* become interchangeable. This section ends with a textual fragment "but I am glad," which not only restores the female voice to the male monster but also gives a lineage to Patchwork Girl. By juxtaposing different quotations from the two texts and by filling in the in-between spaces in each lexia, Jackson is able to subvert Patchwork Girl's freakishness and the monster's monstrous self-image. The attributes that were used to brand them as freaks or outcasts become in fact a source of their empowerment.

The journey from "her" to "hercut, " from the outermost to the innermost narrative, is not a simple jump. It involves going through the history of women's tradition that occurs in the second narrative fold "hercut2/journal" and "hercut3/story." If "hercut2" represents the female body/text that is unaware of itself as it emerges out of the dark background of patriarchal discourse, then "hercut3" is its reverse image in which the dark background is illuminated so as to make the female body/text stand out in its own light. To avoid following the dualistic logic of the male discourse by making "hercut3," a polar opposite of "hercut2," an end in itself, the mediating drawing "chimera," strategically placed as the core of "hercut3" in the tree map (and as a subheading of "hercut3" in the chart view and the outline), intervenes as it introduces the concept of "hybridity" in any rethinking of the text of the body or the body of the text.

Chimera, the fire-breathing she-monster in Greek mythology, possesses the head of a lion, the body of a goat, and the tail of a serpent. Chimera is also the name given to an individual or an organ that is constituted of diverse genetic material, especially at a graft site that marks the joining of tissues from two different genetic sources. The use of "cutting," "grafting," and "joining," recurrent throughout the text, suggests that unlike the continuous dominant male tradition, women's tradition, when seen in the historical context, has been discontinuous, assuming different guises and forms, the connection to which can be realized by the woman writer or artist only by a deliberate act of "grafting" of recollected lives. This recollection is not a search for a direct line of descent but rather an unraveling of a patchwork of connections that have been the fabric of the lives of women, both literary and nonliterary, throughout the centuries. Then, again, chimera is also a mental construction that does not have a basis in the real world; it is the stuff of dreams and myths. Literary monsters and hybrids might not be real; they can, however, serve as dream symbols for empowering women's lives in the material world. The multiple refractions that emanate from "hercut2," "hercut3," and "chimera" thus set the stage for the unfolding of the story of the rebirth and metamorphosis of the female monster that exists in time while at the same time marks the temporal process itself of unfolding female creativity.

The fragments under "journal" narrate Mary Shelley's encounter and parting with the monster that she creates. The journal entries end with remorse on her part for being unable to give a part of herself to the monster so that she might have continued through her. The narrative thread in "journal" suggests thus how difficult it was for women in Shelley's time to write as women. Women could write only through suppressing their voice or disguising it as a male voice, which, however, was inadequate to describe their own experiences.

What is dark and left unsaid in "hercut2/journal" becomes light and achieves voice in the mirror figure "hercut3" and the corresponding narrative titled "story." The "story" is then a continuation of the story of the female monster in Shelley's narrative. The first part of this section is titled "M/S" (which could stand for monster/Shelley, Mary/Shelley, Me/

She)—again, a play on the name that Shelley Jackson shares with Mary Shelley. It includes long excerpts from Mary Shelley's *Frankenstein*—the monster's plea to Dr. Frankenstein to create a companion for him, Frankenstein's promise to do so, and then his treacherous decision to destroy the half-finished female monster and disperse her remains in the ocean. The female monster, destroyed in Shelley's text, comes alive in Jackson's text and reflects on her destiny: "I told her to abort me, raze me from her book; I did not want what he wanted. I laughed when my parts lay scattered on the floor, scattered as the bodies from which I had sprung, discontinuous as I myself rejoice to be. I forge my own links, I am building my own monstrous chain, and as time goes on, perhaps it will begin to resemble, rather, a web."[9]

The resurrected female monster encounters Mary Shelley once again, but this time they have a fruitful exchange. A mutual cut and an exchange of words as well as skin takes place before the final parting. This joining with Shelley is experienced by the monster as a "live scar" that marks a parting that, at the same time, "commemorates a joining."[10] The female monster's metamorphosis goes through several stages (severance, seafaring, seance, falling apart, and rethinking) involving a journey across the sea in disguise, encountering a woman in male disguise, forgetting her past history, hearing ghost voices, remembering her scars, leaving behind the male-defined "feminine identity," and rethinking nomadic identity. At each stage of the metamorphosis, there are doubles, hybrids, and other monsters that are multiple reflections of the female monster.

Marking "phrenology/body of the text" and "hercut4/graveyard" as the third fold, the narrative moves forward and outward. In "phrenology," the exposed head reveals different areas of the brain responsible for different memories or thoughts out of which emerges the coherent "body of the text." Similarly "hercut4" displays the fragmented body out of which arises the body of the modern female monster as revealed in the textual segments "graveyard." The left and the right sides of the third narrative fold that describe the constitution of the body of the modern monster and the modern monster's narrative set in motion another set of relations, which mark the moments of dispersal and reintegration or deterritorialization and reterritoralization that accompany the constitution of female artistic subjectivity as it continually negotiates difference with the dominant tradition.

In an interesting narrative mutation, the section "graveyard" that appears under "hercut4" in the story space map is included under "hercut3" in the chart view, the tree map, and the outline. This suggests that the assembling of the body of the female monster is common to the second fold that narrates the rebirth of Shelley's female monster as well as the third fold that includes the modern monster's narrative. The common birth brings the two narratives together, turning the modern monster into a symbol of the female subjectivity that emerges out of the fragments of the discontinuous women's tradition.

In "graveyard," the monster names her body parts, attributing each to a different woman from the past. The body parts are thus linked to stories of ordinary women who had no opportunity to become extraordinary as their creative spirit wasted away under the burden of social conditions that shaped their lives. The body of the modern monster is thus composed of the clear and calm eyes of the unknown village historian Tituba, the lips of laughing Margaret, the tongue of talkative Susannah, the sharp nose of Geneva, the promising ears of Flora, the trunk of dancing Angela, the finger of nameless scholar Livia, the right leg of unwed and not-yet-crazy Jennifer, the left leg of adventure-starved Jane, the heart of Agatha, the strong foot of Bronwyn, and so on. The body of the modern monster also contains the liver of a man named Roderick. The playful insertion of a male story in a string of women's stories is meant to draw the reader's attention to the fact that the focus in Jackson's text is not to create an essentialist category called "woman" but rather to lay bare the historical shaping of the female subjectivity.

Whereas the body of the modern monster is primarily made out of the fragments from the bodies of women from the past, her narrative in "body of the text," occupying both the periphery and the center of the text at the same time, reflects on the multiplicity of the female subjectivity as well as hybrid nature of all texts. The modern monster describes herself as a double agent who is both whole and dispersed. She incorporates within herself personalities or memories of fictional, real, or disguised women. The women of the past "draw together, bound by a hidden figure that traverses them all."[11] Writing in between the lines of the male discourse, she describes herself as a whole with "haze around the edges." She urges the reader to "come closer, come even closer: if you touch [her], your flesh is mixed with [hers], and if you pull away, you may take some of [her] with you and leave a token behind."[12] Reinscribing the female subjectivity becomes "a matter of redrawing an outline. Snaking through the space between two lives to wrap a line around some third figure."[13] The third figure emerges as the present and the past, now and then, and the dominant and the suppressed traditions come together in the interaction of the reader-writer and the text. Instead of binary opposites becoming subsumed into a unitary synthesis in search of the same masculine model of subjectivity, the encounter is marked by the emergence of a new model of subjectivity that exists in difference, without conforming to any essentialist descriptions of one or the other.

Without making any claims to originality, Jackson's text is then a continuation of Mary Shelley's story, which is every woman's story both literally and metaphorically. Jackson turns Shelley's text inside out by making what is invisible, voiceless, and undefined in the latter into what is visible, voiced, and definable in its indefiniteness. By giving voice to Shelley's monsters, she reinscribes her own monstrous artistic self that finds its continuity with her foremother even as it goes beyond her in self-understanding.

Jackson achieves fragmentation to open in-between spaces by using a patchwork as the thematic as well as the structural principle of her text. As Deleuze and Guattari point out, a patchwork represents a smooth space that has no center because "its basic motif 'block' is composed of a single element; the recurrence of this element frees uniquely rhythmic values distinct from the harmonies of embroidery (in particular in 'crazy' patchwork, which fits together pieces of varying size, shape, and color and plays on the *texture* of the fabrics)."[14] Through its "amorphous collection of juxtaposed pieces that can be joined together in an infinite number of ways, we see that patchwork is literally a Riemannian space or vice versa. The smooth space of patchwork is adequate to demonstrate that 'smooth' space does not mean homogenous, quite the contrary; it is an amorphous, nonformal space prefiguring op art."[15]

In Jackson's text, the textual patchwork creates just this smooth space where individual textual units achieve significance both in themselves and in relation to one another. The visual and textual units become the blocks of the electronic patchwork, even as the thematic focus itself revolves around the exploration of the patchwork subjectivity. Another refraction of the patchwork connects it to the women's tradition that originated in needle and thread. The meaning of the text emerges from relationships that are amorphous, coming into play and dissolving as readers thread their way through the fragmented textual landscape. Readers of the text have an experience very similar to that of the narrator or text ("body of the text"), who/which describes herself/itself as "a discontinuous trace, a dotted line" that is "a potential line, an indication of the way out of two dimensions. . . . Because it is a potential line, it folds/unfolds the imagination in one move. . . . A dotted line demonstrates; even what is discontinuous and in pieces can blaze a trail."[16] In reading Jackson's text, narrative trajectories create a fabric of interrelations that produce a continuous variation of form and meaning in both texts, the body of the text as well as the text of the (female) body. The creation of gaps and in-between spaces allows for an interaction that is liberating both for the writer who must struggle to find her voice as a woman and for the reader who must genuinely engage her text.

As shown above, Jackson's *Patchwork Girl* opens up the patriarchal cultural text from within, showing how it is comprised of heterogeneous discourses that speak through the gaps and the interstitial spaces. Through recursive narration that unfolds as a series of reinscriptions, Jackson reveals the complex discursive matrix that has shaped women's subjectivity for centuries. The electronic media allow her to create a dynamic text that becomes a theater of interaction and enaction that joins the past and the future in the present moment of the reader's experience. Jackson's text, like other hypertext narratives, is performative in nature and writes itself as it is read.

Hypertextual aesthetic is rooted in active and interactive reading, like oral storytelling. Multilinear narratives can be regarded as a return to oral storytelling, which Walter Benjamin reminds us "permits that slow piling one on top of the other of thin, transparent layers which constitutes the most appropriate picture of the way in which the perfect narrative is revealed through the layers of a variety of retellings."[17] Benjamin's lamentation about the death of storytelling in the age of information finds its apotheosis in the birth of hypertext narratives, which continue the tradition of oral storytelling once again.

NOTES

1. N. Katherine Hayles, "The Materiality of Informatics," *Configurations* 1, no. 1 (Winter 1993): 156.

2. Judith Butler, *Bodies That Matter* (London: Routledge, 1993), 10.

3. Ibid., 7.

4. Judy Malloy, *its name was Penelope* (Cambridge, MA: Eastgate Systems, 1993).

5. Shelley Jackson, *Patchwork Girl* (Cambridge, MA: Eastgate Systems, 1995).

6. Mary Shelley, "Introduction," *Mary Shelley Frankenstein,* ed. Johanna M. Smith (Boston: St. Martin's Press, 1992), 23.

7. Mary Wollstonecraft Shelley, *Frankenstein or, The Modern Prometheus* (Los Angeles: University of California Press, 1984), 237.

8. Sandra M. Gilbert and Susan Gubar, *The Madwoman in the Attic: The Woman Writer and the Nineteenth-Century Literary Imagination* (New Haven: Yale University Press, 1984), 235.

9. Jackson, *Patchwork Girl.*

10. Ibid.

11. Ibid.

12. Ibid.

13. Ibid.

14. Gilles Deleuze and Felix Guatarri, *A Thousand Plateaus: Capitalism and Schizophrenia,* trans. Brian Massumi (Minneapolis: University of Minnesota Press, 1994), 476.

15. Ibid., 477.

16. Jackson, *Patchwork Girl.*

17. Walter Benjamin, *Illuminations: Essays and Reflections,* ed. Hannah Arendt (New York: Schocken Books, 1968), 93.

33

Brazilian Counterparts: Old Histories and New Designs

Simone Osthoff

> The crew would respect me more when I wore slacks than when I dressed in skirts. I think they felt diminished by being directed by a woman. Therefore, I had to wear pants. It was very unpleasant. Not the slacks, naturally, but the fact that I had to look like a man to be able to command.
>
> —GILDA DE ABREU, DIRECTOR OF THE FEATURE FILM *O ÉBRIO* (THE DRUNKARD) (1946) STARRING HER HUSBAND, POPULAR SINGER VICENTE CELESTINO

> From the start I decided that I was not going to become a man in the workplace, that I was not going to speak or act like a man. I was going to continue to use mini skirts, and that everything was going to function from this premise. It worked surprisingly well, and with time I have learned how to speak in the workplace in a more personal way.
>
> —SANDRA KOGUT, VIDEO, FILM, AND TV DIRECTOR (1995)

Brazilian women artists, defying popular media images—of tanned sensual Latinas in minuscule bikinis along tropical beaches—have often been more influential locally than their European and North American counterparts. Modernism in the visual arts, for instance, was introduced in São Paulo by two women: Anita Malfati's 1917 exhibit marks the beginning of modernism, and Tarsila do Amaral's anthropophagic paintings from the 1920s, in dialog with literature, synthesized the ideals of cultural cannibalism for decades to come. Notwithstanding inequities along gender, class, and racial lines that persisted throughout the twentieth century, women artists in Brazil have managed to continuously expand their influence nationally and internationally, despite their general distaste for organized feminist movements. The dynamics of gender politics in Brazil, as in the rest of Latin America, is more subtle and less confrontational than in the United States. It is rooted in a Catholic heritage that syncretized with different indigenous and African traditions and that contrasts sharply with the North American Protestant experience. And yet the attention to the politics of gender, race, and class that is prevalent in the United States is not uncommon in Brazilian academic circles.[1] Among women artists in Brazil, however, the general consensus is to avoid the "ghetto" of feminine aesthetics. Iole de Freitas voiced a common sentiment: "I do not know why no one searches for what is masculine in a man's work."[2] Women artists in Brazil have rather preferred to focus on less divisive interests such as the second-citizen status that Latin American artists, unfortunately, still have in the international art scene.[3] Lygia Pape expresses this indignation: "I find it alarming that today an exhibit can still be titled *Latin American Art.* It is discriminatory as well as very reductive!"[4]

The history of art and technology in Brazil began to be written only recently, through the examination of works that have been relegated to oblivion and whose value, in most cases, is only now becoming apparent—regardless of gender.[5] Among the pioneers are three radical visionaries still waiting to receive further historical recognition: Jocy de Oliveira, Sulamita Mairenes, and Tereza Simões.

Composer, pianist, and author Jocy de Oliveira has worked since the early 1960s in multimedia performances merging electroacoustic music, theater, text, and images.[6] Also working in the 1960s but with a less documented trajectory is São Paulo artist Sulamita Mairenes. By the late 1960s, she was addressing psychological issues in electronic and optical media. In the 1967 Ninth International São Paulo Biennial, Sulamita presented electronic *Parapsychological Objects,* and since 1980 she has worked with holograms.[7]

Light art is one of the sources of electronic art not sufficiently explored in Brazil. Credit for the earliest attempt to use neon as a principal material of a sculpture in the country is usually given to Tereza Simões, who encountered much resistance among critics, curators, and collectors in the early 1970s. Another original work from the early 1970s was developed by Analivia Cordeiro. Seeking the interaction of dance with computers and video she developed a software (Nota-Ana) that codified the movements of body and dance into legible signs to facilitate communication between choreographers and dancers.[8] Experiments with film and video, within the visual arts in the early 1970s in Brazil became known as "almost cinema" in reference to the dialog between the visual arts and the cinematic languages.[9] A product of the 1960s' emphasis on process over the creation of finite works of art, these early audiovisual forms employed Super-8, 16 millimeter, and video formats to explore and document a number of experimental and conceptual actions, performances, and happenings.

Seeking to create an awareness of the duration of time and immediacy in the visual arts, the 1970s avant-garde artists added to those experimental concerns critical attitudes toward mainstream film. Among others, we have in Rio de Janeiro Lygia Pape, Anna Bella Geiger, Iole de Freitas, and Regina Vater (based in the United States since the early 1970s), all artists who currently are employing new media on occasion; in São Paulo, Regina Silveira was very influential. She works today with large installations of anamorphic perspectives and shadow projections of objects that reach urban scale.

The explosion of independent video makers in Brazil since the 1980s inaugurated a second generation of artists interested in the medium.[10] Sandra Kogut is one of the exponents of this group that brings a new sensibility to electronic media. One of the characteristics of second-generation video art was the subversion of the language of TV from the inside. Kogut started to work with video in Rio in the early 1980s and achieved international recognition with her 1991 *Parabolic People,* a 41-foot collage of footage taken in the streets of Dakar, Moscow, New York, Tokyo, Paris, and Rio de Janeiro. Her working process reflects a way of thinking in layers, which *Parabolic People*'s superb editing successfully explored.

Kogut's first influences came from Japanese artists in Tokyo, references very far removed from her immediate Rio de Janeiro environment. Despite the distance and the technical challenges, she was determined to make things work in her favor: "I had decided early on, in the first place, to think positively, and in second place, to bring these questions, in some way, into my work. If there is a problem, then it must become a subject. And I did this with all the limitations I encounter, even to understand what it was like to live in Rio and who I was. The work was the path to understanding all this. *What do you think people think Brazil is?* (1990, 5 minutes, 30 seconds) was very much just this. How do other people view us, for us to start seeing who we are."[11]

For this generation of Brazilian independent video makers, television was a primary reference, contrary to previous avant-garde artists who saw television—structured by advertising—either as superfluous or as the embodiment of capitalist manipulation and oppression. Kogut's view on TV was shared by many young artists: "If you do something in front of the camera that you would not ordinarily do, you are not being any less yourself. I wanted people to use the camera as I used people. I'm not pretending that I'm heading toward a naturalism that doesn't exist. How much TV do you have in yourself? We are partially made of TV images."

Sherry Turkle in her 1995 book *Life on the Screen: Identity in the Age of the Internet* addressed the issue in relation to cyberspace.[12] Turkle uses the metaphors of cycling through windows on the computer screen to talk about the change in notions from a singular ego to that of multiple distributed identities of citizens of online communities. Radicalizing the fragmentary language of TV and responding to the fast process of globalization, Kogut charged her work with an international flavor. After her experience with public video cabins in the streets in Rio de Janeiro—*Videocabines são Caixas Pretas* (1990), she expanded the project to other cities in the world:

> I wanted to make a work that didn't belong to any country. It would be spoken in many languages. It would not be dubbed; it would not have subtitles in any language. My idea was to do a video just of people speaking all the time, and depending on the country in which it took place the people would see things differently, in many levels. *Parabolic People* is a totally hybrid work, and the subject is identity.

Kogut's fluid notions of identity are, according to her, a Brazilian trademark. She was born in Rio de Janeiro in 1965 and studied in a French school there. She later abandoned her pursuit of a degree in philosophy to work with video. Today she lives and works between Paris and Rio directing movies and innovative TV programs. She is at home in at least three languages—Portuguese, French, and English. After the release of *Parabolic People,* she was asked on a live TV program in Germany about her Brazilian identity. Why wasn't she addressing Brazilian issues such as misery and hunger—images Europeans were familiar with through the Cinema Novo aesthetic? She answered that this apparent lack of nationality and roots—this freedom to mix up everything—was super-Brazilian, she always thought she was being Brazilian, and it is for this very reason that she is Brazilian.

Kogut's aggressive syncretism and ability to work with highly contrasted elements echo the methods of visual artist Hélio Oiticica and the popular music of the tropicalist movement, which drew on a rich diversity of styles. "The key to understanding the movement is syncretism,"[13] explained composer Caetano Veloso, updating a strategy that has been at the center of Brazilian culture since the anthropophagic manifesto of 1928 endorsed a cannibalist cultural strategy—the devouring of international trends while remaining rooted solidly at home—thus transforming the indigenous "barbarian" ritual into an instructive form for modern behavior. In forging a global identity, anthropophagy pays homage to the oral traditions at the roots of Brazilian culture. This cannibalist metaphor continues to be relevant as cultural strategy for a dialogue with the international artistic scene being the theme for the 1998 twenty-fourth International São Paulo Biennial.

In Brazil, video also became part of a number of multimedia performances and installations during the 1990s. Since 1994, multimedia artist Simone Michelin has explored more closely the language of video in installations both within the gallery space and public environments. She juxtaposes private life and public spheres, global and local relations, appropriated media images and personal narratives—usually with herself as subject and often played in reference to art history icons. Such is the case of her 1994 video performance *1′22″ of Glory: Mondrian, Malevitch, Michelin* and also of her video installation *The Bride Descending the Staircase* at Tyler Gallery in Philadelphia (1998). Michelin was born in 1956 in Bento Gonçalves, Rio Grande do Sul, and has been based in Rio de Janeiro for the past fifteen years. She teaches at the Federal University of Rio de Janeiro and at the School of Visual Arts (EAV) and has been in a two-year residency program in New York. Michelin's work is rooted in the experimentalism of the groups Nervo Otico and the cooperative new media group Espaço N.O. from the city of Porto Alegre, capital of the southernmost state of Brazil in the late 1970s.

Michelin's December 1997 multimedia intervention *Polisensory Zone. N.1* was created at the entrance hall of the dean's building at the Federal University of Rio de Janeiro (UFRJ), which also houses the School of Fine Arts and the School of Architecture. This heavily trafficked area was occupied by a broken nineteenth-century bronze statue of a person that was placed on a small four-wheel cart on the floor surrounded by seven TV monitors set at 45-degree angles on the ground. The statue, an allegory of progress, was originally standing in front of the Central do Brasil, the first train station in Rio de Janeiro. Amid electric cables, wooden stools, and campus employees, surprised students crossed the area avoiding the monitors as though in an obstacle course. Three video channels showed images of the university storerooms full of abandoned broken furniture, fragments of Michelin's master thesis dissertation, and commemorative medals featuring Brazilian presidents. With this intervention in the university routine flux, juxtaposing and displacing images, objects, and people, Michelin offered an allegorical reflection about art's place within this federal institution.

An artist, teacher, editor, and veteran guest curator for many international electronic art symposiums and festivals, Diana Domingues has been widely exhibited both nationally and internationally, collaborating often with specialists in various areas. Best known for her interactive environments, Domingues developed a reputation in Brazil outside the Rio de Janeiro–São Paulo corridor. Her 1995 interactive exhibit *Trans-e,*[14] for instance, was made of four large environments where aspects of the interior of the body are revealed in a fluid, pulsating flux, as opposed to the static images of organs in anatomy books. Domingues employed various technologies that interacted with the physical presence of the public: video cameras register the presence of the visitor; amplifiers echoes their voices; liquids move and drip because of the movement of bodies, which registered by infrared sensors give machines organic functions. This journey through the interior of the body also involves images taken with micro cameras. Domingues started to explore new media in the late 1970s, as part of the experimental artist's collective Espaço N.O. in the city of Porto Alegre, the capital of the southernmost state of Brazil. Domingues is a multimedia artist holding a Ph.D. in communication and semiotics from PUC/São Paulo. She teaches at Caxias do Sul University (U.C.S.), where she also coordinates an interinstitutional graduate program in communication and semiotics at the U.C.S. and the PUC/São Paulo. She also coordinates the research group Novas Tecnologias nas Artes Visuais.

Contrasting with the play between presence and absence, physical and virtual spaces in the performances and installations of the artists Simone Michelin and Diana Domingues are the works of Tania Fraga, Rejane Spitz, and Giselle Beiguelman, which focus on virtual worlds and designs.

Creating virtual fields where three-dimensional objects exist, Tania Fraga's virtual worlds are made for the computer screen. Free from physical constraints, Fraga's interactive stereoscopic simulations challenge viewers' perceptions with new forms of geometric performances that are at once mathematically logical and random, constructive yet surprisingly sensual. Interested in the unexplored possibilities of three-dimensional forms in virtual space—free of gravity and of the laws of physics—the artist challenges our "natural" visual perception, reminding viewers that vision is also ideologically and culturally constructed. Our way of representing three-dimensional forms on two-dimensional space is based on notions of depth inherited from Renaissance perspective and reinforced by the lenses of photographic cameras, cinema, and video. It is, of course, one among many ways of representing depth. Fraga's interactive stereoscopic simulations amplify the user's perception of virtual spaces with their intrinsic logic of movement and unexpected behavior. Fraga, who holds a Ph.D. in semiotics from the University of São Paulo and teaches at the University of Brasília, explains: "The 3D worlds are constructed as complex spaces always disclosing unusual perspectives where emptiness and silence have a place. Therefore, the worlds are characterized by their lightness, their multiplicity, and their openness. We move through them, then beyond them, and perceive their depth. We feel there is a place beyond the image that we cannot clearly see. There is no systematic way to manipulate the objects. There is only a set of predefined options to choose from. At the same time, interior and exterior are part of the same continuum that characterizes the virtual space. The meanings do not come from the outside world, but they emerge in a synchronic process in the viewer's mind when interacting within the simulated world."[15]

We tend to ignore the cultural loss involved in technological progress. The destruction behind creation is in part the subject of Rejane Spitz's *Private Domain (please, keep off!)*. Spitz—an artist, curator, art director, one of the digital pioneers in graphic design in Brazil, a consultant for electronic art—is an acute translator between different cultures and between analogic and digital modes of thinking. She started to work with computers in London, where she received an M.A. from the Central School of Art and Design in 1983. She has developed a number of innovative navigational design projects for the Web and CD-ROM. She holds a Ph.D. in education from PUC–Rio (Catholic University) (1993) and is the graduate coordinator of the art department and the electronic art nucleus, also at PUC-Rio. She is a frequent participant in national and international electronic festivals and symposia (Spitz is the South American representative for SIGGRAPH and a member of the editorial committee of *Leonardo*).

Her concern with the sometimes difficult communication between global and local cultures, and especially between digital and oral traditions, inspired Spitz to create her most personal work: *Private Domain (please, keep off!)* (⟨http://omnibus-eye.rtvf.nwu.edu/Homestead/rspitz/desc.html⟩). This Web site is structured around the relationships between automatic teller machines and the colloquial ways of communication in the northeast of Brazil. The avatars Spitz created for the different characters of her virtual domain are the hand-made clay dolls produced in the northeast region. Each avatar has its social position described by poetry and songs and engages in a different type of dialog with the ATMs. The exchanges with the computer terminals reflect the hierarchical social structures of the region with all the subtleties, flavor, and humor of the local language. The gap between this private domain and the objective language of computers is made tangible by the artist, who pays homage to the beauty of human relations, calling attention to their richness and to our loss: "This work is about those empty hands that are on the fringe of the Web. It is about those words that cannot be translated, those emotions that cannot be shared, and those meanings that cannot be understood by people from other cultures. It is about the richness of human beings living in their different realities, with their own system of ideas, concepts, rules, and meanings."[16]

Also working with Web design is Giselle Beiguelman. Born in São Paulo in 1962, she is based there working as a multimedia essayist and Web artist. Beiguelman holds a Ph.D. in history and was a fellow of the VITAE Foundation. She is the author of various works on contemporary history and a member of the Tenth ISEA International Program Committee and has been the Web editor of *Arte/Cidade* (a nonprofit organization devoted to the arts and urbanism) since 1996. She has worked as Web editor for commercial Internet providers and since 1998 has run ⟨www.desvirtual.com⟩, an editorial studio and her cyberbunker, where she bases her personal projects (like *The Book after the Book,* a hypertextual and visual essay where criticism and Web art melt into the context of the Net's reading and writing condition).

New media artists in Brazil today, regardless of gender, face an old dilemma common in postcolonial societies: on the one hand, artists employing the latest technologies need to be close to international markets and venues and familiar with the Western conceptual artistic legacy; on the other hand, they are required to provide some local and "original flavor" (not necessarily alternative visions) to markets tired of the homogenizing effects of globalization and hungry for the exotic.

Since the beginning of modernism, "Europeans borrowed from other cultures, particularly from African art, but were uninterested in what those cultures might in turn borrow from themselves."[17] As a European export, modernism remained a one-way street. Technology followed old colonial routes, becoming in Brazil, at times, a fetish, not unlike African masks for the beginning of cubism. The answer to the polarized attitudes of xenophobia and xenophilia—mystifying either the national "original culture" or finding originality only in foreign products—continues to be cultural cannibalism. This old strategy—employed successfully by Brazilian avant-garde artists throughout the twentieth century to subvert binary neocolonial hierarchies such as original/copy, nature/culture, local/global, human/technological, female/male—is still a viable strategy for artists today.

The ability to make connections and negotiate among heterogeneous histories and practices—public and private, local and global, visible and invisible—does not start with new media, of course. But the quality of the future, from the biophysical to social ecologies, will certainly depend on the ethical implications of present technological and aesthetic choices, enhancing or erasing the potential for personal and collective agency and action.

NOTES

1. Brazilian anthropologist and political scientist Luiz E. Soares has recently called attention to the Brazilian resistance to feminist ideas and theory. Addressing an audience of anthropologists, Soares asked for the exorcization of the skeletons in the closets of master narratives focusing on apparent more trivial and micropolitical issues to arrive at larger political ones. Soares sees feminist theory as posing some fundamental questions about the author and the relations between subjects and their discourse, as opposed to the emphasis on the relation between discourse and their objects. See "Consequencias de uma Antropologia Trivial," paper presented at the ABA (Brazilian Association of Anthropologists), Conference, Vitoria, Brazil, 8 April 1998. Also Luiz E. Soares, "Political Correctness: The Civilizing Process Is Under Way)," paper presented at the International Conference on

Analytic Philosophy and Pragmatism, Federal University of Minas Gerais, Belo Horizonte, 6–8 August 1997. Here Soares works with the data from a Pesquisa Nacional por Amostra de Domicílios (PNAD) 1988 census stating that the percentage of illiterates in the Brazilian population over five years old was 18 percent among the white population and 36.3 percent among the mulatto and black population. In 1988, salaries were shockingly unequal (and continue to be): men's salaries were more than double women's salaries; white men's salaries were more than double the salaries of mulattos and blacks; white women's salaries were more than double the salaries of mulattas and blacks. Therefore, white men earned more than three times the amount that black women earned. Carlos Hasenbalg and Nelson Silva, *Relações Raciais no Brasil Contemporâneo* (Rio de Janeiro: Rio Fundo Editora and IUPERJ, 1992).

2. Iole de Freitas, interview with Ana Maria Machado, September 1987, quoted in *Quase Catálogo 2: Artistas Plásticas no Rio de Janeiro 1975–1985* (Rio de Janeiro: CIEC, 1991).

3. I believe the resistance of Brazilian women artists to discuss gender issues can be at least partially explained by economic and social structures in the country. The overlapping of professional work and housework has been a reality in Brazil for poor women. Middle-class professional women have been able to rely on the institution of the maid, which although undergoing many changes, is still pervasive. Female domestic servants live at the poverty level and have jobs that require them to live on site and work six to seven days a week at extremely long hours. This complex relation between gender and class interests may play some part in explaining why, for many Brazilian women, feminism is seen as a North American product that does not appeal to local men and women.

4. Lygia Pape, interview with Lúcia Carneiro and Ileana Pradilla, *Lygia Pape* (Rio de Janeiro: Nova Aguilar, 1998), 60.

5. The pioneer project being edited by Eduardo Kac is being published by *Leonardo:* "A Radical Intervention: The Brazilian Contribution to the International Movement of Electronic Art." The articles in this series have appeared regularly in *Leonardo* starting with volume 29, number 2. They can also be found on the Web at ⟨http://mitpress.mit.edu/e-journals/Leonardo/isast/spec.projects/brazil.html⟩. In Brazil, with the exception of a few isolated articles by Mario Pedrosa, Walter Zanini among others, the first serious theoretical examinations of electronic art were Arlindo Machado, *Máquina e Imaginário* (São Paulo: Edusp, 1993) and Machado, *A Arte do Vídeo* (São Paulo: Brasiliense, 1988). For a concise account of the history of art and technology in Brazil, see also Walter Zanini, "Primeiros Tempos da Arte/Tecnologia no Brasil," in *A Arte no Século XXI: a humanização das tecnologieas,* ed. Diana Domingues (São Paulo: Fundação Editora da UNESP, 1997), 233–242.

6. Jocy de Oliveira, *Days and Routes through Maps and Scores* (Rio de Janeiro: Record, 1984); de Oliveira, *Apague meu Spotlight* (São Paulo: Massao Ohno Editora, 1961); de Oliveira, *3º Mundo* (São Paulo: Melhoramentos).

7. Frederico Morais, "Oráculos e Holograms no Teatro de objetos de Sulamita Mairenes," *O Globo,* May 23, 1984.

8. Analívia Cordeiro, "Nota-Ana: uma notação-trajetória dos movimentos do corpo humano," thesis, Universidade Estadual de Campinas, Campinas, SP, 1996.

9. *Almost cinema* (quase cinema) is an expression created by Hélio Oiticica in reference to his experiments with sound and moving images. Today it refers to the whole 1970s and 1980s experimental film production in the visual arts. See Ligia Canongia, *Quase Cinema* Caderno de Textos no. 2 (Rio de Janeiro: Funarte, 1981); Water Zanini, "Primeiros Tempos da Art/Tecnologia no Brasil," in *A Arte no Século XXI: a humanização das tecnologieas,* ed. Diana Domingues (São Paulo: Fundação Editora da UNESP, 1997), 232–246.

10. See Arlindo Machado, "Video Art: The Brazilian Adventure," *Leonardo* 29, no. 3 (1996), and also Arlindo Machado, "A experiencia do Video no Brasil," in *Máquina e Imaginário* (São Paulo: EDUSP, 1993).

11. Interview with the author, Rio de Janeiro, 10 August 1995 (translation mine).

12. Sherry Turkle, *Life on the Screen: Identity in the Age of the Internet* (New York: Simon and Schuster, 1995).

13. For an insightful account of the Tropicalist movement by one of its creators, see Caetano Veloso, *Verdade Tropical* (Rio de Janeiro: Companhia das Letras, 1997).

14. See Leonardo Gallery, *Leonardo* 29, no. 4 (1996): 249.

15. Tania Fraga, "Simulações Estereoscópicas Interativas," in *A Arte no Século XXI: a humanização das tecnologieas,* ed. Diana Domingues (São Paulo: Fundação Editora da UNESP, 1997), 117–125 (translation mine), available at ⟨http://www.lsi.usp.br/~tania/⟩.

16. Leonardo Gallery, *Leonardo* 30, no. 4 (1997): 254.

17. Edward Lucie-Smith, *Visual Arts in the 20th Century* (Englewood Cliffs, NJ: Prentice Hall, 1996), 12.

34

Technology Has Forgotten Them: Developing-World Women and New Information Technologies

Martha Burkle Bonecchi

For several years, my work as a communication professional has been challenged by watching life from feminist perspectives. Reflecting on the relationship between women and Information and Communication Technologies (ICTs), I realized that it is time to overcome cultural stereotypes that keep women away from ICT management and to seize these new tools to improve the way they affect our regions.

The ICTs, including mass media communication and the World Wide Net, constitute a social division between the twentieth and the twenty-first centuries. Until not so long ago, social differences were established by access to the consumption of goods—rich versus poor. Material wealth made the division between different social levels. Today, access to ICT constitutes the division between those who have and will have access to the information culture (the new wealth) and those who do not.

The ICT field's interaction with women reestablishes a platform of discussion on gender and social structures, on which we must reflect if we intend to reach conclusions and make concrete proposals. Furthermore, the novelty of these technologies must be examined regarding the impact they will have on women's personal life and work. Moreover, my personal feminist reflections are made from a Mexican and developing-world point of view and lead me to talk about women and ICTs from those perspectives.

At the World Association of Christian Communication (WACC) congress that took place in Mexico in October 1995, the Hindu feminist Kamla Bhasin analyzed the subordinated role of women in the third world: "Women have been the main victims of the mass media. Our dignity has been shattered by media that make use of our bodies and distort them, that transform us into objects and consumers of goods."[1]

A panel entitled "Media, Culture, and Communication Challenges and Opportunities" was held in Beijing in August 1995 at the Forum of Women Nongovernmental Organizations (NGOs). One panelist, the Japanese journalist Yahori Matsu, urged women to design strategies of participation in the media and strategies to face media.

According to the proposal of the Beijing action platform in respect to communication, "The majority of women, especially in underdeveloped countries, do not have effective access to the information highway and, therefore, cannot access new information. Consequently, it is necessary for women to intervene in the decision-making process that affects the development of NCTs in order to fully participate in their expansion and control their influence."

From this point of view, it is necessary to ask the following questions: Do the ICTs change the social position of women? Will they allow developing-world women to take a step toward liberation and emancipation? Will these ICTs contribute to more equality between the the sexes? Are these ICTs a plausible way to communicate in underdeveloped countries? And will they become accessible to their population? In this chapter, I try to answer most of these questions.

GENDER RELATIONS AND NITs: BREAKING BARRIERS

The ICTs have emerged in a culture that could be called *androcentric*—that is, where decisions, relationships, and even the way we understand life are seen from a perspective of masculine domination.[3] This androcentrism that has permeated Western history since its birth helps to explain why women are not involved, for example, in the technology development of household appliances, which are fabricated and designed by men who imagine the needs and preferences of women and create products that respond to stereotypes many times removed from the real needs of women who do household chores.

The birth of ICTs toward the end of the 1970s became relevant in the study of men's and women's environments. It soon became apparent that these analyses centered on the work of men, under the assumption that this work would be the same for everyone. The gender differences between men and women workers were put aside, and concepts like working skills were not taken into account. Gender was not included when the relationship of ICTs and work was analyzed.

Feminist discourse rescued this gender indifference by asking questions about gender involvement in technologies and proposing that ICTs would not have the same effects for men and for women. On the contrary, technology has come to corroborate work divisions among the sexes and, furthermore, gender roles that are determined by relationships, from family to workplace.

Gender relations and technology produce and reproduce status between men and women, between femininity and masculinity.[4] This means that technologies, designed by men, are based on the relationships these men establish with nature and women. In fact, the evidence suggests that the introduction of new technologies to the workplace does not debilitate the sexual division of labor, the work held by men and women, or the social construction of skill in a substantial way.[5] On the contrary, the emergence of new working technologies reproduces existing labor divisions between the sexes, which indicates that the technologies do not modify work assignments by gender. Furthermore, some feminist theories (Cynthia Cockburn, for example) have demonstrated that while new technologies are not a disadvantage for women (in the developed countries, I suggest), men are more favored by technological changes and therefore the gap among genders is widening.[6]

This reflection on women and ICTs leads us to the need to break paradigms. By this, I mean that we will not be able to understand the role played by new technologies in a society where gender differences continue as a model by which social relationships are established. Even though we can see certain changes in this regard, gender continues to be a decisive factor when signaling women's actual state, whether we talk of the first or third world.

I already mentioned the sexual divisions in the workplace, but although the continuing use of technologies has brought a reduction in the time spent doing housekeeping, it has not had the same effect on male participation in housekeeping, which seems to be less connected to technological options than to gender identity.[7]

Many countries already deal with the issue of having women fully participate in the design and usage of new technologies. Women remained outside academia for centuries because its culture excluded them. Not until the 1970s did women become formally active within the academic arena.

In some European countries (such as England and Germany), those who work in academia have been supported by feminist groups and have succeeded in helping women reach certain areas of knowledge that were formerly banned to them. Great Britain, for example, implemented a program called Women into Science and Engineering (WISE) in 1984. This program pursues, among other objectives, the introduction of women to the logic and application of computer technology. Germany, on the other hand, began a public-information service that encouraged young high school women to choose "difficult courses," such as mathematics, chemistry, and physics. This campaign was well publicized, but its results have not been determined.[8]

Nevertheless, in developing-world countries, families with lower incomes and education levels require daughters to work to pay for their brothers' education. This expectation has reinforced, particularly in my country, the subordination of women as "second-class citizens" who must find the learning of domestic tasks to be sufficient education.

If the acquisition and implementation of new technologies are not accompanied by a radical change in social relationships between genders, any skills that women obtain in academic fields will be useless. The feminist lesson regarding this issue refers precisely to the labor relationships established by women every day and the role that technology plays in them. It cannot be understood without understanding other components of social life.[9]

BEIJING: WOMEN AND ICT CHALLENGES

The Fourth World Conference on Women celebrated in 1995 in Beijing highlighted the topic of women and new technologies. This corroborates the importance that the ICTs have achieved in women's everyday lives around the world. One major challenge facing women is ensuring equal access to decision-making positions in media and communication systems as an indispensable tool for guaranteeing equality among genders. The importance of the strategic role of communication and of revindicated democratic and participative practices in media and communication systems was discussed. Also, the Conference reflected on women's right to access information and freely express themselves in various media.[10]

The Beijing Action Platform also stressed the urgent need for further education, training, and employment for women to promote and guarantee equal access to all spheres and levels of the media. Participating in the production, use, and broadcasting of new communication technologies is an urgent desire common to women of different sectors and countries. Communication constitutes for us all an indispensable element of equality among genders.

The Beijing Conference stressed that it is necessary "to stimulate and recognize women's communication webs and support those groups of women that participate in all areas of ICTs, making us aware of its importance and providing support centers that can facilitate their access. It is already an urgent invitation."[11]

Facing the opportunity to access a great deal of information and a channel of interactive communication, a growing number of female organizations represented in Beijing have been motivated into seizing this tool, overcoming in some cases personal resistance and fear.

In Latin America, many of the NGO offices are using this mechanism, and according to information provided by the Agencia Latinoamericana de Informacion (ALAI), different media specialized in women are feeding its webs. Nevertheless, in this part of the world, most women have very low levels of education and live in conditions of extreme poverty, so that they do not even think about technology.[12]

The Beijing Action Platform stated that women in Latin America have not had effective access to the electronic highways of information and therefore cannot create webs that may offer them new information. Socialist feminists have reminded us that production relationships are built both by gender divisions and by class divisions. Cynthia Cockburn, Erik Arnold, and Wendy Faulkner see women's exclusion from technology to be a consequence of work division by gender that has been developed under the capitalist system.[13]

WOMEN IN DEVELOPING-WORLD COUNTRIES: ACCESS TO NEW TECHNOLOGIES, FACING REALITY

Reflecting on women's access to ICTs in the Third World leads us to a complex situation where injustice, poverty, and marginalization together with education and gender situations are intertwined. According to Capra Kamla Bhasin (the Hindu feminist mentioned earlier), most technology is fragmented and inclined to manipulation and control more than cooperation, is autoassertive more than integrated, is centered more than regional, and is applied by individuals who form an elite. As a result, technology has become profoundly antiecological, antisocial, unhealthy, and antihuman.[14] Some communication theorists, such as

Armand Mattelart, even believe that it is necessary to break with technologies and with the democratic utopia that determines access to and use of media.[15]

A UNESCO research study that took place between 1990 and 1991 and was published in 1994 stresses the social and cultural impacts of ICTs in women's concrete activities.[16] The chapter that analyzes the role that women in Latin America play in media emphasizes that ICTs have generated new production, transmission, and consumption processes. Therefore, it is impossible to separate ICTs from the economical and cultural contexts in which these are held. The study analyzes three realities about the extension of ICTs around the world—the nonegalitarian character of its development, the transnational character of its process, and the increase in the concentration phenomena. These three realities reflect ideological, economic, gender, and workplace inequalities.

The research on female employment in mass media and its relationship to the decision-making organism clearly shows that women are underrepresented in both managerial and technical positions in mass media. On the other hand, women are resistant to mass media, an attitude that is reflected basically in the creation of alternate media and in the possibility that women are not seen as stereotypical products. For many researchers in this field, both attitudes would allow more freedom in the women-media relationship that would facilitate the incursion of women in these fields.

The analysis of the relationship between women and the mass media stresses the differences between women producers and women consumers. In many instances, women repeat male ideological models. Many women consumers do not have an urgent need to transform media stereotypes because in many instances their own lives are built around these false and artificial models. These facts make our investigation more complex.

In Africa, for example, the participation of women in media is limited by the lack of quality TV and radio programming, the difficulty of changing women's situations in developing-world countries, and the need for media messages to consciously make sense to African women.

In Egypt, the illiteracy index among women continues to be very elevated. Feminist topics are important in the media, but they still insist on portraying women as housekeepers. Women hold 60 percent of managerial positions in Egyptian radio and TV, a high percentage in comparison with other countries. Nevertheless, this percentage does not mean that these women engage in important ideological transformations. The higher social classes decide on the content of the mass media, and, therefore, their own interests determine content. As a consequence, there is a great distance between these interests and the population's real needs.

The situation is very similar in Latin America. In Bolivia, for example, women are great media consumers; nevertheless, the content shown is far from their everyday reality. Access and consumption continue to be privileges of the elite. Women journalists are not uncommon, but this has not benefitted Bolivian women in their everyday lives.

In Bolivia, an alternative communication project called *Warmin Arupa* is shown on radio, on television, and on video. Among its primary goals is for women to express their own reality and spread their message to other women.

In Chile, radio and television are the most popular media. The underground radio stations created in mainstream areas have interrupted their transmissions in the hope that legislation will permit the regulation of their airwaves. The enrollment of women in communication courses such as journalism, public relations, and publicity has continued to grow; however, the participation of women in managerial positions in the mass media remains scarce. In this Latin American country, in 1991, the first feminist radio station in Latin America was created with the support of the Danish government. This initiative is attempting to change the function of media and the place of women in society.

In Singapore, many women are involved in radio and TV, but their presence is lower than men's. A great number of women work in fields such as journalism and writing rather than in ICTs. Men are trying hard to gain their control of ICTs, while women only see ICTs as complementary accessories to their work.

MEXICO'S INDIAN WOMEN: A REALITY FAR FROM ICTs

As in almost every country in Latin America, the contrast between Mexico's urban citizens and its rural inhabitants is dramatic. Indians' educational levels are very low, and a great number of Indians are unable to read or write. A great number of Mexican women survive day by day, looking for food for their families.[17]

In some cities, women are hired not only for their manual skills but also because they do a job without arguing, asking for fair pay, or complaining about work loads. Women are programmed to obey and are undereducated.

We could talk a lot about the monopoly in TV and its control over Mexican ideology, but a recent study done by the Mexican Radio Institution (IMER) shows that the number of women reporters and researchers is increasing. Nevertheless, the image of women that emerges from their labors is one of a superficial, stereotyped person.

Talking about media and democracy, Mattelart proposes that "on an international level, democracy is in jeopardy. . . . The dramatic reality we have to face is that decisions that affect the future of world communication are being made by small groups that do not pay attention to our real needs."[18]

MEXICAN WOMEN NETWORKING

To create new ways of accessing technology is a must. To help women to improve their educational levels and get better jobs, in Mexico as in other developing-world countries, access to technology has to be cheap. An example of accessible ICTs has been the work of the Association for Progress in Communication (APC), a global communication net that provides low-cost access to the Internet and the World Wide Web for NGOs working on human rights, ecology, social development, women's rights, and peace. Men and women using APC communication services can access much information on these and some other issues, can share their own experiences, and can participate in discussions.[19]

APC has developed training workshops in Africa and Asia for women who want to work with ICTs. In Mexico, APC works through LaNeta, a nonprofit organization that offers some Net services like electronic mail, forums, and workshops. The first goal is to make ICTs accessible to everyone and to make access to new technology a democratic right.

I believe that it is urgent that we work for a fairer distribution of ICT resources for women in the developing world. I recently had the opportunity to interview five women who work as Net trainers, Web masters, and Web information feeders.[20] They are all committed to helping women learn how to use information technologies, to work as a network, to exchange information, and to join forces from every point of the world:

Women do naturally networking, and the making up of nets is precisely this. What we have to do is to improve the networking, to build strategies so that we would be able to be more present, to have more impact all around the world. Mexican women have to be ready to form part of politics in order to be able to face the telecommunications challenge, to build the cyberspace as a place for each and every one of them."[21]

CONCLUSIONS

The feminist discourse on ICTs was developed in Europe and the United States, but since the Beijing resolutions women in the developing world have been vocal about trying to access ICTs.

The new information technologies present different opportunities for women and men because of economic and geographical constraints and the tendency to replicate the established gender division of labor and patterns of work. The question is how to sort out the economic and geographical problems from the gender and culturally related problems. We could begin by exploring the implications of an English-language and Western-dominated Net.

Any analysis of developing-world women and ICT utilization should not neglect issues of poverty and justice. ICTs need to be accessible by all kinds of women around the world. Access to ICTs needs to be cheaper in developing-world countries, and women there need more chances for a free education and for higher expectations for ther lives.

NOTES

1. Kamla Bhasin, speech given at the World Association of Christian Communication (WACC) Congress, Mexico, 1995.

2. Beijing Action Platform Fourth World Conference on Women, Beijing China, 1995, Action for Equality, Development and Peace (1995), ch. 4.

3. Judith Astelarra, "Tecnología y valores," *Telos* 13 (March–May 1988): 95–103.

4. Cynthia Cockburne and Ruza Fürst, eds., *Bringing Technology Home: Gender and Technology in a Changing Europe* (Buckingham England: Open University Press, 1994), 15.

5. Juliet Webster, "What Do We Know about Gender Information Technology at Work?" *European Journal of Women's Studies* 2 (1995): 314.

6. Cynthia Cockburn and Ruza Fürst, eds., "Bringing Technology Home," in *Gender and Technology in a Changing Europe* (Buckingham England: Open University Press, 1994), 13.

7. Cynthia Cockburn and Ruza Fürst, research on this matter, 1994.

8. Liesbet van Zoonen, "Feminist Theory and Information Technology," *Media Culture and Society* 14 (January 1992): 9–29.

9. Webster, "What Do We Know about Gender Information Technology at Work?," 329.

10. Beijing Action Plataform, ch. 4, J.1.

11. Beijing Action Platform, ch. 4, J.239F.

12. According to data obtained from the Ministerio de Asuntos Sociales de Espana and the Facultad Latinoamericana de Ciencias Sociales (FLACSO), published by UNICEF and UNIFEM (Mexico) in 1990, 15 percent of Mexican women are illiterate, which means that one-quarter of the entire population over fifteen years old considered themselves to be illiterate. Illiteracy is concentrated in people who live in rural zones, poor people, and men and women over forty years old. Women's illiteracy rates are clearly above those of men. In 1991, almost 60 percent of the total population over twelve years old had not finished elementary school, 26 percent had reached high school, and only 14 percent reached college or university. This reality was even more dramatic in rural zones: in 1991, three-quarters of the population in suburban areas had not reached elementary school; 19 percent had reached high school, and only 6.2 percent had reached college or university.

13. Webster, "What Do We Know about Gender Information Technology at Work?," 329.

14. Bhasin, WACC, 1995.

15. Armand Mattellart, comments at a conference at the Instituto Francés de América Latina (IFAL), Mexico, 28 April 1996.

16. Silvia Pérez-Vitoria, "Las Mujeres y las Tecnologías de la comunicacion," in *Estudios y documentos de Comunicación de Masas* (UNESCO, 1994).

17. Poverty in Mexico in 1994 affected 40 percent of homes. There has been a very small decline in poverty since then, but this difference is not noticed in rural regions, where poverty remains extreme. In 1992, almost half of Mexican rural families were located below the poverty level, and most of them lived in substandard homes. Teresa Valdez Echenique and Enrique Gomez Moraga, *Mujeres latinoamericanas en cifras* (Ministerio de Asuntos Sociales, Mexico, Instituto de la Mujer, Spain & Flacso, Chile, 1995).

18. Mattellart, conference at IFAL.

19. APC, information brochure, 1995.

20. Erika Smith is an American woman who works for the Women's Program at Laneta in Mexico; Britta Sholtys is a German journalist who works for CIMAC, women journalists who work with the goal of bringing information about women to newspapers; Zaida Rodríguez is a Mexican Webmaster who works for Modemmujer, a NGO inspired by the Beijing project to involve more women in Net usage; Beatriz Cavazos is a cofounder of the project Modemmujer, and Sylvia Cadena is a Colombian woman who works as the Webmaster of Colnodo, an electronic Net for Colombian women.

21. Interview with Erika Smith, 27 March 1998.

REFERENCES

Alberdi, Inés. "Nuevas tecnologías y educación." *Telos* 13 (Madrid 1988): 7–79.

Astelarra, Judith. "Tecnologa y valores." *Telos* 13 (Madrid 1988): 89–94.

Bhasin, Kamla. "Las Mujeres Dinamizando las Comunicaciones—de Bangkok a Pekín." World Association of Christian Communication (WACC) Congress, México, October 1995.

Camps, Victoria. "El Juicio Femenino ante las Nuevas Tecnologías." *Telos* 13 (Madrid 1988): 66–69.

Cockburn, Cynthia, and Ruza Fürstç Dilic, eds. *Bringing Technology Home: Gender and Technology in a Changing Europe.* Buckingham, England: Open University Press, 1994.

Escario, Pilar. "Mujer y Trabajo ante el Cambio Tecnológico: el Riesgo de Segregación sectorial." *Telos* 13, (Madrid 1988): 69–72.

Fitzsimons, Annette. "Women, Power and Technology." In Kathleen Lennon and Margaret Whitford, eds., *Knowing the Difference: Feminist Perspectives in Epistemology.* New York: Routledge, 1994.

Franquet, Rosa. "La mujer en el hogar tecnológico." *Telos* 13 (Madrid 1988): 85–91.

Gender Working Group. United Nations Commission on Science and Technology for Development. *Missing Links: Gender Equity in Science and Technology for Development.*

Lasagni, M. Cristina. "Mujeres y Viejas Tecnologías." *Telos* 13 (Madrid 1988).

Marcelle, Gillian. Strategies for Including a Gender Perspective in African Information and Communications Technologies (ICTs) Policy. *IDRC* (1998).

Mattelart, Armand. "Acceder a los Mass Media no Crea más Democracia." *La Jornada,* 24 April 1996.

Mattelart, Armand. "El Engañoso Concepto de la Globalización Cultural." Paper delivered at a conference at the Instituto Francés de América Latina (IFAL), Mexico, 24 April 1996.

Mies, María. "Towards a Methodology for Feminist Research." In Gloria Bowles and Renate Duelli Klein, eds., *Theories of Women's Studies.* London: Routledge & Kegan, 1983.

Perez-Vitoria, Silvia. *Las Mujeres y las Tecnologías de la Comunicación. Estudios y documentos de Comunicación de Masas.* UNESCO, 1994.

Piva, Paola. "El trabajo a distancia." *Telos* 13 (Madrid 1988): 104–109.

Van Zoonen, Liesbet. "Feminist Theory and Information Technology." *Media Culture and Society* 4, no. 1 (1992).

Wagner, Ina. "Hard Times: The Politics of Women's Work in Computerized Environments." *European Journal of Women's Studies.* London: Sage, 1995.

Wright, Barbara Drygulski, ed. *Women, Work and Technology.* Ann Arbor: University of Michigan Press, 1987.

35

Crossing the Threshold: Examining the Public Space of the Web through *Day Without Art Web Action*

Carol Stakenas

I wish to balance my desires.
—KATHRIN BECKER

I wish that all my friends that have passed on reincarnate as happy people.
—FRANK ANDREWS

Despite the odds, one hopes for—in fact, demands—the future!
—LEON GOLUB

I want money, I want money, I wish I was ten years younger.
—KATHE BURKHART

The act of wishing, especially as a communal ritual, can start a subtle, yet powerful transitional moment for an individual. In the instant of vocalizing a wish, your imagination pushes possibility to the front of consciousness. To wish, or desire, becomes especially resonant against the broad continuum of how the HIV/AIDS pandemic continues to effect our lives and our culture from the promising news of a possible vaccine to the ominous certainty that this fatal virus is continuing to spread. Love, peace, health, hope, sleep, communication, home, money, lust, and happiness: what do you wish for?

DAY WITHOUT ART, NOT A DAY WITHOUT ACTION

Day With(out) Art (DWA), held annually on December 1, provides a moment to mobilize the arts community, both nationally and internationally, and to increase awareness of HIV/AIDS and its devastating impact. Held in conjunction with World AIDS Day since 1989, this observance has evolved over time to become a proactive moment, including exhibitions, events, public interventions, and programs that bring focus to our society living with HIV/AIDS. It is an occasion that emphasizes the important role the arts play in understanding the impact of this disease while also continuing the ritual shrouding of artworks and the closing of exhibitions to remember those we have lost and the devastation our field has sustained.

Since 1995, I have annually coordinated Creative Time's *DWA Web Action* to facilitate the coming together of individuals and organizations to defy the geographical boundaries and unite publicly around the grave impact of the AIDS pandemic. It is somewhat daunting to create a project in response to the AIDS crisis now that we are solidly planted in the second decade of living under the shadow of this disease. Recent developments in drug-treatment programs hallowed by the media, the demographic shift in the populations that are being ravaged (the statistics concerning rate of infection for drug users and minorities is alarming), and our lack of connection to the global impact of AIDS have numbed many Americans' sense of urgency. And yet family members, lovers, and coworkers continue to get sick. Perhaps you cross the threshold into positive. Now the computer power button is pushed, the modem sings, and then the Internet opens as a fertile terrain to address your needs to search for critical medical information, seek out services, find community—to become an electronic activist.

DWA WEB ACTION STARTS FROM HOME BASE

DWA Web Action emerged through my experience as a Creative Time staff member. When I joined the organization, Anne Pasternak, the executive director, was eager for the organization to address the emerging public space of the Internet, especially the Web. Since I wanted a chance to combine my artistic practice with my position, this new frontier was extremely appealing.

Creative Time commissions and presents artists who work in the public realm and who challenge notions of what art is and can be. It seeks new strategies that can help expand and strengthen artistic expression and contribute to larger dialogs on cultural production. Throughout its twenty-five-year history, this not-for-profit arts organization has worked with artists from all disciplines in an amazing array of public spaces. For Creative Time, the artwork and the public are inextricably linked.

One of the most obvious aspects of the Web that encourages an interpretation of public space in this electronic landscape is the ease of hypertext linking from one Web site to another. By focusing on this invisible passing from site to site, sense of location and domain shifts into a tracery of networks. From its inception, *DWA Web Action* has been a community-based endeavor that relies on the cooperation of many individuals to reveal an aspect of this electronic matrix—a network of activism already present in the arts community.

Employing practical coalition-building tactics to generate participation, I crafted a simple e-mail invitation in which sites were invited to participate by featuring the Day Without Art logo on their home pages, which was then linked to the *DWA Web Action* site. I targeted both arts- and AIDS-related sites and any e-mail pals I could find. The hope was to heighten visibility by benefiting from individuals' online presence as well as organizations' extension into this virtual realm.

In designing the project, we kept the requirements for participation minimal so that Web designers of all skill levels would be able to join. Since I had just learned my first HTML tags, the baseline was not hard to find. At the same time, we wanted to encourage more elaborate interpretations. Although many participants simply added the linked DWA icon to their home pages, several sites employed a more radical approach that echoed the classic DWA shrouding ritual: they disabled entry to their Web sites and displayed the linked DWA logo on a black background for the entire day. Other sites opted for an elegant solution that placed the DWA logo on a special buffer page that held users' attention for a moment to consider the special focus of the day before allowing them to "pass through" to the site. A list of all sites participating in the project was posted to link back to each site. Although a list did not conjure a visual sense of this network, it was a first step in representing the coalition that formed for *DWA Web Action.*

In the first year, the Web appealed to me as an effective research tool. With 14.4 modems as our initial standard, elaborate graphics created too much lag, and I really wanted to move users through the *DWA Web Action* into other sites. To that end, we posted a resources section that featured a list of information links connected to medical resources, specific services for artists, youth outreach information, and links to advocacy sites and other information- and service-rich community organizations.

THE WORLD IS GETTING EMPTY . . .

In addition to activating a network of consciousness, *DWA Web Action* featured artwork as a way to create a moment of pause for viewers as they passed through the site. The first artwork used was Visual AIDS *BROADSIDE* (1993) by John Giorno. *BROADSIDE* is a project in which text, image, and information artworks are made available as camera-ready mechanicals for easy dissemination. Each broadside focuses on the effect of AIDS on a specific community.

For the initial year, artnetweb (⟨http://artnetweb.com⟩) agreed to host the site, and their artist-in-residence, G. H. Hovagimyan, collaborated on the site presentation. G. H. evolved the online experience by using a meta tag refresh, which created a performative reading of the piece. Once a user came to the site through the DWA link, G. H. used visual cues in the *BROADSIDE* to segment the phrasing of Giorno's words. The resulting series of five pages are automatically loaded one by one, offering the viewer a moment of contemplation. In addition to this performative moment, *DWA Web Action 1995* offered Giorno's piece as a full page to download (figure 35.1). Three newly commissioned works by artists William Cullum, Roberto Juarez, and Steed Taylor were also made available in digital form.

TRIPLE COCKTAIL

For *DWA Web Action 1996,* I collaborated with artist C. B. Cooke. At that time, the new triple-drug therapy (the most common combination of drugs was AZT, 3TC, and a protease inhibitor) was yielding promising results. Yet our optimism was laced with ambivalence. Drug-combination treatment does not work for everyone. For those that it can help, how will they get access to these costly drugs? For HIV/AIDS service organizations, this new protocol means exploding budgets to serve an ever-increasing client base. For young people, still one of the fastest-growing populations infected by HIV, could the hype surrounding this "magic bullet" create a false impression that a safety net is available if you get infected? As a response to these pressing concerns, C. B. created a stunning Web animation, *VisionNeverstopCrying,* that wove together three cycling poems revealing his personal anger, frustration, and even love in response to how AIDS has impacted his life. Rhythmic flashes of pharmaceutical logos punctuate the ebb and flow of his words.

Users who arrive at the site are required to pass through HIV-infected T-cells to enter the site to reinforce a sense of personal contact with AIDS. Then in a triple combination of our own design, C. B. and I developed an interface for the *DWA Web Action 1996* site that offered three areas of exploration. In addition to the Resources and Participants sections, we added an Actions area. Unlike the 1995 site, which simply listed and linked the Web sites that participated, this new area allowed any online visitor to contribute to the site.

In a collective posting area entitled "What are three things you could not live without?" users were invited to engage in a delicate word play to respond to the question. After filling out a preliminary form, each person's three words were posted at the top of three columns. As the list grew, the entries, which ranged from lighthearted and cliché "things" to more

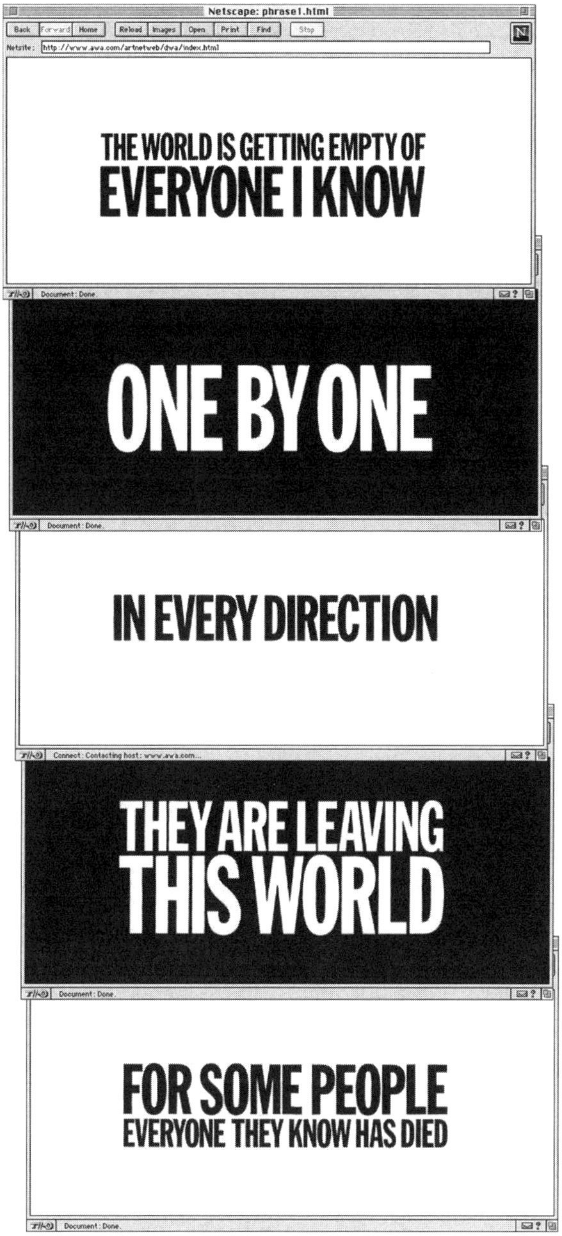

Figure 35.1

DWA Web Action, 1995, collaboration with G. H. Hovagimyan. Using a meta tag refresh created a performative reading of *A Visual AIDS BROADSIDE* by John Giorno, 1993.

arresting comments, began to offer a collective list that could be read in various ways. Totemic profiles created by each column encourage the eye to link the words vertically as well as horizontally. The repetition of popular entries—*love,* for instance—appeared in all three columns repeatedly, offering a rhythm and shape to this collective action. In addition to creating space for active users, this section also provided a link to a memorial list as a reminder of the growing community of those who have been lost.

As I began to do research to update the resources links for this second observance, I found an extraordinary increase in sites offering information and range of topics and services on HIV/AIDS accessible via the Web. I also noticed innovative Web designs and an increasing sophistication in the visual language being used to engage visitors. Most noticeable were the many jazzy, animated GIFs fluttering about.

There was also a palpable sense of territory being created and claimed as more and more fellow arts organizations and museums posted Web sites and as new entities were dedicated to online artistic practices. As in the previous year, I began disseminating invitations primarily to fellow colleagues in the arts community but hoped the resonance of *DWA Web Action* would extend to the broader public that was rapidly expanding online. Eventually, a "buzz" began to travel, and to my delight, as participation grew, the number of individual Web sites that joined the observance increased. (In fact, the West Hollywood neighborhood of the GeoCities server annually yields an enthusiastic enclave of participants.) Local community-based service organizations and activists groups such as Housing Works, Gay Men's Health Crisis, ACT UP, and GLAAD also joined the observance.

At this time, Creative Time made the commitment to build our own Web site and take a domain name (⟨http://www.creativetime.org⟩). Now that we had a way to track activity on the site, we were surprised to learn that over 4,000 visitors had passed through our site that first week in December.

THE WISH MACHINE

Day Without Art Web Action 1997 was created in conjunction with artist Chrysanne Stathacos. This online project was originally inspired by Stathacos's *The Wish Machine* project, which Creative Time presented in New York City's Grand Central Terminal (December 1997 to January 1998). Inspired by a wishing tree that the artist saw in India, this customized vending machine offered urban commuters the opportunity to make a wish, focus on a vintage daguerreotype portrait, and smell the essence of a plant that evokes the essence of the wish—extending the wishing ritual into a multisensory experience. Chrysanne was already interested in associating *The Wish Machine* with DWA, and it was immediately apparent that we could create an online version that would have its own integrity.

Using the vending machine to shape the interface (figure 35.2), I worked with Paul Kontonis of Silvershock and Manos Megagiannis of TeKnowledge Industries. Using a Java applet, the project offered a place for personal and collective action and contemplation. Instead of hard currency, the user was invited to submit a wish to activate the powerful energy of imagination and hope. Unlike the machine in Grand Central, which encouraged a silent and private ritual, the on-line version gave users a place to publicly confess a desire.

In addition to the more ambitious Java scripting, *DWA Web Action 1997* used a leaf image to create "fields" in the Participants section. This image was developed from the wishing tree motif that Stathacos used in *The Wish Machine.* Arranged alphabetically, each group of links nested in a virtual hedge. Our intent in constructing these fields, instead of a traditional list, was to offer an extremely low-band-width solution that could present a simple but more visual sense of this network of consciousness. Although I can intuit the nature of this matrix, I am still searching for ways to visualize a user's sense of place within the geography of this electronic network.

SILENT ORPHEUS **AND** ***TIME CAPSULE***

So often, *Internet* and *World Wide Web* are considered to be synonymous, when in fact the Web is only one protocol on the Internet. I wanted to extend *DWA Web Action* onto another protocol to break out of my Web biases, so I invited the Plaintext Players to present *Silent Orpheus* (1997), a live Internet performance. This performance group is pioneering a new form of drama that is intrinsic to cyberspace. It uses the Internet's multiuser environments known as MOOs. Performances are based on written scenarios that are a unique hybrid of fiction, theater, poetry, and verbal improvisation that is shaped and encouraged by the director, Antoinette LaFarge. Past Plaintext performances such as *Little Hamlet* (1995) or *The White Whale* (1997) reveal the group's interest in the dynamic

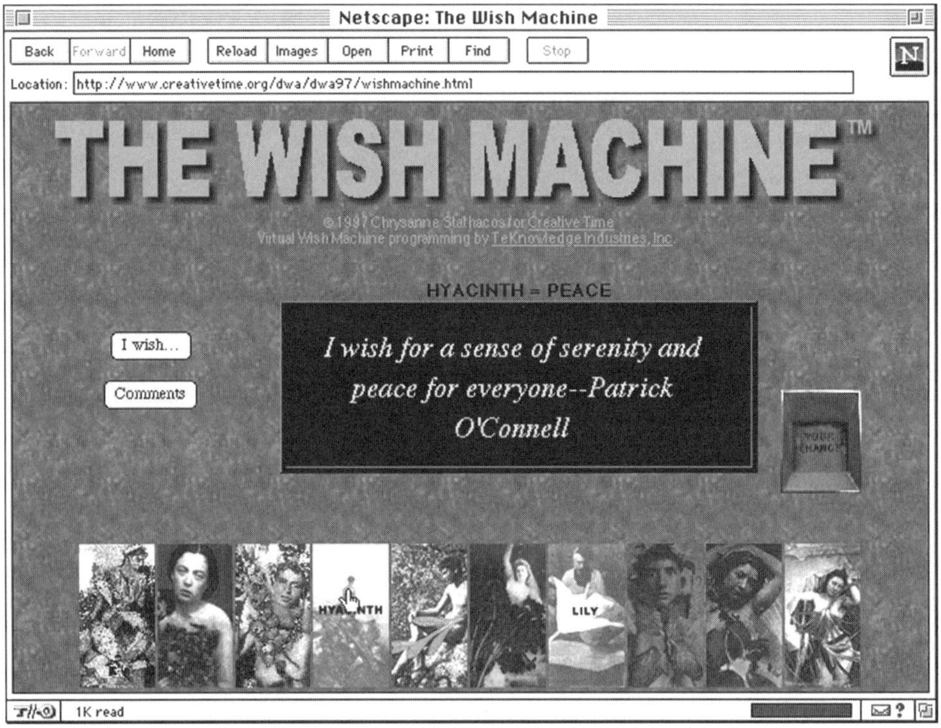

Figure 35.2

DWA Web Action, 1997, collaboration with Chyrsanne Stathacos. This quote from *The Wish Machine* posting area is by Patrick O'Connell, the founding executive director of Visual AIDS and a cofounder of Day With(out) Art.

retelling of classic literary narratives with a soap operatic, improvisational flair. For our event, they retold the story of Orpheus focusing on the bleak time when the legendary singer and poet must descend to hell to bring his love, Eurydice, back from the dead. During Orpheus's absence, music and poetry vanish from earth. The intertwined themes of art, death, and resurrection of this classic Greek myth were especially poignant.

As with other Plaintext Player performances, *Silent Orpheus* was viewable to audiences online via Telnet on the virtual world known as ID MOO, which is hosted by the School of Visual Arts, New York. We also arranged offline access coast to coast with simultaneous live projections at the Institute of Cultural Inquiry (Santa Monica, California), Harvard University (Boston), and Thundergulch (New York).

In cooperation with ArtAIDS (United Kingdom) and Visual AIDS, Creative Time also offered access to *Time Capsule* (1997). Facilitated by artist Ming Wei Lee, this project asked individuals to submit text and image offerings to preserve for posterity personal accounts of the AIDS pandemic impact at this specific moment. A simple e-mail ftp process was set up for online submissions. To extend the range of participation, in-person contributions were scanned or digitally photographed for *The Time Capsule* at the Edward John Noble Education Center at the Museum of Modern Art for the first two days in December 1997. Individuals brought their small objects, photographs, memories, and statements about the AIDS pandemic. The virtual contents were "launched" on January 30, 1998, and will not be accessible until the unlocking on 1 December 2002.

It should be quite telling to reminisce on our present-day reality in the years to come. How will we use technology and its networking capabilities to change and evolve the way we construct community, interact with others, educate ourselves, do business, make art? We have seen these capabilities already start to challenge the way we define these activities. Through the development of *DWA Web Action*, I have come to appreciate the value of working collaboratively—with the artists I have specifically mentioned and also with an online community whose voice grows stronger and stronger as we continue to activate this network and find our bearings in an electronic landscape.

Creative Time continued to commission DWA Web Actions *through December 1, 2000. Please visit ⟨http://www.creativetime.org⟩ to explore the complete archives.*

36

Contested Zones: Futurity and Technological Art

Zoë Sofia

Futures are contested zones, and the language we use to talk about them is possibly more important than we usually realize. To speak with apparent certainty of things that will or are bound to happen is to use the "collapsed future tense" and to risk reproducing a form of technological determinism that pictures technology as an autonomous entity evolving under its own momentum, independent of human decisions, actions, or motives that could be contested from a variety of perspectives in a variety of languages. Can a future already known in advance even be considered a future? Is it not simply a projection of what Heidegger describes as an "entrapping securing" of a world and its inhabitants made available as resources and ordered to serve precalculated outcomes?[1]

Perhaps since Marx and certainly since the rise of the political struggles of feminists, indigenous peoples, and ethnic minorities, it has become increasingly accepted that there is no one "history" but that histories are narratives told from particular standpoints and with characteristic blind spots. The same relativistic perspective is not usually applied to futurity, which (in Western culture, anyway) is almost always invoked in the singular—the future—and assumed to be a destiny shared by all. But just as we have learned to interrogate "history," so can we interrogate accounts of the future: Whose future gets to be the future? Whose visions are named as realistic and attainable, and whose are deferred as impractical and utopian? Who gets empowered and legitimated by such language? Who is ignored?

Linguistic habits such as the collapsed or singular future arise partly as ideological effects of late capitalism, reflecting the sense that technologies "arrive" or "impact" on ordinary people at speeds and from directions beyond our control. Yet at stake is not merely the possibility of gaining control over this or that technology, for, as philosopher of technology Don Ihde has pointed out, so coextensive are technology and culture that the question of how to gain control over technology is really a question of how to "control" a whole culture.[2] Thus, questions concerning technological futurity can be translated into questions about who—or what—are the agents of cultural change.

Originally published in *Leonardo* 29, no. 1 (1996).

Western technologies have long been caught up in the progress narrative overtly promoted as icons of modernity, newness, and futurity. At different historical periods, artifacts such as the clock, the steam engine, and, more recently, the computer have been taken up as metaphors defining humanness and culture.[3] Even at the supposedly postmodern moment, faith in the progress narrative persists in the form of technological neophilia (the love of the new). Advertising rhetoric that describes new products as "revolutionary" can reinforce the idea that technological change automatically brings about social "revolutions," ending domination rather than (as is more usually the case) allowing it to alter its sites and forms. The historical awareness that decisions about technologies are made by particular social actors on the basis of particular interests—often business and military ones—is frequently eclipsed by a technological determinist perspective emphasizing machines as agents of social and historical change, with their own evolutionary powers. Successive generations of ever "smarter" tools are heralded as manifestations of the future today, to which "we" can do nothing but acquiesce.

Collapsed futurity is also discernible on the post-1960s left, which has successively lost faith in socialism and utopian thought. In a 1982 essay, Marxist cultural critic Fredric Jameson proclaimed that "the past is dead" and that while "the future . . . may still be alive in some heroic collectivities on the Earth's surface, it is for us either irrelevant or unthinkable."[4] But the "us" for whom the future is a nonissue is not completely inclusive: Jameson himself cites utopian feminist writers (including Le Guin, Russ, and Delany) as counterexamples. My suggestion is that although past and future may be of little interest to disenchanted veterans of the new left, both history and futurity are very much alive and contested by members of the newer social movements (such as feminism, environmentalism, and land rights).

THE POLITICS OF NEOPHILIA

I am interested in becoming more discerning of differences in attitudes to newness and futurity in discussions of technological arts. Understandably, many arts practitioners and critics are excited about the possibilities of electronic media to open up new forms of art, authorship, and sponsorship. But do these new technologies have to be taken up by artists with the same neophilic spirit that pervades the capitalist and military matrix in which they were promoted and developed? The visual arts and crafts are potentially important sites for exploring techniques and technologies in relative (though certainly not absolute) freedom from hegemonic technological thrusts toward a premapped future; craft forms from antiquity can happily coexist with industrial and digital processes. Here, interest in "the new" can be satisfied not only by the development of new media, techniques, or technologies but also through stylistic variations within or innovative combinations of existing media.

The positing of the new as a good in itself has political and aesthetic implications within discussions of electronic arts. Writing with reference to Latin America, Maria Fernandez criticizes the assumption that advanced Western technologies will inevitably and incontestably diffuse throughout developing countries, supposedly due to the rapidity and autonomy of technological development itself or peoples' willful abandonment of their own cultural traditions in favor of imported state-of-the-art equipment.[5] To artists working in countries that cannot constantly upgrade computer equipment (let alone produce a stable electricity supply to run it), a fetishization of scientific and technological aesthetics and artworks produced on the latest high-end machines could be seen as another instance of a European or U.S. preoccupation being proclaimed as a "universal aesthetic" and forming yet another canon that favors artists who are white and male—once again "re-establishing the aesthetic superiority of wealthy nations over poor countries."[6] Recalling the lessons of modernism, Fernandez argues that computer arts likewise "are products of specific political and economic environments" and therefore "should not become the norm by which other forms are judged."[7]

One alternative to the progress narrative and neophilia is the nonteleological perspective described by design theorist Tony Fry, who recognizes that technologies presently coexist in many different forms, not all of them new:

> From the nonteleological perspective, the new is neither celebrated nor lamented as the end of the old (which either continues or returns, but not necessarily in its original form). Certainly, qualities of "the new" can be brought to an ethical evaluation of object, process, or use. This cannot be done, however, on the basis of values wherein the new is posited as good in itself. Ethics themselves do not

exist at the end of a road of moral development. They are also ensnared in a Web of difference.[8]

One example of artwork that makes use of new media without diminishing the value of previously existing forms is Acha Debelah's *Digital Painting,* which includes scanned images of local artifacts and patterns of traditional Ethiopian design as well as scanned images of pictures he has painted. The new is here incorporated as part of a technocultural assemblage that includes handmade work and invokes strong senses of tradition, origin, and heritage. Debelah expresses the view that electronic arts, far from making history and tradition irrelevant, "can breathe new life" into them.[9]

Differences in attitudes toward the new may be discerned across gender lines within technology-rich countries such as Australia. Several subjects interviewed so far for Virginia Barratt's and my study of Australian women electronic artists[10] have compared their own styles of relating to new equipment with those assumed to be normal in technoculture, where a mania for the newest and latest is conventionally associated with the "hard edge" of discovery and invention. Moira Corby, who describes herself as "fully devoted to very high-end Silicon Graphics workstations and what they can do for me as an artist," nevertheless represents an alternative approach to the equipment.[11] Referring to the difficulties in learning from technical experts who attempt to help solve problems by taking over the equipment and doing the work themselves, Corby says:

> And of course you don't learn that way. You have to be shown, and you have to do it yourself, in the different steps. . . . I've learned by myself. I've got the catalog, and learned, through trial and error, trial and error. . . . With the boy's thing, it's always, "Oh, here's the newest latest software. Let's do what we can do with it and then throw it out and get the newest, latest software." There's this big competition thing happening, whereas that doesn't happen among the women I know. We like to spend a bit more time on one thing and really explore it. You come up with things that so-called experts on the machine don't even know you could do. That's happened to me several times.[12]

In other words, discoveries can arise out of careful and thorough exploration, not just from the speedy conquest and instant trashing of the new. Isabelle Delmotte is a Sydney-based artist who taught herself to use a

Silicon Graphics workstation to make computer animations that explore changes in perception and consciousness during and after epileptic seizures. Delmotte argues that women artists' access to expensive high-end equipment is not necessarily as big a problem as it might seem, since the mania for upgrading makes available good secondhand machines whose full technical and artistic potentials have yet to be thoroughly explored; the high-end secondhand equipment market represents an excellent source to be tapped by women (and other) artists.[13]

Nola Farman, who makes interactive installations for galleries and public spaces, is also critical of technological neophilia within the electronic arts. In her view, technological developments in this century have proceeded faster than our coming to grips with their implications, and artists still have a lot to do in helping culture "digest" and work through earlier technologies—even things as humble as the light switch. Works of artistic and cultural merit can be produced with equipment "off the shelf of a good secondhand electronics store."[14]

An electronic artwork concerned with the collapse of futurity in an explicitly critical rather than in an unwittingly hegemonic way is the installation *Carsick,* by Nola Farman, with Helen Britton, Brad Clinch, and Anna Gibbs, which was exhibited at the Art Gallery of Western Australia in 1992 and at the Performance Space in Sydney in 1994 (figure 36.1). It comprises sections of old cars pulled from the bush, worked over, and coupled with various interactive and noninteractive sound and video devices. The artists state their concern with "the social place of cars within Western culture as well as their approaching obsolescence" and counterpoise this general scenario against the ways that the car works at the individual level as "a nomadic utopic space offering a rich complex of possibilities for the owner/driver."[15]

The car has been an icon of progress and futurity, promising individuals tremendous powers of speed, mobility, and freedom through the mastery of a vehicle frequently experienced as a prosthetic extension of the body and/an object for ego identification. *Carsick* literally explodes the smooth self-contained ego body of the car and its utopic womblike enclosing space into a set of partial objects. As Farman explains, this work explores the disintegrating postcar as "a failed prosthesis." The installation's title refers to cars that are sick (rusted beyond repair) and to the idea of being "love sick" for cars (having a sick love of cars and the perverse eroticism of automobility). Car sickness also is nausea induced by car

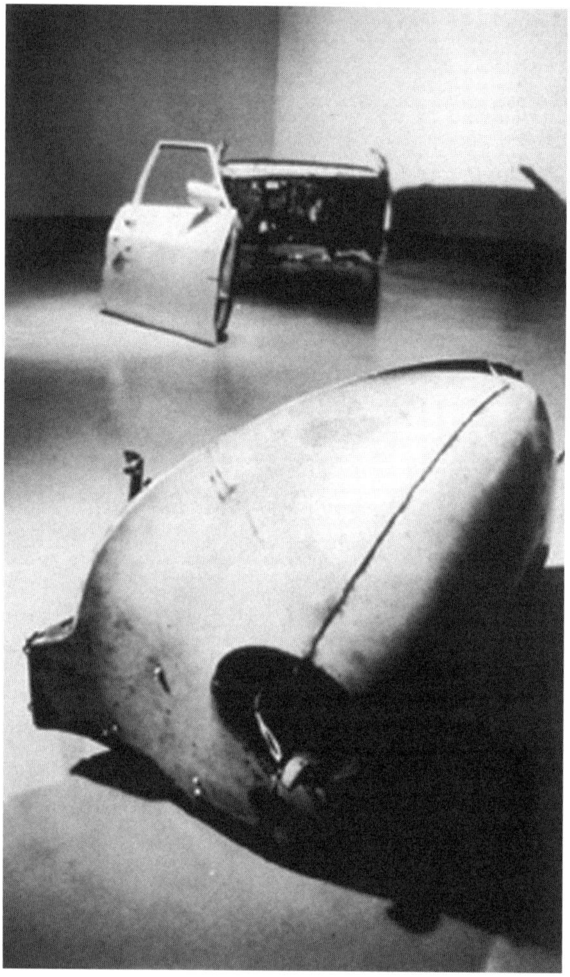

Figure 36.1
Nola Farman, Helen Britton, Brad Clinch, and Anna Gibbs, *Carsick,* 1992, mixed-media installation with rusted car parts, video, audio, and various suveillance technologies. Fragmented car bodies evoke "the aftermath" of car culture as a failed dystopia.

travel; the utopian promise of a nomadic trajectory is spoiled by vomit on the curbside—the abject product of the body's disorientation. Sounds of road crashes are heard when the visitor approaches the fender part of installation, reminding us that this dystopian tendency culminates in blood when an accident crashes through the limits of car and human bodies alike.

Instead of euphorically collapsing a predicted future of human obsolescence onto the present, *Carsick* explores the impending redundancy and dysphoria of a once-futuristic technology and the blood sacrifices that continue to be made for its sake. The installation invokes for me a temporal sense of "the aftermath," a potentially, if perversely, utopian time or space from which to imagine surviving the florescence of disaster and getting on with the clean-up operations. It prompts us to realize we are already living in the aftermath, the earth-sickness caused by cars; we have to start cleaning up now. In *Carsick,* the aftermath of car culture is signaled by the rusted and failed prostheses, as well as by the phoenixlike presence of electric and electronic technologies arising within and around them. Some of these media—the rear-view mirror video with its images of Lassie and roadside farewells, the dashboard radio that constantly switches between various sounds and frequencies "as if searching through memory for something lost and possibly irrecoverable"[16]—are deployed for evocations of memory and nostalgia, reminding us that even our electronic future technologies will be caught within human purposes, memories, and meanings.

CONTESTED FUTURES

Artists working in digital media express a diversity of attitudes toward technology, newness, and futurity. But to the extent that discourses on electronic arts are pervaded by hegemonic neophilic attitudes, this diversity can go unacknowledged. Arguably something like this is going on in the 1993 video documentary *Artists in Cyberculture,* which shows interviews and artworks of some artists at the 1992 Third International Symposium on Electronic Arts (TISEA) in Sydney.[17] This informative text presents a diversity of artists' works and views but subsumes this diversity within a textual framing that reiterates pseudo-universal sensibilities of neophilia and collapsed futurity. A written quote chosen to open the documentary and thus authoritatively address the viewer states, "We move under the shadow of our machines . . . for our time is over and their time has begun."[18]

This statement of collapsed futurity—"our time is over"—pictures machines as beyond human control or responsibility and implies that the human body is redundant. It resonates with the ideas of Hans Moravec and other "tomorrow makers"[19] that rapid advances in information technologies and artificial intelligence are making human bodies outmoded and that the future of humanity will see a "downloading" of intelligence and reproductive powers from human bodies into high technologies that might live a remote-control or extraterrestrial existence,[20] while human bodies disappear.[21] In the documentary, this perspective is immediately reinforced by a cut to the Australian Stelarc, perhaps "the first cyborg artist,"[22] who is shown questioning whether in this high-tech age "male-female intercourse is the best way to prolong the species." Known for his interest in body obsolescence and his "shamanistic" and increasingly biomedical arts practice, Stelarc has elsewhere stated: "The body must burst from its biological, cultural, and planetary containment. The significance of technology may be that it culminates in an alien awareness—one that is POSTHISTORIC, TRANS-HUMAN, and even EXTRATERRESTRIAL."[23]

Another articulation of collapsed futurity in the documentary occurs in the later segment, where artist Jon McCormack (whose own works involve explorations of computer "A-life" and ecological themes)[24] discusses the rapid pace of technological change compared to human adaptation and claims "we have stopped evolving."

At the beginning and end of the TISEA documentary, a robotic voice proclaims, "It is the law of nature . . . the strong survive." This Darwinian idea, long applied to capitalist ventures, is extended to imply the survival of machines as the fittest inheritors of human evolutionary powers. For those who have "stopped evolving," it is no longer a question of gaining mastery over the machines or of worrying about who owns or controls them but simply a matter of masochistically surrendering to *their* control, being content to live in their shadow or even, as in cyberpunk and hacker fantasies, of "jacking into cyberspace"—of letting oneself be engulfed by the oceanic matrix. There is a chillingly profound passivity and a regressive disclaiming of responsibility in these framing quotations, which imply that machines are something like an emergent Fourth Reich to which "we" must inevitably, even enthusiastically, submit. As Maria Fernandez points out, "Anything that is presented as inevitable, predestined, or part of the natural order of things resists contestation."[25] She goes on to cite Walter Benjamin's views that what most corrupted the German working class was "the notion that it was moving with the current . . . the illusion that the factory work which was supposed to tend towards technological progress constituted a political achievement."[26] I would suggest that like earlier futurist celebrations of the machine's aesthetics and powers, attitudes of neophilia and collapsed futurity associated with new technologies could also play into a form of fascist politics in which people relinquish social responsibility in submission to the incontestable progress of "stronger" and "higher" machinic powers.

Who are the "we" who have stopped evolving? And who are the "our" whose times are over? What of those whose time is just beginning (or re-beginning)? Referring to modernity's uneasiness about the links between machines and women, seen as at once automatic and yet not entirely controllable emergent intelligences,[27] British cultural theorist Sadie Plant has suggested that, in the information age, "women's emergence is man's emergency" and that while white men have dominated and "re-membered" history as his-story, "the future is unmanned"[28]—that is, neither dead nor collapsed—but animated by other dynamic agents, including women and machines. From the perspective of "cyberfeminism" (a Haraway-inspired term coined simultaneously in 1991 by Plant in England and the feminist art group VNS Matrix in Australia), the question is not one of dominance and control or of submission and surrender to machines; instead it is one of exploring alliances, affinities, and coevolutionary possibilities, especially between women and technology. The hope is that despite their patrilineage, new technologies may be effectively deployed in feminist projects to "morph" the premapped future, mutate its preset algorithms and codes, and bring about a social reality that is not just a speedier, higher version of the high-tech today.

(TECHNO)BODY POLITICS

The Australian cyberfeminist art group VNS Matrix, comprising Virginia Barratt, Francesca da Rimini, Julianne Pierce, and Josephine Starrs,[29] deploys vivid physical imagery to actively contest the masculinist rationales of high-technology and dominant visions of the future—the "contested zone"—ironically appropriating the hacker metaphor in artworks while developing a computer video game called *All New Gen.* The enemies are Big Daddy Mainframe, a logo-headed suit, and Circuit Boy, "a dangerous techno-bimbo" imaged as a counter to the usual mechanical woman fembot visions, whose penis morphs into a mobile phone in the work's recent computer animation version. The player measures his or her energy in stores of G-slime and is aligned with a feminine-coded cyberspace inhabited by the amorphous *All New Gen* (variously described as a hostile mist, intelligent slime, and so on) and receives assistance from the DNA Sluts, described as "mutant sheroes" (figure 36.2).

One of the alternatives to neofuturist submission to dominant technoculture pursued by VNS Matrix is active viral identification—imaginatively entering technology and the corporate system as a microentity that can penetrate and corrupt the data banks of Big Daddy Mainframe. Unlike male hackers with their Oedipal fantasies of using viruses and data piracy to take over and control the place of Big Daddy,[30] VNS Matrix highlights the infective principle: the virus becomes a metaphor of a political process as the spread of feminist consciousness causes permanent disruption to organs and functions of the male-controlled corporate technobodies.

Technology, futurity, the matrix, and the body are all contested zones for VNS Matrix, whose works aim to physicalize technology and insist on the sensual and erotic character of the human-technology interface. Donna Haraway's "Cyborg Manifesto,"[31] with its vision of an ironic cyborg politics based on affinities and partial identifications rather than identities and essences, is an important reference for the group's *Cyberfeminist Manifesto for the Twenty-first Century.* The work has appeared in various forms and sites, including as a billboard at the Tin Sheds Gallery in Sydney and at the Chicago ACM SIGGRAPH convention (1992). Resonant with the viral metaphor, the woman here appears as a phallic and penetrative force capable of moving in the cyberspace matrix, seeking pleasure and knowledge. Developing Haraway's insights into the perverse and embodied character of cyborg technics, the artists articulate erotic and physical metaphors of the technobody, acknowledging possibilities

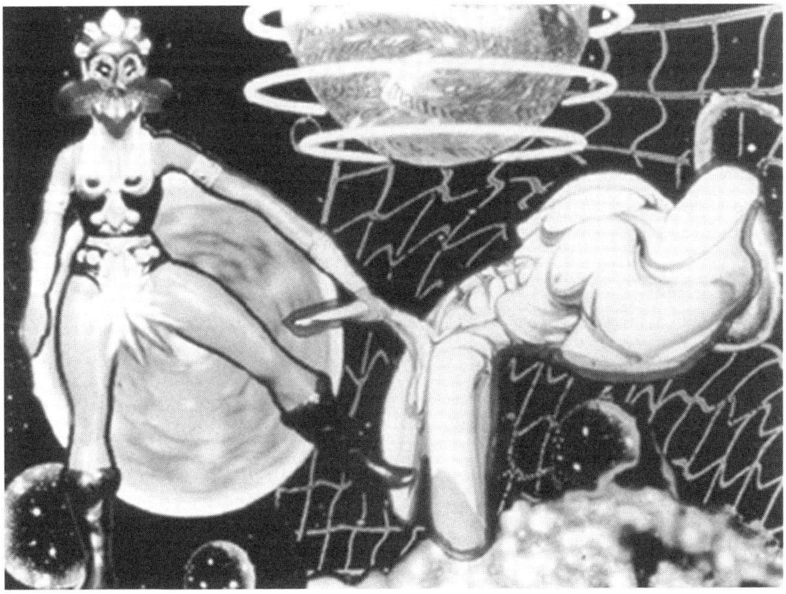

Figure 36.2

VNS Matrix, *All New Gen,* 1994, computer-game project. One of the DNA sluts — the "mutant sheroes" who keep players that bond with them stocked with G-slime — shown wrangling with the chrome-plated technobimbo Circuit Boy, whose penis is morphing into a mobile phone.

for a creative woman-centered technophilia: *the clitoris is a direct line to the matrix.*

For VNS Matrix and several other Australian women artists, Stelarc's extremist statements about body redundancy and the quest for transcendence of the body through technology have helped clarify a position from which to differ.[32] Although conceived as part of a rigorous personal artistic inquiry into the limits of the body and its technological prostheses or substitutes, Stelarc's works and statements are readily interpretable in relation to Christian, metaphysical, and technoscientific desires to escape what cyberpunks call the "meat" of the body. His practices have included various suspensions by hooks, a robotic "third arm" wired up as a prosthetic double of his own, and the insertion of various biomonitoring devices and the amplification and display of their signals to create a sound and light show in which his internal body functions become exteriorized as environment, a technoscape that can include virtual elements (such as a virtual reality hand or a whole body).[33] His recent biomedical performance sculpture, involving a specially made endoscopic device inserted (with much discomfort) into the stomach, further extends his interest in implanting the body with technology: "The body is no longer a mere container of a self; it now contains a sculpture. . . . The body . . . is seen as physiologically and metaphysically hollow."[34]

While invasive biomedical art practice is occasionally pursued by women (such as the French performance artist Orlan, who is remaking her face and body in a documented series of artistic surgical events), more women seem to be interested in less invasive ways of interfacing with technologies to animate environments. Anne Marsh compares Stelarc to Laurie Anderson, whose performances also involve visions of "a future where mind and body are enmeshed in technology" but who tends more to play with questions of (especially gender) identity and who wears technology "on her skin."[35] One might also compare Stelarc's technologically invaded body with the very much intact body of the Sydney performance artist Anna Sabiel in her works. Wearing an abseiling harness, Sabiel attaches herself via piano wire and a system of pulleys and counterweights to one or more 44 gallon drums and, sometimes, ladders. She moves against these weights and in turn causes them to move through her own body motions. The amplified tinkling and occasional scraping sounds of objects and water inside these drums provide a synchronized audio accompaniment to the movements, which explore themes of physical memory—the knowledge and feelings held and remembered by the body.

Sabiel contrasts Stelarc's insertion of biomechanical technologies into a body he wants to abandon with her own aim of achieving a holistic being in the body, explicitly motivated by a "social responsibility" to counter what she considers his "depressing" masculinist visions of the high-tech future and "hollow, hardening, dehydrating" bodies.[36] For Stelarc, the body-technology interface is a ruptured boundary that flows in both directions, whereas for Sabiel, the body-technology interface is surface-to-surface: the harness is donned like a garment that preserves the integrity of the body inside it. This and the overall arachnid effect of the performance reminds me of the movie *Aliens,*[37] where Ripley does battle with the monstrous alien queen from within the Powerloader, a robotic electromechanical exoskeleton: the body remains intact and surrounded by the technology. Sabiel's body movements animate a dynamic system in which what is externalized is not a spectacle extracted from the "meat" of the body by biomedical equipment but a dimension of physical interiority almost unthinkable within Western metaphysics—namely, "physical memory," the body's own historical consciousness, its kinesthetic knowings.

For many women artists working with digital media, the body's physicality is not transcended or obsolesced by technology; rather, it is a source of poetic efforts to at once use and counteract the machine's own anti-body logics by using it as a medium to explore organic or visceral forms. Women artists interviewed for *Artists in Cyberculture* included German artist Ulrike Gabriel discussing her computer-mediated installation *Breath* (1992), where images and sounds changed according to a breath monitor worn by the participant. Gabriel made the point that the work was not transcendent but was tied to the specific limitations of the body's energies and the aesthetic parameters coded into the machine: "You can't get out of your body; it can't get out of the [aesthetic] system, but still a lot can happen." New York artist Patricia Search, whose longstanding love of math made her at home with computer forms and logic, spoke of her desire to enrich her computer-based art with "color, form, and light" to keep it from "being too rigid," while Canadian Char Davies rejected the term *computer artist* and described her computer-generated works as organic and textured forms exploring inner mystical experiences and the natural world.

Whereas Davies and Search counter the computer's biases with an aesthetic of organic beauty, some Australian women artists have adopted an (anti)aesthetic of viscera and abjection. Melbourne-based computer image maker Linda Dement has articulated her desire to "put some guts into the machine" and describes her enjoyment of the way computer scanning allows the use of objects as "paint" in combination with "inward-looking" autobiographical narrative explorations of "brutality, violence, madness" as well as beauty, desire and pleasure, exemplified in her work *Typhoid Mary* (1993).[38]

Dement's images often feature detached organs, anatomical dissections, and medical images that are arranged in formal compositions with a strong decorative emphasis on layered images and textured surfaces. In a catalog note for the *Tekno Viscera* exhibition of women's electronic art and performance in Brisbane 1993, Jo Frare and Vicky Sowry vividly summed up the feminist interest of "putting some guts into the machine":

> Given the obsession of many new technology applications (for example, video games and VR systems) with the female body, it is not surprising that feminist artists are rewiring the network so that these technologies become receptacles for abject female excess—for that which is subversive in gendered difference. The precision and order of these contemporary entertainment and/or representational technologies are being challenged with the intromission of viscera and fluid into what has previously been perceived as a hermetic, dry, closed circuit.[39]

Dement's recent CD-ROM *Cyberflesh Girlmonster* (Perspecta, Art Gallery of New South Wales, 1995) includes some anatomical and violent imagery, along with fragments of written text, but features whimsical animated monsters formed by grafted-together scanned images of body parts, enthusiastically donated by women at the 1994 Adelaide Festival Artists' Week.

The thematics of "putting guts into the machine" and a similarly ironic play with the fragmentation of women's bodies into fetishized parts can be discerned in the interactive mixed-media installation *Mapping E-motion* (1993) by Sydney artist Sarah Waterson. It consists of a number of hanging rectangular perspex plates on which are mounted latex casts of different breasts. A strong sense of "interface" is conveyed by the latex itself, which is shaped by a surface-to-surface interaction between skin and rubber and which is described by the artist as "a semi-permeable membrane between reaction and sensation." Wires run between the latex breasts, which respond to different degrees and speed of visitors' movement in the gallery space by pulsating, their nipples becoming erect. In the 1994 version, the breasts also emitted electronic "chirps." The artist's catalog notes begin:

> We are all dislocated organs in a mapped symbolic state open to operation open to social and philosophic interpretation and reading. We are already mapped we are constantly written we are informed we are reproduced in our most predictable and deducible form. Technology and the button/switch manipulation proved this and requires recruits.[40]

The installation provides an alternative mapping: we are inscribed into this space by an unpredictable and (on first experience anyway) a nondeducible form, for it may take a while to realize that our movements stimulate the breasts, that we are caught in an intangible "interactive interface." In this way *Mapping E-motion* not only experiments with an alternative to "button/switch" interfaces; it also successfully realizes the artist's stated aim of playing with the idea of pheromones—olfactory hormones that may allow us to be erotically stimulated across a distance. But it is able only to represent the idea of pheromones, a simulation of the interactions between erotically charged bodies; the perspex mounted breasts have no insides that might feel surging and erectile sensations. There is, however, the perverse possibility that some visitors will get stimulated by the simulation and that their own body parts move in "sympathetic resonance" with the latex models.

The catalog notes state that "*Mapping E-motion* is about the impossibilities of electronics in simulating what is usually not seen or felt—the internal landscape of being stimulated by another." And a few lines later, we read: "*Mapping E-motion* explores the possibility of using technology to simulate erotic pathways."

The contradiction between these statements about the possibilities and impossibilities of electronics to simulate the internal landscape is not a product of careless writing but precisely expresses that generative tension between technologic and bodily interiority whose exploration comprises the poetic essence of this intriguing and gently ironic artwork. *Mapping E-motion* fully acknowledges the contemporary technological trajectories that entrap us as recruits to "button/switch manipulations" but nevertheless responds to this situation in ways that are uninterpretable within regimes of artistic production and reception that valorize newness, highness, futurity, and sexual/physical redundancy.

My conclusion then is simple: theorists, critics, art lovers, and art makers need to learn to listen to the diversity of voices and visions expressed in technological media and develop more appropriate frameworks in which to appreciate artistic works that do not merely reproduce or celebrate machine logics (such as algorithms creating electronic wallpaper) but actively challenge and pervert them and the futures they imply. Otherwise, we risk not only misunderstanding specific artworks but also reducing their potential effectiveness as alternatives to the specters of monolithic futurelessness and a posthuman world. In such a world, art itself could lose all meaning, for poetic strivings to remember and embody past experiences, to critically reflect on present situations, and to shape imagined and future worlds would all be eclipsed by the overshadowing machines to which our historically transformative and evolutionary powers would have been ceded. I hope that this chapter, by focusing interest on works by contemporary women technological artists, contributes to their success in shaping alternative futures that do not simply intensify the powers of the already strong but enlarge the influence of the values and interests of those not satisfied by the pursuit of the new as a good—or even a god—in itself.

ACKNOWLEDGMENTS

This chapter draws on collaborative research with Virginia Barratt on Double Agents, a study of women technological artists currently underway with gratefully acknowledged assistance from the Australian Network for Arts and Technology (ANAT), the Australia Council, and Murdoch University. Barratt conducted the interviews with Moira Corby and Anna Sabiel quoted here. I thank the various artists for spending time answering our questions and for providing documentation of their work. The Institute of Modern Art in Brisbane gave me the opportunity to view several of the artworks discussed here at the *Tekno Viscera* event in October 1993. Versions of this chapter have been presented at the Future Languages Day of the Adelaide Festival Artists Week (February 1994) and under the title "Intersections, Interdictions, Interfacings: Women, Technology, Art, Philosophy" at the West Australian Art Gallery (November 1993) for the Jillian Bradshaw Memorial Lecture series. A portion of that talk dealing with Joan Brassil, Joyce Hinterding, and Sarah Waterson has been published in Nicholas Zurbrugg, ed., *Electronic Arts in Australia,* a special issue of the journal *Continuum* 8, no. 1 (1994), that is recommended for readers wanting to survey a full range of Australian technological arts.

NOTES

1. Martin Heidegger, *The Question Concerning Technology and Other Essays,* trans. W. Lovitt (New York: Harper & Row, 1977), 27.

2. Don Ihde, *Technology and the Lifeworld: From Garden to Earth* (Bloomington: Indiana University Press, 1990), 140.

3. J. Bolter, *David Turing's Man: Western Culture in the Computer Age* (Chapel Hill: University of North Carolina Press, 1984), ch. 2.

4. Fredric Jameson, "Progress versus Utopia; or, Can We Imagine the Future?," *Science Fiction Studies* 9 (1982): 152.

5. Maria Fernandez, "Technological Diffusion and the Construction of a Universal Aesthetic (discussed with reference to Latin America)," paper presented at the Adelaide Festival Artists Week, *Broadsheet* 23, no. 2 (1994): 3–6.

6. Ibid.

7. Ibid.

8. Tony Fry, "Art Byting the Dust: Some Considerations on Time, Economy, and Cultural Practices of Postmodernity," in Philip Hayward, ed., *Technology, Culture and Creativity in the Late Twentieth Century* (London: Libbey, 1990), 164.

9. Acha Debelah, interview, in Jonathan Cohen, director, *Artists in Cyberculture* (Sydney: Cracked Metal Productions, 1993; distributed by Ronin Films).

10. Virginia Barratt is an electronic artist, performer, curator, and member of the feminist art group VNS Matrix. We are jointly conducting an interview-based study of women electronic artists and their relationships to technologies (see acknowledgments).

11. Moira Corby, interview with Virginia Barratt for the Double Agents project, December 1993; see also "*My Memory Your Past*: Interview with Moira Corby," *Mesh: Journal of the Modern Image Makers Association* 2 (Summer 1993): 5–7.

12. Corby, interviewed with Barratt.

13. Isabelle Delmotte, personal communication, December 1994.

14. Nola Farman, interview with the author, November 1992.

15. Nola Farman, Helen Britton, Brad Clinch, and Anna Gibbs, artists' statement accompanying the installation *Carsick,* 1992.

16. Ibid.

17. TISEA was the third symposia in the series coordinated by the Inter-Society for the Electronic Arts (ISEA), a membership organization. The symposium series began in 1985 and now takes place on an annual basis. For more information, contact ISEA, P.O. Box 8656, 3009 AR Rotterdam, The Netherlands, e-mail: ⟨isea//mbr.frg.eur.nl⟩. A short history of ISEA can be found in Wim van der Plas, "ISEA: An Introduction," *Leonardo* 28, no. 4 (1995): 288.

18. Cohen, *Artists in Cyberculture.*

19. Grant Fjermedal, *The Tomorrow Makers: A Brave New World of Living-Brain Machines* (New York: Macmillan, 1986).

20. See Jean-François Lyotard, "Can Thought Go on without a Body?," trans. B. Boone and L. Hildreth, *Discourse* 11, no. 1 (1988): 74–87.

21. Arthur Kroker and Marilouise Kroker, "Body Digest: Theses on the Disappearing Body in the Hyper-Modern Condition," *Body Digest* issue of *Canadian Journal of Social and Political Theory* 11, nos. 1–2 (1987): i–xvi.

22. Anne Marsh, "Bad Futures: Performing the Obsolete Body," in N. Zurbrugg, ed., *Electronic Arts in Australia,* issue of *Continuum: The Australian Journal of Media and Culture* 8, no. 1 (1994): 281.

23. Ibid., 284.

24. Jon McCormack, "Post-Visual," *Photofile* 42 (June 1994): 46–48; Bill Seaman, "The Emergence of New Electronic Forms in Australian Art: Rodney Berry, John Colette, Linda Dement, Philip George, Joyce Hinterding, Jon McCormack, Stelarc, VNS Matrix," in Zurbrugg, *Electronic Arts in Australia,* esp. 359–360.

25. Fernandez, "Technological Diffusion," 4.

26. Ibid., 4, quoting Walter Benjamin, "Theses on the Philosophy of History."

27. See also Andreas Huyssen, "The Vamp and the Machine: Technology and Sexuality in Fritz Lang's *Metropolis,*" *New German Critique,* nos. 24–25 (1982): 221–237; Judith Halberstam, "Automating Gender: Postmodern Feminism in the Age of the Intelligent Machine," *Feminist Studies* 17, no. 3 (1991): 439–459; and Claudia Springer, "The Pleasure of the Interface," *Screen* 32, no. 3 (1991): 303–323.

28. Sadie Plant, "Cybernetic Hookers," paper presented at the Adelaide Festival Artist's Week, reprinted in *Australian Network for Art and Technology Newsletter* (April–May 1994): 4–8.

29. For discussions of VNS Matrix and other Australian women technological artists, see Glenda Nalder, "Under the VR Spell? Subverting America's Masculinist Global Hologram," *Eyeline* 21 (Autumn 1993): 20–22; see also Nicola Teffer, "Body with Organs," Margriet Bonnin, "Zero 1: Technology and Future Art," and Cath Kenneally, "All New Gen," all in the *Natural/Unnatural* issue of *Photofile* 42 (June 1994); Jyanni Steffensen, "Gamegirls: Working with New Imaging Technologies," *Mesh: Journal of the Modern Image Makers Association* 3 (Autumn 1994): 8–11; Bernadette Flynn, "Woman/Machine Relationships: Investigating the Body within Cyberculture," *Body's Image* issue of *Media Information Australia* 72 (May 1994): 11–19; and "VNS Matrix and Virginia Barratt Interviewed by Bernadette Flynn," in Zurbrugg, *Electronic Arts in Australia,* 419–432.

30. Andrew Ross, "Hacking Away at the Counterculture," in C. Penley and Andrew Ross, eds., *Technoculture* (Minneapolis: University of Minnesota Press, 1991), esp. 107–120.

31. Donna Haraway, "A Cyborg Manifesto: Science, Technology, and Socialist Feminism in the Late Twentieth Century," in *Simians, Cyborgs, and Women: The Reinvention of Nature* (New York: Routledge, 1991): 149–181.

32. VNS Matrix interview in Zurbrugg, *Electronic Arts in Australia,* 429–430.

33. See Marsh, "Bad Futures"; "Stelarc Interviewed by Martin Thomas: 'Just Beaut to Have Three Hands,'" in Zurbrugg, *Electronic Arts in Australia,* 377–395.

34. "Stelarc Interviewed by Martin Thomas," 388. For more on technophagy, see Dale Nason and Troy Innocent, "Cyber Dada Manifesto," *Leonardo* 24, no. 4 (1991): 489; David Cox, "Nothing If Not Full-On: Notes on the S-Thetix of Cyber Dada Films," *Mesh: Journal of the Modern Image Makers Association* 3 (Autumn 1994): 16–20; and for a more general account, Margaret Morse, "What Do Cyborgs Eat? Oral Logic in an Information Society," *Discourse* 16, no. 3 (1994): 86–121.

35. Marsh, "Bad Futures," 288.

36. Anna Sabiel, interview with Virginia Barratt, December 1993.

37. James Cameron, director, *Aliens* (Twentieth Century Fox, 1986).

38. Linda Dement, in Virginia Barratt, ed., *Tekno Viscera,* exhibition catalog (Brisbane: Institute of Modern Art, 1993), n. 8; see also "Linda Dement Interviewed by Glenda Nalder," in Zurbrugg, *Electronic Arts in Australia,* 166–177.

39. Jo Frare and Vicky Sowry, "Intromission," in Barratt, *Tekno Viscera,* 5–6.

40. Sarah Waterson, note in Barratt, *Tekno Viscera,* 11.

Appendix: Listing of Web Site Contents

A Web site associated with this book—located at ⟨http://mitpress2.mit.edu/e-journals/LEA/wow⟩, with a mirror site at ⟨hittp://www.judymalloy.net/newmedia⟩, supplements this book with papers and short statements by women whose work is core to the field, including updates to chapters in this book, as well as work not covered in this book.

INTRODUCTION

Judy Malloy, Switched On: A Female Interface to New Media

DISCUSSION

GENID/NEME, Gender & Identity in New Media, an online conference: Participants include Robert Atkins, Virginia Barratt, Annick Bureaud, Cynthia DuVal, Marjorie Franklin, Carolyn Guertin, Patricia Kim, Tina LaPorta, Jacalyn Lopez Garcia, Amanda McDonald, J. Dawn Mercedes, Karen O'Rourke, Christiane Paul, Andrea Polli, Frank Popper, Christiane Robbins, Cynthia Beth Rubin, Anne-Marie Schleiner, Carolyn P. Speranza, Rejane Spitz, Helen Thorington, Joan Truckenbrod, and Lorri Ann Two Bulls, among others.

ARTISTS' PAPERS AND STATEMENTS

Lutz Bacher, Huge Uterus
Tim Collins and Reiko Goto, Urban Reclamation: Place, Value, Use: The Nine Mile Run Project
Abbe Don, We Make Memories
Marjorie Franklin, Digital Blood
JoAnn Gillerman, The Sun Drops Its Torch; EROS INterACTive; AnArchy Partycam
Lucia Grossberger-Morales, Sangre Boliviana
Carolyn Guyer, Quibbling, a hyperfiction
Helen Mayer Harrison and Newton Harrison, new work
Lynn Hershman, new work
Kathy Rae Huffman and Eva Wohlgemuth, Face Settings—An Updated Chronology
Mary Jean Kenton, Engineers' Notebooks: The Geometry of Color

Laurie Lundquist, Lagoon Project: San Francisco Exploratorium
Muriel Magenta, Virtual Justice
Judy Malloy, OK Research/OK Genetic Engineering/Bad Information, Information Art Defines Technology
Cathy Marshall, Subverting the Link
Pauline Oliveros, Continued work with the Expanded Instrument System
Ann Powers, North Water World
Aviva Rahmani: Ghost Nets: The Medicine Wheel Garden
Sonya Rapoport, new work
Sara Roberts, Early Programming: An Interactive Installation
Jill Scott, Machine Dreams
Deborah Whitman, Deus ex Machina/Closet of Angels
Bibliography

The Web site will continue to be updated with new works by artists in this book, as well as papers by artists not covered in the book.

Contributors

Rebecca Allen
UCLA School of the Arts and Architecture
Center for Digital Arts
1300 Dickson Art Center
Los Angeles, CA 90095-1615
⟨rallen@arts.ucla.edu⟩

Linda Austin
1731 SW Taylor #3
Portland, OR 97205
⟨lindapaustin@earthlink.net⟩

Judith Barry
Box 708
Knickerbocker Station
New York, NY 10002
⟨j.barry@pop.mindspring.com⟩

Pat Bentson
Leonardo
425 Market Street, 2nd floor
San Francisco, CA
94105
⟨pbentson@sfsu.edu⟩

Dara Birnbaum
140 Thompson Street, Apt. 3A
New York, NY 10012
⟨dara@spacelab.net⟩

Kathy Brew
41 West 28th Street
New York, NY 10001
⟨KBrew66933@aol.com⟩

Martha Burkle Bonecchi
Monterrey Institute of Technologies, campus Guadalajara
Av. General Ramón Corona 2514
Col. Nuevo México
Zapopan, Jalisco, C.P. 45140
Mexico
⟨mm_burkle@infosel.net.mx⟩

Anna Couey
1937A Carelton
Berkeley, CA 94704
⟨couey@well.com⟩

Donna J. Cox
NCSA
605 East Springfield
Champaign, IL 61820
⟨cox@ncsa.uiuc.edu⟩

Char Davies
Immersence
3837 rue de Bullion
Montreal, Quebec
H2W 2E2 Canada
⟨char@immersence.com⟩

Diane Fenster
287 Reichling Avenue
Pacifica, CA 94044
⟨diane@dianefenster.com⟩

Monika Fleischmann
Frauenhofer Institute for Media-communication
Mars Exploratory Lab
D-53754 Sankt Augustin
Germany
⟨fleischmann@gmd.de⟩

Jennifer Hall and Blythe Hazen
Do While Studio
122 South Street
Boston, MA 02111
⟨jenhall@massart.edu⟩,
⟨bhazen@montserrat.edu⟩

Jo Hanson
201 Buchanan Street
San Francisco, CA 94102

Helen Mayer Harrison and Newton Harrison
P.O. Box 446
Del Mar, CA
92014
⟨harrstudio@aol.com⟩

Agnes Hegedüs
Karl Seckinger Strasse 40A
76229 Karlsruhe
Germany
⟨agnes-hegedus@t-online.de⟩

Lynn Hershman
Hotwire Productions
147 10th Street
San Francisco, CA
94103
⟨hotwirelh@aol.com⟩

Kathy Rae Huffman
Director of Visual Arts
Cornerhouse
70 Oxford Street
Manchester M1 5NH
United Kingdom
⟨huffman@vgtv.com⟩

Erkki Huhtamo
Linnankatu 47 A3
20100 Turku
Finland

Joan Jonas
112 Mercer Street
New York, NY 10012
⟨lunardogs@earthlink.net⟩

Brenda Laurel
14004 Madrone Place
Los Gatos, CA 95033
⟨blaurel@tauzero.com⟩

Cécile Le Prado
30 rue de la Folie Mericourt
75011 Paris
France
⟨leprado@ircam.fr⟩

Judy Malloy
5306 RidgeView #5
El Sobrante, CA 94803
⟨judymalloy@judymalloy.net⟩

Margaret Morse
1230 Colusa Avenue
Berkeley, CA 94707-2709
⟨memorse@attbi.com⟩

Jaishree K. Odin
Liberal Studies Program
University of Hawaii at Manoa
2500 Dole St.
Krauss Hall 116
Honolulu, HA 96822
⟨odin@hawaii.edu⟩

Pauline Oliveros
156 Hunter Street
Kingston, NY 12401
⟨pauline@deeplistening.org⟩

Simone Osthoff
School of Visual Arts
Pennsylvania State University
210 Patterson Building
University Park, PA 16802
⟨sosthoff@psu.edu⟩

Nancy Paterson
39 Rhinestone Drive
Toronto, ON
M9C 3W8 Canada
⟨nancy.paterson@senecac.on.ca⟩

Sheila Pinkel
210 North Avenue 66
Los Angeles, CA 90042
⟨SPinkel@aol.com⟩

Patric D. Prince
2340 Caminito Cala
Del Mar, CA 92014
⟨patric@siggraph.org⟩

Celia Rabinovitch
684 Corbett Street, Apt. 6
San Francisco, CA 94114
⟨moodindigo9@juno.com⟩

Sonya Rapoport
6 Hillcrest Court
Berkeley, CA 94705
⟨sonyarap@lmi.net⟩

Leslie Ross
520 East 14th Street, #35
New York, NY 10009
⟨rossbsn@tellurian.com⟩

Valerie Soe
262 Gates Street
San Francisco, CA 94110
⟨vsoe@sirius.com⟩

Zoe Sofia
Institute for Cultural Research
and School of Humanities
University of Western Sydney
P.O. Box 1797
Penrith South Dc N.S.W. 1797
Australia
⟨z.sofoulis@uws.edu.au⟩

Carol Stakenas
Creative Time
307 Seventh Avenue, Suite 1907
⟨carols@creativetime.org⟩

Steina
RR 6, Box 100
Santa Fe, NM 87501
⟨woodyv@santafe.edu⟩

Allucquère Rosanne Stone
Advanced Communications Technology Lab (ACTLab)
Department of Radio-Television-Film
University of Texas at Austin
1 University Station A0800
Austin, TX 78712-0108
⟨sandy@sandystone.com⟩

Dawn Stoppiello and Mark Coniglio
Troika Ranch
321 Graham Ave. #4R
Brooklyn, NY 11211
⟨dawn@troikaranch.com⟩

Wolfgang Strauss
Frauenhofer Institute for Media-communication
Mars Exploratory Lab
Schloss Birlinghaven
D-53754 Sankt Augustin
Germany
⟨strauss@imk.fraunhofer.de⟩

Nell Tenhaaf
20 Brockton Avenue, #2
Toronto, ON
M6K 155 Canada
⟨tenhaaf@yorku.ca⟩

Eva Wohlgemuth
Kriehubergasse 5
Vienna, A 1050 Austria
⟨evasys@thing.net⟩

Pamela Z
540 Alabama Street, Studio 213
San Francisco, CA 94110
⟨pamelaz@pamelaz.com⟩

Index